Springer Theses

Recognizing Outstanding Ph.D. Research

For further volumes:
http://www.springer.com/series/8790

Aims and Scope

The series "Springer Theses" brings together a selection of the very best Ph.D. theses from around the world and across the physical sciences. Nominated and endorsed by two recognized specialists, each published volume has been selected for its scientific excellence and the high impact of its contents for the pertinent field of research. For greater accessibility to non-specialists, the published versions include an extended introduction, as well as a foreword by the student's supervisor explaining the special relevance of the work for the field. As a whole, the series will provide a valuable resource both for newcomers to the research fields described, and for other scientists seeking detailed background information on special questions. Finally, it provides an accredited documentation of the valuable contributions made by today's younger generation of scientists.

Theses are accepted into the series by invited nomination only and must fulfill all of the following criteria

- They must be written in good English.
- The topic of should fall within the confines of Chemistry, Physics and related interdisciplinary fields such as Materials, Nanoscience, Chemical Engineering, Complex Systems and Biophysics.
- The work reported in the thesis must represent a significant scientific advance.
- If the thesis includes previously published material, permission to reproduce this must be gained from the respective copyright holder.
- They must have been examined and passed during the 12 months prior to nomination.
- Each thesis should include a foreword by the supervisor outlining the significance of its content.
- The theses should have a clearly defined structure including an introduction accessible to scientists not expert in that particular field.

Philipp Hövel

Control of Complex Nonlinear Systems with Delay

Doctoral Thesis accepted by
Technische Universität Berlin, Germany

Springer

Author
Dr. Philipp Hövel
Institut für Theoretische Physik
Technische Universität Berlin
Hardenbergstr. 36
10623 Berlin
Germany
e-mail: phoevel@physik.tu-berlin.de

Supervisor
Prof. Dr. Eckehard Schöll, PhD
Institut für Theoretische Physik
Technische Universität Berlin
Hardenbergstr. 36
10623 Berlin
Germany
e-mail: schoell@physik.tu-berlin.de

ISSN 2190-5053 e-ISSN 2190-5061

ISBN 978-3-642-14109-6 e-ISBN 978-3-642-14110-2

DOI 10.1007/978-3-642-14110-2

Springer Heidelberg Dordrecht London New York

Library of Congress Control Number: 2010932329

Cover design: eStudio Calamar, Berlin/Figueres

Printed on acid-free paper

Springer is part of Springer Science+Business Media (www.springer.com)

Ich bin von je der Ordnung Freund gewesen.[1]

(Johann Wolfgang von Goethe, Faust I)

[1] I was a friend of order always.

Contributions to Publications

Es gehört ein wenig
kleinlicher Geist dazu,
immer Ordnung zu halten.[2]

(Ernst Ludwig Kirchner)

This thesis contributed considerably to the following publications:

Hövel P, Schöll E (2005) Control of unstable steady states by time-delayed feedback methods. Phys Rev E 72:046203.

Schikora S, Hövel P, Wünsche HJ, Schöll E, Henneberger F (2006) All optical noninvasive control of unstable steady states in a semiconductor laser. Phys Rev Lett 97:213902.

Yanchuk S, Wolfrum M, Hövel P, Schöll E (2006) Control of unstable steady states by long delay feedback. Phys Rev E 74:026201.

Choe C-U, Flunkert V, Hövel P, Benner H, Schöll E (2007) Conversion of stability in systems close to a Hopf bifurcation by time-delayed coupling. Phys Rev E 75:046206.

Fiedler B, Flunkert V, Georgi M, Hövel P, Schöll E (2007) Refuting the odd number limitation of time-delayed feedback control. Phys Rev Lett 98:114101.

Dahms T, Hövel P, Schöll E (2007) Control of unstable steady states by extended time-delayed feedback. Phys Rev E 76:056201.

[2] Only a small-minded person keeps order at all times.

Just W, Fiedler B, Flunkert V, Georgi M, Hövel P, Schöll E (2007) Beyond odd number limitation: a bifurcation analysis of time-delayed feedback control. Phys Rev E 76:026210.

Hövel P, Dahlem MA, Schöll E (2007) Synchronization of noise-induced oscillations by time-delayed feedback. In: Proc. 19th Internat. Conf. on Noise and Fluctuations (ICNF-2007), American Institute of Physics, College Park, Maryland 20740-3843.

Schöll E, Hizanidis J, Hövel P, Stegemann G (2007) Pattern formation in semiconductors under the influence of time-delayed feedback control and noise. In: Schimansky-Geier L, Fiedler B, Kurths J, Schöll E (eds) Analysis and control of complex nonlinear processes in physics, chemistry and biology. World Scientific, Singapore, pp 135–183.

Blyuss KB, Kyrychko YN, Hövel P, Schöll E (2008) Control of unstable steady states in neutral time-delayed systems. Eur Phys J B 65:571.

Dahms T, Hövel P, Schöll E (2008) Stabilizing continuous-wave output in semiconductor lasers by time-delayed feedback. Phys Rev E 78:056213.

Fiedler B, Yanchuk S, Flunkert V, Hövel P, Wünsche HJ, Schöll E (2008) Delay stabilization of rotating waves near fold bifurcation and application to all-optical control of a semiconductor laser. Phys Rev E 77:066207.

Guzenko P, Hövel P, Flunkert V, Fradkov AL, Schöll E (2008) Adaptive tuning of feedback gain in time-delayed feedback control. In: Fradkov A, Andrievsky B (eds) Proc. 6th EUROMECH Nonlinear Dynamics Conference (ENOC-2008), IPACS Open Access Library http://lib.physcon.ru (e-Library of the International Physics and Control Society).

Fiedler B, Flunkert V, Georgi M, Hövel P, Schöll E (2008) Delay stabilization of rotating waves without odd number limitation, in Reviews of nonlinear dynamics and complexity. In: Schuster HG (ed) Annual reviews of nonlinear dynamics and complexity, Vol 1. Wiley-VCH, Weinheim.

Fiedler B, Flunkert V, Georgi M, Hövel P, Schöll E (2008) Beyond the odd number limitation of time-delayed feedback control. In: Schöll E, Schuster HG (eds) Handbook of Chaos Control. Wiley-VCH, Weinheim, 2008, second completely revised and enlarged edition.

Kyrychko YN, Blyuss KB, Hövel P, Schöll E (2009) Asymptotic properties of the spectrum of neutral delay differential equations. Dyn Syst 24:361.

Schöll E, Hiller G, Hövel P, Dahlem MA (2009) Time-delayed feedback in neurosystems. Phil Trans R Soc A 367:1079.

Schöll E, Hövel P, Flunkert V, Dahlem MA (2009) Time-delayed feedback control: from simple models to lasers and neural systems. In: Atay FM (ed) Complex Time-Delay Systems. Springer, Berlin.

Hövel P, Dahlem MA, Schöll E (2010) Control of synchronization in coupled neural systems by time-delayed feedback. Int J Bifur Chaos 20:3.

Supervisor's Foreword

A central issue in applied nonlinear science is the deliberate design and control of temporal, spatial, and spatio-temporal states with desired properties. The number of publications in this field has been steadily growing since the first pionieering papers appeared two decades ago, and has reached almost 1,000 each year. Nowadays the original notion of *chaos* control has been extended to a much wider class of problems involving the stabilization of unstable periodic or stationary states in complex nonlinear dynamic systems. A particularly simple and efficient control concept is provided by time-delayed feedback, where the control signal is constructed from some time-delayed output variable of the system.

This Thesis deals with delay effects in nonlinear systems which are ubiquitous in various fields of physics, chemistry, biology, engineering, and even in social and economic systems. They may arise due to processing and latency times or the finite propagation speed of information between the constituents of a complex system, for instance in electronic or optical systems or neural networks. Time delay has two complementary, counterintuitive and almost contradicting facets. On the one hand, delay is able to induce instabilities, bifurcations of periodic and more complicated orbits, multi-stability and chaotic motion. On the other hand, delay can suppress instabilities, stabilize unstable stationary or periodic states and may control complex chaotic dynamics. This thesis deals with both aspects, and presents novel fundamental results on the controllability of nonlinear dynamics by time-delayed feedback, as well as applications to lasers, hybrid-mechanical systems, and coupled neural systems.

The theoretical analysis in this thesis uses simple generic models—normal forms—which are universally applicable, but still allow for an analytic treatment. Among the most interesting results are the refutation of the alleged odd-number limitation of time-delayed feedback control, which was erroneously believed for a long time, and new insights into delay control of stochastic synchronization of neuronal activity which might lead to novel therapeutic approaches, for instance, the suppression of pathological states of synchronous firing of neurons, which occur in Parkinson's disease or epilepsy. In further work, one might apply time-delayed feedback control to suppress excitation waves that occur in the brain during migraine or stroke, or use it to control autonomous robots that adaptively select different walking patterns and react to changing environment.

Berlin, May 2010 Eckehard Schöll

Acknowledgments

I took off a coat
to feel lighter
then added it to my pack.

(Bashou)

Above all, I thank Prof. Dr. Eckehard Schöll, PhD. for the chance of exploring the field of nonlinear systems as part of his group. His supervision, support, and experience was a constant source of inspiration and motivation. Especially, I enjoyed the organization and the directing of the group expressed, for instance, in the weekly Jour Fixe, continuous support during the publication process, and his encouragement to frequent visits of international conferences which provided many contacts to researchers from around the world.

Furthermore, I owe many thanks to the fellow members of the Schöll group. They are all responsible for the comfortable and friendly atmosphere which accompanied the official Jour Fixe as well as the unofficial gatherings around the espresso machine. Especially, I thank Valentin Flunkert, Thomas Dahms, and Robert Hinz for the thought exchange and vivid discussions during their time as master and Ph.D. students in the Schöll group as well as Dr. Johanne Hizanidis and Niels Majer who shared an office with me for many years. I am also indebted to Dr. Markus A. Dahlem for his extensive knowledge in the area of biological and neural systems and discussions beyond science.

I am also grateful for various discussions with many scientists working in the area of nonlinear dynamics. Without claim of completeness, special thanks are due to Dr. Andreas Amman, Dr. Yuliya N. Kyrychko, Dr. Konstantin B. Blyuss, Dr. Hans-Jürgen "Ede" Wünsche, Sylvia Schikora, as well as numerous guests to the Schöll group who contributed to this thesis in many ways.

This work was supported by Deutsche Forschungsgemeinschaft in the framework of the collaborative research center 555 Complex Nonlinear Processes[3]. In addition, I acknowledge support by Deutscher Akademischer Austauschdienst[4] for parts of Chapter 6 about excitable, neural systems in two ways: I am grateful for a short term scholarship which gave me the opportunity to visit the group of Prof. Dr. Kazuyuki Aihara at the University of Tokyo[5] in 2007. In addition, I was lucky to participate in the Research Internships in Science and Engineering 2008 program (RISE)[6] by which Sarang A. Shah (Georgia Institute of Technology)[7] spend three months with the Schöll group in 2008.

Last, but certainly not least, I thank my parents and sister Maike for their never ending support under all circumstances and finally my girlfriend Christine for her love and care. This is for you.

Berlin, May 2010 Philipp Hövel

---

3 http://sfb555.physik.hu-berlin.de

4 http://www.daad.de

5 http://www.aihara.jst.go.jp

6 http://www.daad.de/rise

7 http://www.gatech.edu

Contents

Chapter 1
Introduction

Gebraucht der Zeit,
sie geht so schnell von hinnen,
doch Ordnung lehrt Euch Zeit gewinnen.[1]

(Johann Wolfgang von Goethe, Faust I)

Control of a specific state is an important and challenging task that is present in a wide range of different systems. To mention two examples that have nothing in common on the first glance think of the children's play of stick balancing on a finger tip and the treatment of symptoms of neurological diseases. Although they seem to be very different, both of these two examples deal with fundamental questions which are addressed by control theory.

In the first case, a child tries to keep a stick in its upright position by adequate repositioning of the arm and finger. Thereby, it acquires motor skills such as coordination and balance. From the perspective of control theory, the movement of the child's arm and finger leads to a change of stability of the unstable fixed point correspond to the stick in upright position.

The control goal of the second case is opposite nature. In order to treat symptoms of neurological diseases like Parkinson's tremor, it is most desirable to destabilize the stable, but pathological state. This state can be related to unwanted neural synchrony in the brain.

Although, these examples stem from a completely different background, on can achieve either control goal by application of an adequate control method, for instance, time-delayed feedback control [1]. This control method compares the current state of a system to its time-delayed version. It generates a control force from this difference which is applied to the system via a feedback loop.

Coming back to the examples above, a robot will be able to automatically perform the stabilization task of stick balancing if it acts according this this control

[1] Use time wisely for it runs so fast. But order teaches you to save time.

P. Hövel, *Control of Complex Nonlinear Systems with Delay*,
Springer Theses, DOI: 10.1007/978-3-642-14110-2_1,

scheme. Applied to the neurological example, one could aim destroying the pathological synchrony by the same method. Once the symptoms are treated, the neurons might even recover and form new connections. Thus, control results eventually in a therapeutic effect.

Of course, it is still a long way to achieve this ambitious goal. Intermediate steps, however, are the development of a deeper understanding how and for which type of systems control meads to the desired outcome.

Over the past decade control of unstable states has evolved into a central issue in applied nonlinear science. See, for instance, recent reviews [2–6]. This area of research has various aspects and diverse applications: In the context of stabilization of unstable periodic orbits embedded in a strange attractor of deterministic, chaotic systems, this is generally referred to as chaos control. Control can also concern the stabilization of unstable steady states or the coherence and timescales in stochastic systems.

Various methods of control, going well beyond the classical control theory [7–9], have been developed since the pioneering, seminal work of Ott et al. [10] in which they demonstrated that small time-dependent changes in the control parameters of a nonlinear system can turn a previously chaotic trajectory into a stable periodic motion. As mentioned above, one scheme where the control force is constructed from the difference of the present state of a system to the time-delayed value [1] has turned out to be very robust, universal to apply, and easy to implement experimentally. It is known as time-delayed feedback control, time-delay autosynchronization, or Pyragas control named after its inventor Kestutis Pyragas.

Time-delayed feedback control has been used in a large variety of systems which will be pointed out in the following overview. In general, applications can be found in physics, chemistry, biology, medicine, and engineering [4, 11–13], in purely temporal dynamics as well as in spatially extended systems [3, 14–29].

Next to the easy realization of time-delayed feedback, the control method has another advantage: It is noninvasive in the sense that the control force vanishes if the target state under control is reached. This can be achieved by proper tuning of the control parameter, if, for instance for periodic orbits, the time delay is equal to the period of the desired orbit. Thus, the control method does not require a reference system and needs only a minimum of a priori knowledge. This enables stabilization of ultra-fast systems which are found, e.g., in optics and electronics [30–48]. Note that time-delayed feedback can also create additional delay-induced modes [29, 47, 49] which may result in multistability [50].

Another area of research, where time-delayed feedback has been successfully implemented, concerns nonlinear charge transport in semiconductor nanostructures [14, 16, 51–55]. These might consist of only a few quantum wells as in the case of the double-barrier resonant tunneling diode [18, 28, 56–59] or extended devices like a superlattice [19, 54, 60, 61].

Moreover, time-delayed feedback control has recently been shown to be applicable also to noise-induced oscillations and patterns [62–65]. This is an interesting observation in the context of ongoing research on the constructive influence of noise in nonlinear systems [66–71].

In order to finish this overview of potential systems, where time-delayed feedback control is useful, let me mention also the field of neurodynamics or reaction–diffusion systems of activator-inhibitor type [72–75]. On one hand, time-delayed effects are present in these excitable systems as part of the underlying dynamics and occur, for instance, in the interplay between vascular and neural components in the brain [26, 76, 77] which leads to global and local delays, respectively [78]. On the other hand, time-delayed feedback can be used for the therapeutic treatment via an external stimulus in Parkinson's disease, epilepsy, migraine, and stroke [26, 79–83].

In this context, the concept of synchronization is fundamental [84, 85] as it describes how single elements like neurons cooperate in a large network, e.g, the brain, by adjusting the individual dynamics. Next to synchronization in a network, the question of functionality is of increasing interest [86] and the influence of delay in the network dynamics with respect to this issue is a topic of ongoing research [87, 88].

This summary demonstrates that time-delayed feedback is a powerful tool with many possible applications. Besides the implementation in an experiment or numerical simulations, it is also of importance to understand the underlying control mechanisms. This will be investigated in the present thesis, where time-delayed feedback is applied to different generic models. These models serve as normal forms for fixed points, periodic orbits, or excitable dynamics. Next to the mathematical background of delay-differential equations, I will also consider possible extensions in connection to experiments such as additional filtering or control loop latency.

This thesis is organized as follows: Chapter 2 introduces the control scheme. It contains various extensions of the original controller as well as modifications important in experimental realizations. In addition, Chap. 2 provides a review of fundamental techniques such as linear stability analysis and can be seen as the central element which connects all subsequent discussions. See also Fig. 1.1 for

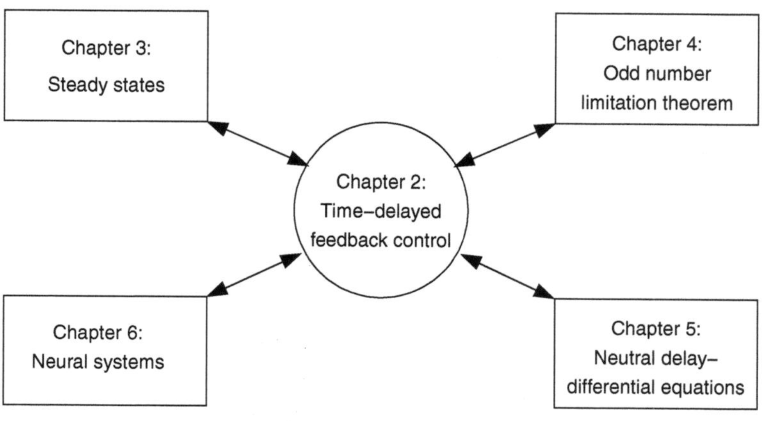

Fig. 1.1 Schematic structure of the thesis

Table 1.1 Notation of frequently used variables and parameters

Chapter 2	
$\mathbf{x}(t)$	State of the system at time t, $\mathbf{x}(t) \in \mathbb{R}^n$
$\mathbf{f}$	Dynamic function $\mathbf{f} : \mathbb{R}^n \to \mathbb{R}^n$, $\mathbf{f} : \mathbf{x} \mapsto \mathbf{f}(\mathbf{x})$
$\mathbf{g}$	Coupling function $\mathbf{g} : \mathbb{R}^n \to \mathbb{R}^m$, $\mathbf{g} : \mathbf{x} \mapsto \mathbf{s} = \mathbf{g}(\mathbf{x})$
$\mathbf{s}(t)$	Control signal at time t: $\mathbf{s}(t) = \mathbf{g}(\mathbf{x}(\mathrm{t})) \in \mathbb{R}^{\mathrm{n}}$
τ	Time delay
$\mathbf{F}(t)$	Control force $\mathbf{F} : \mathbb{R}^m \to \mathbb{R}^m$, $\mathbf{F} : \mathbf{s} \mapsto \mathbf{s}(t) - \mathbf{s}(t-\tau)$
K	Feedback gain
$\mathbf{h}$	Transducer function $\mathbf{h} : \mathbb{R}^m \to \mathbb{R}^n$, $\mathbf{h} : \mathbf{F} \mapsto \mathbf{h}(K\mathbf{F})$
K	Feedback gain
$\zeta(t)$	Periodic orbit $\zeta(t) \in \mathbb{R}^n$
$\mathbf{x}^*$	Steady state, $\mathbf{x}^* \in \mathbb{R}^n$
R	Memory parameter, $R \in [-1, 1]$
$\mathbf{A}$	$n \times n$ coupling matrix
φ	Feedback phase
δ	Latency time
$\bar{\mathbf{s}}(t)$	Filtered version of $\mathbf{s}(t)$
α	Cutoff frequency
$\hat{\mathbf{s}}(\omega)$	Fourier transform of $\mathbf{s}(t)$, $\hat{\mathbf{s}}(\omega) = \mathcal{F}[\mathbf{s}(t)]$
Δx	Spatial delay
$\kappa(x)$	Integral kernel
$\boldsymbol{\delta x}(t)$	Small deviation from $\mathbf{x}(t)$
$\mathbf{u}(t)$	Floquet mode, $\mathbf{u}(t) = \mathbf{u}(t-\tau)$
Λ	Floquet exponent or eigenvalue, $\Lambda \in \mathbb{C}$
p	Real part of the Floquet exponent, $\mathrm{Re}(\Lambda) = p$
q	Imaginary part of the Floquet exponent, $\mathrm{Im}(\Lambda) = q$
$\mathbf{Id}$	$n \times n$ identity matrix
$\Phi(t)$	$n \times n$ fundamental matrix
$T(\omega)$	Transfer function
Additional variables of Chap. 3	
Λ_0	Uncontrolled Floquet exponent or eigenvalue, $\Lambda_0 \in \mathbb{C}$
λ	Real part of the uncontrolled eigenvalue, $\mathrm{Re}(\Lambda_0) = \lambda$
ω	Imaginary part of the uncontrolled eigenvalue, $\mathrm{Im}(\Lambda_0) = \omega$
W	Lambert function, $W : \mathbb{C} \to \mathbb{C}$, $W : z \mapsto W(z)$
T_0	Intrinsic period, $T_0 = 2\pi/\omega$
T_{trans}	Transient time
ψ	Control phase
Additional variables of Chap. 4	
$\mathbf{\Phi}$	Fundamental matrix of the uncontrolled system
$\mathbf{\Psi}$	Fundamental matrix of the system with control
μ	Floquet multiplier, $\mu = \exp(\Lambda\tau) \in \mathbb{C}$
β	Phase of the feedback gain K
b_0	Amplitude of the feedback gain K
Additional variables of Chap. 5	
M, m, l	System parameters
z_M, z, θ	System variables

(continued)

Table 1.1 (continued)

$\mathcal{L}$	Lagrange function
τ_1, τ_2	Time delays
Additional variables of Chap. 6	
$u(t)$	Activator variable
$v(t)$	Inhibitor variable
ε	Timescale separation, $\varepsilon \ll 1$
a	Threshold parameter
D	Noise intensity
C	Coupling strength
$\xi(t)$	Gaussian white noise
T_{ISI}	Interspike interval
$\langle T_{\mathrm{ISI}} \rangle$	Average interspike interval
$p(T_{\mathrm{ISI}})$	Interspike interval distribution
T_Σ	Interspike interval calculated from the summarized variable $u_\Sigma = u_1 + u_2$
$S(\omega)$	Power spectrum
$\Psi(s)$	Autocorrelation function
t_{cor}	Correlation time
$\varphi(t)$	Phase variable
$\Delta\varphi(t)$	Phase difference
γ	Phase synchronization index
δ	Average length of the phase synchronization intervals
τ_K	Time delay of the self-coupling
$\mathbf{G}$	Adjacency matrix

Most of variables and parameters are introduced in Chap. 2, but will also be used in the subsequent chapters

the structure of this thesis. Chapters 3–6 are devoted to the application of time-delayed feedback control to different classes of dynamic systems. Chapter 3 presents the control of steady states subject to time-delayed feedback control in various modifications and coupling schemes. Chapter 4 deals with the stabilization of periodic orbits in a subcritical Hopf bifurcation which provides a counterexample to the often invoked odd number limitation theorem. Chapter 5 discusses time-delayed feedback control applied to neutral delay-differential equations which have an intrinsic time delay in the highest derivative. The subject of Chap. 6 is the application of the time-delayed feedback scheme to excitable neural systems where the control method influences the cooperative dynamics of coupled elements. The resulting synchronization effects are also important in the context of larger networks. Finally, Chap. 5 concludes with a summary and an outlook. As a reference for the notation, see Table 1.1 which lists the frequently used variables and parameters.

Throughout all subsequent chapters, the analysis of the control mechanisms makes use of different techniques ranging from bifurcation analysis, linear stability analysis, and Floquet theory to time series analysis, for instance, on the basis of the power spectrum.

To summarize this introduction, I will demonstrate in the following that the time-delayed feedback scheme provides a powerful tool to control the dynamics of a given system and can be used especially for the stabilization of unstable states.

References

1. Pyragas K (1992) Continuous control of chaos by self-controlling feedback. Phys Lett A 170:421
2. Bär M, Engel H, Schöll E, Torcini A (eds) (2004) Trends in pattern formation: stability, control and fluctuations, special issue of Physica D, vol 199 (no. 1, 2). Elsevier, Amsterdam, pp 1–277
3. Schöll E, Hizanidis J, Hövel P, Stegemann G (2007) Pattern formation in semiconductors under the influence of time-delayed feedback control and noise. In: Schimansky-Geier L, Fiedler B, Kurths J, Schöll E (eds) Analysis and control of complex nonlinear processes in physics, chemistry and biology. World Scientific, Singapore, pp 135–183
4. Schöll E, Schuster HG (eds) (2008) Handbook of chaos control. Wiley-VCH, Weinheim (second completely revised and enlarged edition)
5. Schimansky-Geier L, Fiedler B, Kurths J, Schöll E (eds) (2007) Analysis and control of complex nonlinear processes in physics, chemistry and biology. World Scientific, Singapore
6. Schöll E, Hövel P, Flunkert V, Dahlem MA (2010) Time-delayed feedback control: from simple models to lasers and neural systems. In: Atay FM (eds) Complex time-delay systems: theory and applications. Springer, Berlin
7. Nijmeijer H, Schaft AVD (1996) Nonlinear dynamical control systems, 3rd edn. Springer, New York
8. Ogata K (1997) Modern control engineering. Prentice-Hall, New York
9. Fradkov AL, Miroshnik IV, Nikiforov VO (1999) Nonlinear and adaptive control of complex systems. Kluwer, Dordrecht
10. Ott E, Grebogi C, Yorke JA (1990) Controlling chaos. Phys Rev Lett 64:1196
11. Gauthier DJ (2003) Resource letter: controlling chaos. Am J Phys 71:750
12. Pyragas K (2006) Delayed feedback control of chaos. Philos Trans R Soc A 364:2309
13. Sieber J, Krauskopf B (2007) Control based bifurcation analysis for experiments. Nonlinear Dyn 51:365
14. Franceschini G, Bose S, Schöll E (1999) Control of chaotic spatiotemporal spiking by time-delay autosynchronisation. Phys Rev E 60:5426
15. Kim M, Bertram M, Pollmann M, von Oertzen A, Mikhailov AS, Rotermund HH, Ertl G (2001) Controlling chemical turbulence by global delayed feedback: Pattern formation in catalytic CO oxidation on Pt(110). Science 292:1357
16. Beck O, Amann A, Schöll E, Socolar JES, Just W (2002) Comparison of time-delayed feedback schemes for spatio-temporal control of chaos in a reaction-diffusion system with global coupling. Phys Rev E 66:016213
17. Baba N, Amann A, Schöll E, Just W (2002) Giant improvement of time-delayed feedback control by spatio-temporal filtering. Phys Rev Lett 89:074101
18. Unkelbach J, Amann A, Just W, Schöll E (2003) Time-delay autosynchronization of the spatiotemporal dynamics in resonant tunneling diodes. Phys Rev E 68:026204
19. Schlesner J, Amann A, Janson NB, Just W, Schöll E (2003) Self-stabilization of high frequency oscillations in semiconductor superlattices by time-delay autosynchronization. Phys Rev E 68:066208
20. Beta C, Bertram M, Mikhailov AS, Rotermund HH, Ertl G (2003) Controlling turbulence in a surface chemical reaction by time-delay autosynchronization. Phys Rev E 67:046224
21. Beta C, Mikhailov AS (2004) Controlling spatiotemporal chaos in oscillatory reaction–diffusion systems by time-delay autosynchronization. Phys D 199:173

22. Montgomery KA, Silber M (2004) Feedback control of travelling wave solutions of the complex Ginzburg–Landau equation. Nonlinearity 17:2225
23. Postlethwaite CM, Silber M (2007) Stabilizing unstable periodic orbits in the Lorenz equations using time-delayed feedback control. Phys Rev E 76:056214
24. Ahlborn A, Parlitz U (2007) Controlling spatiotemporal chaos using multiple delays. Phys Rev E 75:65202
25. Ahlborn A, Parlitz U (2008) Control and synchronization of spatiotemporal chaos. Phys Rev E 77:016201
26. Dahlem MA, Schneider FM, Schöll E (2008) Failure of feedback as a putative common mechanism of spreading depolarizations in migraine and stroke. Chaos 18:026110
27. Kyrychko YN, Blyuss KB, Hogan SJ, Schöll E (2009) Control of spatio-temporal patterns in the Gray–Scott model. Chaos 19:043126
28. Majer N, Schöll E (2009) Resonant control of stochastic spatio-temporal dynamics in a tunnel diode by multiple time delayed feedback. Phys Rev E 79:011109
29. Kehrt M, Hövel P, Flunkert V, Dahlem MA, Rodin P, Schöll E (2009) Stabilization of complex spatio-temporal dynamics near a subcritical Hopf bifurcation by time-delayed feedback. Eur Phys J B 68:557
30. Gauthier DJ, Sukow DW, Concannon HM, Socolar JES (1994) Stabilizing unstable periodic orbits in a fast diode resonator using continuous time-delay autosynchronization. Phys Rev E 50:2343
31. Socolar JES, Sukow DW, Gauthier DJ (1994) Stabilizing unstable periodic orbits in fast dynamical systems. Phys Rev E 50:3245
32. Sukow DW, Bleich ME, Gauthier DJ, Socolar JES (1997) Controlling chaos in a fast diode resonator using time-delay autosynchronisation: experimental observations and theoretical analysis. Chaos 7:560
33. Wieczorek S, Krauskopf B, Lenstra D (1999) Unifying view of bifurcations in a semiconductor laser subject to optical injection. Opt Commun 172:279
34. Krauskopf B, Lenstra D (eds) (2000) Fundamental issues of nonlinear laser dynamics. AIP Conference Proceedings 548. American Institute of Physics, Melville
35. Blakely JN, Illing L, Gauthier DJ (2004) Controling fast chaos in delay dynamical systems. Phys Rev Lett 92:193901
36. Wieczorek S, Krauskopf B, Simpson T, Lenstra D (2005) The dynamical complexity of optically injected semiconductor lasers. Phys Rep 416:1
37. Erzgräber H, Krauskopf B, Lenstra D, Fischer APA, Vemuri G (2006) Frequency versus relaxation oscillations in a semiconductor laser with coherent filtered optical feedback. Phys Rev E 73:055201(R)
38. Illing L, Gauthier DJ (2006) Ultra-high-frequency chaos in a time-delay electronic device with band-limited feedback. Chaos 16:033119
39. Schikora S, Hövel P, Wünsche HJ, Schöll E, Henneberger F (2006) All-optical noninvasive control of unstable steady states in a semiconductor laser. Phys Rev Lett 97:213902
40. Green K, Krauskopf B (2006) Mode structure of a semiconductor laser subject to filtered optical feedback. Opt Commun 258:243
41. Erzgräber H, Krauskopf B, Lenstra D (2007) Bifurcation analysis of a semiconductor laser with filtered optical feedback. SIAM J Appl Dyn Syst 6:1
42. Erzgräber H, Lenstra D, Krauskopf B, Fischer APA, Vemuri G (2007) Feedback phase sensitivity of a semiconductor laser subject to filtered optical feedback: experiment and theory. Phys Rev E 76:026212
43. Erzgräber H, Krauskopf B (2007) Dynamics of a filtered-feedback laser: influence of the filter width. Opt Lett 32:2441
44. Dahms T, Hövel P, Schöll E (2007) Control of unstable steady states by extended time-delayed feedback. Phys Rev E 76:056201
45. Wünsche HJ, Schikora S, Henneberger F (2008) Noninvasive control of semiconductor lasers by delayed optical feedback. In: Schöll E, Schuster HG (eds) Handbook of chaos control. Wiley-VCH, Weinheim (second completely revised and enlarged edition)

46. Flunkert V, Schöll E (2007) Suppressing noise-induced intensity pulsations in semiconductor lasers by means of time-delayed feedback. Phys Rev E 76:066202
47. Fiedler B, Yanchuk S, Flunkert V, Hövel P, Wünsche HJ, Schöll E (2008) Delay stabilization of rotating waves near fold bifurcation and application to all-optical control of a semiconductor laser. Phys Rev E 77:066207
48. Flunkert V, D'Huys O, Danckaert J, Fischer I, Schöll E (2009) Bubbling in delay-coupled lasers. Phys Rev E 79:065201 (R)
49. Balanov AG, Janson NB, Schöll E (2005) Delayed feedback control of chaos: bifurcation analysis. Phys Rev E 71:016222
50. Hizanidis J, Aust R, Schöll E (2008) Delay-induced multistability near a global bifurcation. Int J Bifur Chaos 18:1759
51. Reznik D, Schöll E (1993) Oscillation modes, transient chaos and its control in a modulation-doped semiconductor double-heterostructure. Z Phys B 91:309
52. Schöll E, Pyragas K (1993) Tunable semiconductor oscillator based on self-control of chaos in the dynamic Hall effect. Europhys Lett 24:159
53. Schöll E, Pyragas K, Cooper D, Döttling R (1994) Tuning of semiconductor oscillators by chaos control. Semicond Sci Technol 9:559
54. Schöll E (2004) Pattern formation in semiconductors: control of spatio-temporal dynamics. Ann Phys (Leipzig) 13:403. Special topic issue edited by Friedrich R, Kuhn T, Linz S
55. Schöll E (2008) Delayed feedback control of chaotic spatio-temporal patterns in semiconductor nanostructures. In: Schöll E, Schuster HG (eds) Handbook of chaos control, Chap 24. Wiley-VCH, Weinheim, pp 533–558 (second completely revised and enlarged edition)
56. Stegemann G, Balanov AG, Schöll E (2005) Noise-induced pattern formation in a semiconductor nanostructure. Phys Rev E 71:016221
57. Stegemann G, Balanov AG, Schöll E (2006) Delayed feedback control of stochastic spatiotemporal dynamics in a resonant tunneling diode. Phys Rev E 73:016203
58. Stegemann G, Schöll E (2007) Two-dimensional spatiotemporal pattern formation in the double-barrier resonant tunneling diode. New J Phys 9:55
59. Schöll E, Majer N, Stegemann G (2008) Extended time delayed feedback control of stochastic dynamics in a resonant tunneling diode. Phys Stat Sol C 5:194
60. Hizanidis J, Balanov AG, Amann A, Schöll E (2006) Noise-induced oscillations and their control in semiconductor superlattices. Int J Bifur Chaos 16:1701
61. Hizanidis J, Balanov AG, Amann A, Schöll E (2006) Noise-induced front motion: signature of a global bifurcation. Phys Rev Lett 96:244104
62. Janson NB, Balanov AG, Schöll E (2004) Delayed feedback as a means of control of noise-induced motion. Phys Rev Lett 93:010601
63. Balanov AG, Janson NB, Schöll E (2004) Control of noise-induced oscillations by delayed feedback. Phys D 199:1
64. Pomplun J, Amann A, Schöll E (2005) Mean field approximation of time-delayed feedback control of noise-induced oscillations in the Van der Pol system. Europhys Lett 71:366
65. Janson NB, Balanov AG, Schöll E (2008) Control of noise-induced dynamics. In: Schöll E, Schuster HG (eds) Handbook of chaos control, Chap. 11. Wiley-VCH, Weinheim, pp 223–274 (second completely revised and enlarged edition)
66. Hu G, Ditzinger T, Ning CZ, Haken H (1993) Stochastic resonance without external periodic force. Phys Rev Lett 71:807
67. Pikovsky AS, Kurths J (1997) Coherence resonance in a noise-driven excitable system. Phys Rev Lett 78:775
68. García-Ojalvo J, Sancho JM (1999) Noise in spatially extended systems. Springer, New York
69. Masoller C (2002) Noise-induced resonance in delayed feedback systems. Phys Rev Lett 88:034102
70. Lindner B, García-Ojalvo J, Neiman A, Schimansky-Geier L (2004) Effects of noise in excitable systems. Phys Rep 392:321

71. Sagués F, Sancho JM, García-Ojalvo J (2007) Spatiotemporal order out of noise. Rev Mod Phys 79:829
72. Just W, Bose M, Bose S, Engel H, Schöll E (2001) Spatio-temporal dynamics near a supercritical Turing–Hopf bifurcation in a two-dimensional reaction-diffusion system. Phys Rev E 64:026219
73. Schlesner J, Zykov V, Engel H, Schöll E (2006) Stabilization of unstable rigid rotation of spiral waves in excitable media. Phys Rev E 74:046215
74. Balanov AG, Beato V, Janson NB, Engel H, Schöll E (2006) Delayed feedback control of noise-induced patterns in excitable media. Phys Rev E 74:016214
75. Dahlem MA, Schneider FM, Schöll E (2008) Efficient control of transient wave forms to prevent spreading depolarizations. J. Theor Biol 251:202
76. Stepan G (2009) Delay effects in brain dynamics. Philos Trans R Soc A 367:1059
77. Stepan G (2009) Delay effects in the human sensory system during balancing. Philos Trans R Soc A 367:1195
78. Jirsa VK (2009) Neural field dynamics with local and global connectivity and time delay. Philos Trans R Soc A 367:1131
79. Schiff SJ, Jerger K, Duong DH, Chang T, Spano ML, Ditto WL (1994) Controlling chaos in the brain. Nature (London) 370:615
80. Tass PA, Rosenblum MG, Weule J, Kurths J, Pikovsky AS, Volkmann J, Schnitzler A, Freund HJ (1998) Detection of n:m phase locking from noisy data: application to magnetoencephalography. Phys Rev Lett 81:3291
81. Grosse P, Cassidy MJ, Freund HJ (2002) EEG-EMG, MEG-EMG and EMG-EMG frequency analysis: physiological principles and clinical applications. Clin Neurophysiol 113:1523
82. Rosenblum MG, Pikovsky AS (2004) Delayed feedback control of collective synchrony: an approach to suppression of pathological brain rhythms. Phys Rev E 70:041904
83. Barnikol UB, Popovych OV, Hauptmann C, Sturm V, Freund HJ, Tass PA (2008) Tremor entrainment by patterned low-frequency stimulation. Philos Trans R Soc A 366:3545
84. Rosenblum MG, Pikovsky AS, Kurths J (2001) Synchronization—a universal concept in nonlinear sciences. Cambridge University Press, Cambridge
85. Balanov AG, Janson NB, Postnov DE, Sosnovtseva OV (2009) Synchronization: from simple to complex. Springer, Berlin
86. Newman MEJ (2003) The structure and function of complex networks. SIAM Rev 45:167
87. Coombes S, Laing C (2009) Delays in activity-based neural networks. Philos Trans R Soc A 367:1117
88. Ermentrout GB, Ko TW (2009) Delays and weakly coupled neuronal oscillators. Philos Trans R Soc A 367:1097

Chapter 2
Time-Delayed Feedback Control

Chaos is found in greatest abundance wherever order is being sought. Chaos always defeats order because it is better organized.

Terry Pratchett

In the seminal work by Ott et al. [1], they demonstrated that small time-dependent changes of a parameter in a deterministic chaotic system can lead to periodic motion. Their findings are beyond classical control theory [2–5] and opened the field of chaos control which has become an aspect of increasing interest in nonlinear science [6, 7].

An especially powerful control scheme was introduced by Pyragas [8]. It is called time-delayed feedback control or time-delay autosynchronization and constructs a control force from the difference of the present state of a given system to its delayed value, i.e., $\mathbf{s}(t) - \mathbf{s}(t - \tau)$. For proper choices of the time delay τ, the control force vanishes if the state to be stabilized is reached. Thus, the method is noninvasive. This feedback scheme is easy to implement in an experimental setup and numerical calculation. It is capable of stabilizing fixed points as well as periodic orbits even if the dynamics are very fast. Furthermore, the Pyragas scheme has no need for a reference system since it generates the control force from information of the system itself. Also from a mathematical perspective it is an appealing instrument as the corresponding equations fall in the class of delay differential equations.

This chapter provides a summary of the time-delayed feedback scheme which is investigated in the subsequent chapters of this thesis and includes basic concepts for its analysis. Thus, it can be seen as the central node in this thesis and connects the other parts, where time-delayed feedback is applied to different classes of dynamic systems. The chapter is organized as follows: In Sect. 2.1, I will introduce the general concept of time-delayed feedback control starting with the original work by Pyragas [8]. Section 2.2 is devoted to extended time-delayed feedback invented by Socolar et al. [9]. This is an extension of the Pyragas scheme which will be used frequently in the subsequent chapters. Sections 2.3 and 2.4

P. Hövel, *Control of Complex Nonlinear Systems with Delay*,
Springer Theses, DOI: 10.1007/978-3-642-14110-2_2,

cover special realizations and further extensions of time-delayed feedback control. These include different coupling schemes, control loop latency, filtering, and nonlocal feedback. Section 2.5 describes the concept of linear stability analysis in the presence of time delay. This technique will be used several times in this thesis. Section 2.6 deals with the formalism of transfer functions and provides an additional perspective on the control mechanism. Finally, Sect. 2.7 concludes this chapter with an intermediate summary.

2.1 Control Method

In this section, I will discuss the time-delayed feedback method in its original form introduced by Pyragas [8]. The focus at this point is not the application of the control scheme to a specific system, but its introduction and general properties. The application to various classes of models will be the subject of the subsequent chapters.

Consider the following, general dynamic system

$$\frac{d}{dt}\mathbf{x}(t) = \mathbf{f}(\mathbf{x}(t)), \tag{2.1.1}$$

where $\mathbf{x}$ denotes a state vector of the n-dimensional state space, i.e., $\mathbf{x} \in \mathbb{R}^n$ and $\mathbf{f}$ is a function $\mathbf{f} : \mathbb{R}^n \to \mathbb{R}^n$ with $\mathbf{f} : \mathbf{x} \mapsto \mathbf{f}(\mathbf{x})$. In the following, I will present the above mentioned control method called time-delayed feedback control in a general notation. This method is also called time-delay autosynchronization or Pyragas control [8].

Figure 2.1 depicts a schematic diagram of the time-delayed feedback loop. The control parameters are given by the time delay τ, the feedback gain K, and in an extension of the Pyragas method the memory parameter R [9]. The red (dashed) color shows this extension of the original Pyragas control including multiple delays which will be discussed in Sect. 2.2. First, let me introduce the Pyragas control scheme.

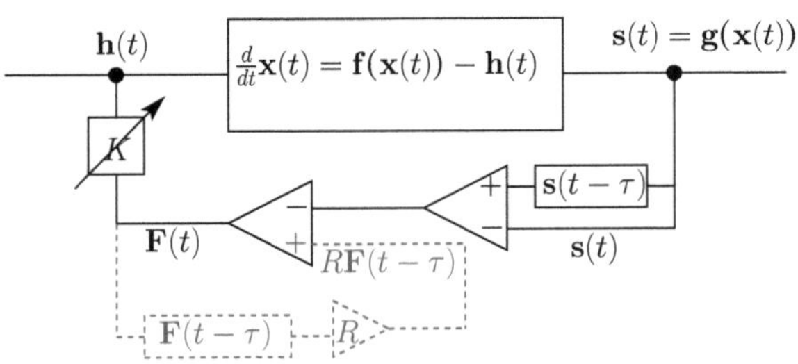

Fig. 2.1 Diagram of the time-delayed feedback control method. $\mathbf{x}(t)$ denotes the state of the system at time t, $\mathbf{s}(t)$ the control signal, i.e., some component of $\mathbf{x}(t)$ measured by $\mathbf{g}(\mathbf{x}(t))$, and $\mathbf{F}(t)$ is the control force. The real constants τ, K, and R denote the time delay, the feedback gain, and the memory parameter, respectively. The transducer function $\mathbf{h}(t)$ describes the coupling of $\mathbf{F}$ to the dynamic system $\mathbf{x}$. The extension of the original time-delayed feedback [8] as introduced by Socolar et al. (see Ref. [9]) is shown in *red* (*dashed*) color and discussed in Sect. 2.2

From the state vector $\mathbf{x} \in \mathbb{R}^n$, one can calculate a control signal $\mathbf{s} \in \mathbb{R}^m$ via a function $\mathbf{g} : \mathbb{R}^n \to \mathbb{R}^m$ with $\mathbf{g} : \mathbf{x} \mapsto \mathbf{s} = \mathbf{g}(\mathbf{x})$, which measures the state $\mathbf{x}$ to create a control signal in the m-dimensional signal space. This control signal could be, for instance, a single component of the state vector $\mathbf{x}$. The crucial part of the Pyragas control is to generate a control force $\mathbf{F}$ that consists of the difference between the current signal $\mathbf{s}(t)$ and a time-delayed counterpart $\mathbf{s}(t - \tau)$, i.e., $\mathbf{F} : \mathbb{R}^m \to \mathbb{R}^m$ with $\mathbf{F} : \mathbf{s} \mapsto \mathbf{s}(t) - \mathbf{s}(t - \tau)$. This control force is further multiplied by a control gain $K \in \mathbb{R}$. The application procedure of $\mathbf{F}$ to the dynamic system $\mathbf{x}$ is then specified by a transducer function $\mathbf{h} : \mathbb{R}^m \to \mathbb{R}^n$ with $\mathbf{h} : \mathbf{F} \mapsto \mathbf{h}(K\mathbf{F})$.

In a special realization of time-delayed feedback called diagonal control, for instance, the composition of the functions $\mathbf{g}$ and $\mathbf{h}$ becomes the identity, i.e., $\mathbf{h} \circ \mathbf{g} : \mathbb{R}^n \to \mathbb{R}^m \to \mathbb{R}^n$ with $\mathbf{h} \circ \mathbf{g} : \mathbf{x} \mapsto (\mathbf{h} \circ \mathbf{g})(\mathbf{x}) = \mathbf{h}(\mathbf{g}(\mathbf{x})) = \mathbf{x}$. Thus, for diagonal control the control force applied to the ith component of the system consists of the time-delayed difference of the same component. To summarize, the controlled system can be written as

$$\frac{d}{dt}\mathbf{x}(t) = \mathbf{f}(\mathbf{x}(t)) - \mathbf{h}[K\mathbf{F}(\mathbf{s}(t))] \tag{2.1.2a}$$

$$= \mathbf{f}(\mathbf{x}(t)) - \mathbf{h}[K(\mathbf{s}(t) - \mathbf{s}(t - \tau))] \tag{2.1.2b}$$

$$= \mathbf{f}(\mathbf{x}(t)) - \mathbf{h}[K\{\mathbf{g}(\mathbf{x}(t)) - \mathbf{g}(\mathbf{x}(t - \tau))\}]. \tag{2.1.2c}$$

Note that the difference term in the argument of $\mathbf{h}$ guarantees the noninvasive property of the control scheme which will be discussed next.

In order to investigate some properties of the control method, assume that the system $\mathbf{x}$ exhibits an unstable periodic orbit $\zeta(t)$ with period T which is meant to be stabilized via the Pyragas method. By choosing the time delay as $\tau = T$, the feedback method becomes noninvasive because the control force vanishes if the orbit ζ is stabilized: $\zeta(t) = \zeta(t - \tau) = \zeta(t - T)$. Let me stress the important feature once more: Only a minimum knowledge of the system $\mathbf{x}$ is required. The sole quantity of the system that needs to be known is the period T of the periodic orbit which determines the choice of the time delay. Methods to calculate the period of the target orbit a priori will be mentioned later in this section. Note that the exact knowledge of the unstable periodic orbit ζ is not necessary. This has the important consequence that invariant solutions of the uncontrolled system's equation persist unperturbed even in the presence of time-delayed feedback control. Only the neighborhood of the orbit is altered such that the dynamics converges to the target state.

The noninvasive property also holds if the desired state of the system is not a periodic orbit but a steady state $\mathbf{x}^*$ [10]. In the latter case, the control force also vanishes after reaching the stabilized state: $\mathbf{x}^*(t) \equiv \mathbf{x}^*(t - \tau)$. Optimal choice of the time delay are not so obvious as in the case of periodic orbits. This will be the topic of Chap. 3.

Instead of time-delayed feedback, it is tempting to use proportional feedback, where the control force $\mathbf{F}$ is given by the difference of the current state to the target state, for instance, a periodic orbit $\zeta(t)$

$$\mathbf{F}(t) = \mathbf{g}(\mathbf{x}(t)) - \mathbf{g}(\zeta(t)). \tag{2.1.3}$$

In order to apply this method, the orbit must be known a priori, e.g., reconstructed from experimental data or from a reference system, which eventually turns out to be a complicated process or is numerically expensive.

As mentioned above, the only quantity that needs to be known from the system is the period is the target orbit. There are various methods to calculate this period which then yields a promising choice of the time delay in the control loop. For instance, the feedback scheme itself can be used as a detection tool [11]. This might be interesting in an experimental setup, when a mathematical model in terms of differential equations is not at hand. In order to use the control method as a detection tool, the time delay needs to be varied. For the detection goal, the resonance behavior in terms of a vanishing control force **F** yields the desired period. Note that the occurrence of this resonance can be very sensitive with respect to small changes of the time delay [12–15].

One can also determine the period by optimization of a performance function which enables detection of unstable periodic orbits embedded in a strange attractor [16–18]. See also a comment on Ref. [16] and its reply [19, 20]. Alternatively, one can set up a few coupled equations relating the induced period, the true period, and the mismatched time delay by repeated application of an analytic approximation formula [21, 22] or by explicitly computing the unstable periodic orbit using a damped Newton solver [23].

Time-delayed feedback can also be used to explore a bifurcation diagram. Imagine a periodic which loses its stability as a bifurcation parameter is varied. This can happen, for instance, via a period-doubling cascade which subsequently leads to chaos. Without the control method, only the stable states would be visible in the bifurcation diagram. Applying time-delayed feedback could eventually stabilize a periodic orbit an area of the parameter space in which it would be unstable otherwise. This way, the periodic orbit can be tracked beyond the bifurcation point. Note, however, that the control parameters might need to be adjusted to follow the orbit in its previous unstable regime in the bifurcation parameter varies. This can be done by continuous change of the control parameters and this adjustment can even be performed automatically as was shown for periodic orbits in discrete as well as continuous systems [24, 25]. Let me stress that the target which is subject to the tracking procedure involving time-delayed feedback can also be a steady state as has been demonstrated numerically in Ref. [26] and experimental in an electrochemical system [27]. In addition, tracking by time-delayed feedback can also be used in spatio-temporal systems [28, 29].

To summarize, the main advantages of time-delayed feedback are the minimum knowledge of the investigated system and no need of a reference signal. In fact, the time-delayed feedback method generates the reference signal from the delayed time series of the system under control.

Another advantage of the control method is its easy experimental implementation. The control force can be realized, for instance, in a laser experiment all-optically by an external resonator, where the propagation of the electric field in the cavity results in a time delay [30, 31], or opto-electronically by an additional electronic delay line [32, 33]. This enables stabilization of systems with fast dynamics. An additional electronic control loop with delay can be used for the control of electronic systems such as fast diode resonators [34, 35].

The easy experimental implementation shows that time-delayed feedback is superior to other control schemes. Consider, for instance, the famous OGY method named after its inventors Ott, Grebogi, and Yorke, which triggered the field of chaos control [1]. In their method, they showed how small perturbations of accessible parameters lead to a stabilization of periodic orbits in a chaotic system and thus turn previously chaotic motion into a stable periodic behavior. These perturbations are calculated such that the trajectory is pushed towards the desired orbit once it enters a neighborhood of this state. However, the difficulty is that one needs to calculate the unstable directions of the target state which in principle can be done by delay embedding, but they involve often time-consuming calculations. See Ref. [1]. These a priori calculations are not necessary for time-delayed feedback control.

Before discussing various extensions and modifications in the following sections, let me stress again that time-delayed feedback has been successfully employed in the context of chaos control. For a recent review including both basic aspects and applications see Ref. [7]. The applications cover various fields of research ranging from chaos communication, optics, electronic systems, chemical reactions, biology, and engineering.

2.2 Extended Time-Delayed Feedback

Socolar et al. [9] introduced an extension of the Pyragas method by taking states into account which are delayed by integer multiples of a basic time delay τ. This method is known as extended time-delayed feedback control or extended time-delay autosynchronization. Calculating the difference between two states which are one time unit τ apart yields the following control force [9, 36, 37]

$$\mathbf{F}(t) = \sum_{n=0}^{\infty} R^n [\mathbf{s}(t - n\tau) - \mathbf{s}(t - (n+1)\tau)] \tag{2.2.1a}$$

$$= \left[\mathbf{s}(t) - (1 - R) \sum_{n=1}^{\infty} R^{n-1} \mathbf{s}(t - n\tau) \right] \tag{2.2.1b}$$

$$= [\mathbf{s}(t) - \mathbf{s}(t - \tau)] + R\mathbf{F}(t - \tau). \tag{2.2.1c}$$

The absolute value of the real constant R is smaller than unity, i.e., $|R| < 1$, such that it can be interpreted as a memory parameter that weights information of states further in the past. Note that the case $R = 0$ recovers the original Pyragas control scheme (2.1.2b). The equivalence of the three forms in (2.2.1a)–(2.2.1c) is shown in the following. They can be done by reordering of the infinite series. First, I consider the derivation from (2.2.1a) to (2.2.1b):

$$\mathbf{F}(t) = \sum_{n=0}^{\infty} R^n[\mathbf{s}(t-n\tau) - \mathbf{s}(t-(n+1)\tau)] \tag{2.2.2a}$$

$$= \sum_{k=0}^{\infty} R^k\mathbf{s}(t-k\tau) - \sum_{m=0}^{\infty} R^m\mathbf{s}(t-(m+1)\tau) \tag{2.2.2b}$$

$$= \sum_{k=0}^{\infty} R^k\mathbf{s}(t-k\tau) - \sum_{m=1}^{\infty} R^{m-1}\mathbf{s}(t-m\tau) \tag{2.2.2c}$$

$$= \mathbf{s}(t) + \sum_{k=1}^{\infty} R^k\mathbf{s}(t-k\tau) - \sum_{m=1}^{\infty} R^{m-1}\mathbf{s}(t-m\tau) \tag{2.2.2d}$$

$$= \mathbf{s}(t) - \sum_{m=1}^{\infty} R^{m-1}\mathbf{s}(t-m\tau) - (-R)\sum_{k=1}^{\infty} R^{k-1}\mathbf{s}(t-k\tau) \tag{2.2.2e}$$

$$= \mathbf{s}(t) - (1-R)\sum_{n=1}^{\infty} R^{n-1}\mathbf{s}(t-n\tau). \tag{2.2.2f}$$

A similar derivation yields the third recursive form (2.2.1c) of the extended time-delayed feedback control force:

$$\mathbf{F}(t) = \sum_{n=0}^{\infty} R^n[\mathbf{s}(t-n\tau) - \mathbf{s}(t-(n+1)\tau)] \tag{2.2.3a}$$

$$= \underbrace{[\mathbf{s}(t) - \mathbf{s}(t-\tau)] - [\mathbf{s}(t) - \mathbf{s}(t-\tau)]}_{=0} \tag{2.2.3b}$$

$$+ \sum_{n=0}^{\infty} R^n[\mathbf{s}(t-n\tau) - \mathbf{s}(t-(n+1)\tau)] \tag{2.2.3c}$$

$$= \mathbf{s}(t) - \mathbf{s}(t-\tau) + \sum_{n=1}^{\infty} R^n[\mathbf{s}(t-n\tau) - \mathbf{s}(t-(n+1)\tau)] \tag{2.2.3d}$$

$$= \mathbf{s}(t) - \mathbf{s}(t-\tau) + \sum_{n=0}^{\infty} R^{n+1}[\mathbf{s}(t-(n+1)\tau) - \mathbf{s}(t-(n+2)\tau)]$$

$$= [\mathbf{s}(t) - \mathbf{s}(t-\tau)] + R\underbrace{\sum_{n=0}^{\infty} R^n[\mathbf{s}(t-n\tau-\tau) - \mathbf{s}(t-(n+1)\tau-\tau)]}_{=\mathbf{F}(t-\tau)} \tag{2.2.3e}$$

$$= [\mathbf{s}(t) - \mathbf{s}(t-\tau)] + R\mathbf{F}(t-\tau).$$

Although the first form of the extended time-delayed feedback force (2.2.1a) can be seen as an analogy of the Pyragas control, this form is not feasible for numerical implementation because it requires to store information of all states in the past. However, there is also the equivalent recursive form given in (2.2.1c) which involves next to the time-delayed control signal $\mathbf{s}(t-\tau)$ the delayed version of the control force $\mathbf{F}(t-\tau)$ itself. This form becomes more suitable for an experiment. The extension of the original time-delayed feedback scheme is depicted schematically in Fig. 2.1, where the red dashed color displays the additional recursion component according to (2.2.1c). In an all-optical experimental setup, the feedback scheme in its extended form can be realized by a Fabry–Perot resonator as will be discussed in Sect. 3.5.3.

It is worth noting that, similar to time-delayed feedback, the extended version possesses the noninvasive property. The control force vanishes if the target state, e.g., periodic orbit or steady state, is stabilized. Thus, the target states are invariant solutions of the uncontrolled system which persist unperturbed in the presence of (extended) time-delayed feedback. The delayed feedback can also induce additional solutions which are not solutions of the uncontrolled system. For these delay-induced states, the control scheme is invasive and the control force does not vanish. They are important if the corresponding modes become dominant and are used in Chap. 4 in an exchange of stability with the orbit to be stabilized.

So far, nothing was said about the specific choice of the coupling function $\mathbf{g}$ which extracts information of the system to generate a control signal and the transducer function $\mathbf{h}$ which determines the realization of the feedback scheme. This topic will be discussed in the next section.

2.3 Coupling Schemes

In the framework of, for instance, neural systems of 2-variable activator–inhibitor type as it will be investigated in Chap. 6, it is of crucial importance to carefully distinguish between different coupling schemes. First of all, there is a difference by generating the control signal form the inhibitor or the activator variable. In addition, the application of the control force to the system leaves again the choice of coupling to the activator or the inhibitor. Depending on the specific realization of the coupling, one can expect different responses of the system to the control scheme [38, 39].

In order to discuss the roles of the coupling function $\mathbf{g}$ and the transducer function $\mathbf{h}$, it is convenient to rewrite (2.1.2b) of the controlled system in vector form:

$$\frac{d}{dt}\mathbf{x}(t) = \mathbf{f}(\mathbf{x}(t)) - \mathbf{h}[K(\mathbf{s}(t) - \mathbf{s}(t-\tau))] \qquad (2.3.1a)$$

$$\begin{pmatrix} \dot{x}_1(t) \\ \vdots \\ \dot{x}_n(t) \end{pmatrix} = \begin{pmatrix} f_1(x_1(t),\ldots,x_n(t)) \\ \vdots \\ f_n(x_1(t),\ldots,x_n(t)) \end{pmatrix} - \begin{pmatrix} h_1[K(s_1(t)-s_1(t-\tau)),\ldots,K(s_m(t)-s_m(t-\tau))] \\ \vdots \\ h_n[K(s_1(t)-s_1(t-\tau)),\ldots,K(s_m(t)-s_m(t-\tau))] \end{pmatrix}, \tag{2.3.1b}$$

where $x_i(t)$ denotes the ith component of the state vector $\mathbf{x}(t) = (x_i(t),\ldots,x_i(t))^T$ and similar notations hold for **f**, **g**, **h**, and **s**, where the function **g** is given by

$$\mathbf{g}(\mathbf{x}(t)) = \begin{pmatrix} g_1(x_1(t),\ldots,x_n(t)) \\ \vdots \\ g_m(x_1(t),\ldots,x_n(t)) \end{pmatrix} \tag{2.3.2}$$

with $g_i : \mathbb{R}^m \to \mathbb{R},\ g_i : x_1,\ldots,x_n \mapsto s_i = g_i(x_1,\ldots,x_n)$ for each vector element $i = 1, \ldots, m$. A similar notation can be applied to Eqs. 2.1.2a and 2.1.2c as well. For a schematic diagram of the control method see Fig. 2.1. This notation seems lengthy, but it can be shortened assuming that **g** and **h** are linear functions which can be written as matrices with proper dimensions, i.e., **g** becomes a $n \times m$ matrix and **h** turns into a $m \times n$ matrix. Then, (2.3.1b) can be rewritten as

$$\begin{pmatrix} \dot{x}_1(t) \\ \vdots \\ \dot{x}_n(t) \end{pmatrix} = \begin{pmatrix} f_1(x_1(t),\ldots,x_n(t)) \\ \vdots \\ f_n(x_1(t),\ldots,x_n(t)) \end{pmatrix} - K \begin{pmatrix} h_{11} & \cdots & h_{1m} \\ \vdots & \ddots & \vdots \\ h_{n1} & \cdots & h_{nm} \end{pmatrix} \times \begin{pmatrix} g_{11} & \cdots & g_{1n} \\ \vdots & \ddots & \vdots \\ g_{m1} & \cdots & g_{mn} \end{pmatrix} \begin{pmatrix} x_1(t)-x_1(t-\tau) \\ \vdots \\ x_n(t)-x_n(t-\tau) \end{pmatrix} \tag{2.3.3a}$$

such that the equation for the ith component of the state vector **x** becomes

$$\frac{dx_i(t)}{dt} = f_i(x_1(t),\ldots,x_n(t)) - K\sum_{j=1}^{n} \mathbf{A}_{ij}[x_j(t) - x_j(t-\tau)] \tag{2.3.4a}$$

$$\frac{d}{dt}\mathbf{x}(t) = \mathbf{f}(\mathbf{x}(t)) - K\mathbf{A}[\mathbf{x}(t) - \mathbf{x}(t-\tau)], \tag{2.3.4b}$$

where the elements of the $n \times n$ coupling matrix **A** are given by

$$\mathbf{A}_{ij} = \sum_{k=1}^{m} h^{ik} g_{kj}. \tag{2.3.5}$$

This matrix selects which components of **x** are used for construction of the control signal and specifies the application of the control force back to the system.

The function $\mathbf{g} = (g_1, \ldots, g_m)^T$ determines which components of the system **x** are measured to generate the difference to the time-delayed signal. In the simplest case, g_i extracts the ith component of **x**:

$$g_i(x_1(t), \ldots, x_j(t), \ldots, x_n(t)) = x_i(t)\delta_{i,j} \qquad \forall i,j = 1, \ldots, n. \tag{2.3.6}$$

If $h_i = KF_i$ also selects the ith component, then this coupling scheme is called diagonal.

The control signal $\mathbf{s} = (s_1, \ldots, s_m)^T$ can also contain global information of the system. In this case of global control, **g** yields, for instance, the calculation of the average of the system components or mean field:

$$g_i(x_1(t), \ldots, x_n(t)) = \frac{1}{n}\sum_{k=1}^{n} x_k(t) \qquad \forall i = 1, \ldots, n. \tag{2.3.7}$$

This kind of control signal is often used in the context of networks, where a measurement of a single node is not possible. Global delayed feedback is applied, for instance, to coupled phase oscillators of Kuramoto type [40, 41], Hindmarsh-Rose neurons [42], and limit cycle oscillators [43, 44]. It is also of importance for the class of spatio-temporal systems, when a spatial variable is not accessible, but an overall current turns out to be a convenient choice for the control signal [15, 28, 45–56]. In these spatio-temporal systems, the coupling function **g** involves a spatial average which can be realized by an integral, for instance, in one-dimensional systems:

$$\mathbf{g}(\mathbf{x}(x,t)) = \int_0^L \mathbf{x}(x,t)\, dx, \tag{2.3.8}$$

where L denotes a constant length which determines the spatial dimension of the system. Different local and global coupling schemes were systematically compared in Refs. [28, 45].

The function **g** can also act as a differential operator if the derivative of **x** is accessible in an experiment. In Chap. 5, for instance, the second derivative enters the control signal and thus, **g** reads:

$$g_i(x_1(t), \ldots, x_j(t), \ldots, x_n(t)) = \ddot{x}_i(t)\delta_{i,j} \qquad \forall i,j = 1, \ldots, n. \tag{2.3.9}$$

Other choices of **g** will be discussed later in this section and in the subsequent chapters for various classes of dynamic systems.

While the function **g** specifies the generation of the control signal **s**, the transducer function $\mathbf{h} = (h_1, \ldots, h_n)^T$ with $h_i : \mathbb{R}^m \to \mathbb{R},\ \ h_i : s_1, \ldots, s_m \mapsto h_i(s_1, \ldots, s_m)$ ($i = 1, \ldots, m$) determines the component of the system to which the control force is applied. In optical system, some components of the system's state **x** describes the electric field. In this context, **h** takes into account changes of the polarization axis. This will be discussed in detail in Sect. 3.5, where an additional phase parameter comes into play [30, 57, 58].

In the field of networks, when $x_i(t)$ describes the dynamics of the ith node and can be understood as a vector quantity itself, prominent choices of the function **h** include all-to-all coupling and nearest-neighbor coupling. In the first case, all components of **h** are identical. In the latter case, **h** connects only nodes which are next to each other in the network. In a one-dimensional ring configuration, for instance, the function **h** can be written as

$$h_i[K\mathbf{F}(\mathbf{s}(t))] = K(s_j(t) - s_j(t-\tau))\big(\delta_{i-1,j} + \delta_{i+1,j}\big), \qquad (2.3.10)$$

where s_{n+1} and s_0 are identified with s_1 and s_n by periodic boundary condition, respectively. In terms of a coupling matrix **A** as in Eqs. 2.3.4, Equation 2.3.10 corresponds to a matrix with only secondary diagonal entries, i.e., $a_{i,i\pm 1} = 1$ for all $i = 1, ..., n$. The two choices of all-to-all coupling and nearest-neighbor coupling serve also as paradigmatic case for global and local control.

In the linear case of **g** and **h**, the two functions could be merged into a single control function. If **h** becomes nonlinear, however, this is not possible anymore because of the noninvasive property and hence, the main advantage of time-delayed feedback would be lost.

Before turning towards extensions of time-delayed feedback, let me discuss further examples of **g** and **h**. Next to the already mentioned case, where g_i extracts the ith component of the system, **g** can also be used to construct the control signal from a single component only, i.e., **g** is given for fixed $i \in 1, ..., n$ by

$$g_i(x_1(t), \ldots, x_j(t), \ldots, x_n(t)) = x_j(t) \quad \text{and} \quad g_k \equiv 0 \quad \text{for} \quad k \neq i. \qquad (2.3.11)$$

As an example of this choice, see Ref. [59], where the chaotic Rössler system (with real constants a, b, and μ) is subject to time-delayed feedback in the following realization

$$\begin{pmatrix} \dot{x}_1(t) \\ \dot{x}_2(t) \\ \dot{x}_3(t) \end{pmatrix} = \begin{pmatrix} -x_2(t) - x_3(t) \\ x_1(t) + a x_2(t) \\ b + x_3(t)(x_1(t) - \mu) \end{pmatrix} - K \underbrace{\begin{pmatrix} 1 & 0 & 0 \\ 0 & 0 & 0 \\ 0 & 0 & 0 \end{pmatrix}}_{=\mathbf{A}} \begin{pmatrix} x_1(t) - x_1(t-\tau) \\ x_2(t) - x_2(t-\tau) \\ x_3(t) - x_3(t-\tau) \end{pmatrix}. \qquad (2.3.12)$$

One can see the composition $\mathbf{h} \circ \mathbf{g}$ yields a 3×3 matrix **A** with only one non-zero element.

Another special realization that will become important and act as a reference case in the following chapters is called diagonal control. In this choice, the control force applied to the ith component of the system consists only of information of the same component and thus, the composition of the functions $\mathbf{g}$ and $\mathbf{h}$, i.e., the coupling matrix $\mathbf{A}$, is the identity. Then, the controlled system can be simplified to

$$\frac{d}{dt}\mathbf{x}(t) = \mathbf{f}(\mathbf{x}(t)) - K\mathbf{A}[\mathbf{x}(t) - \mathbf{x}(t-\tau)] \tag{2.3.13a}$$

$$\Leftrightarrow \begin{pmatrix} \dot{x}_1(t) \\ \vdots \\ \dot{x}_n(t) \end{pmatrix} = \begin{pmatrix} f_1(x_1(t),\ldots,x_n(t)) \\ \vdots \\ f_n(x_1(t),\ldots,x_n(t)) \end{pmatrix} - K \begin{pmatrix} 1 & & \mathbf{0} \\ & \ddots & \\ \mathbf{0} & & 1 \end{pmatrix} \begin{pmatrix} x_1(t) - x_1(t-\tau) \\ \vdots \\ x_n(t) - x_n(t-\tau) \end{pmatrix} \tag{2.3.13b}$$

$$= \begin{pmatrix} f_1(x_1(t),\ldots,x_n(t)) \\ \vdots \\ f_n(x_1(t),\ldots,x_n(t)) \end{pmatrix} - K \begin{pmatrix} x_1(t) - x_1(t-\tau) \\ \vdots \\ x_n(t) - x_n(t-\tau) \end{pmatrix}.$$

The example of the Rössler system given above can be seen as a simplification of diagonal control because only a single diagonal entry of the coupling matrix is non-zero.

In optical systems as mentioned above, the polarization leads to a phase-dependent coupling. This can be realized by the introduction of a control phase φ and the controlled system is then given by

$$\begin{pmatrix} \dot{x}_1(t) \\ \dot{x}_2(t) \\ \vdots \\ \dot{x}_n(t) \end{pmatrix} = \begin{pmatrix} f_1(x_1(t),\ldots,x_n(t)) \\ f_2(x_1(t),\ldots,x_n(t)) \\ \vdots \\ f_n(x_1(t),\ldots,x_n(t)) \end{pmatrix} - K \begin{pmatrix} \cos\varphi & \sin\varphi & \mathbf{0} & \ldots & \mathbf{0} \\ -\sin\varphi & \cos\varphi & \mathbf{0} & \ddots & \vdots \\ \mathbf{0} & \mathbf{0} & \mathbf{0} & \ddots & \vdots \\ \vdots & \ddots & \ddots & \ddots & \vdots \\ \mathbf{0} & \ldots & \ldots & \ldots & \mathbf{0} \end{pmatrix} \begin{pmatrix} x_1(t) - x_1(t-\tau) \\ x_2(t) - x_2(t-\tau) \\ \vdots \\ x_n(t) - x_n(t-\tau) \end{pmatrix}, \tag{2.3.14}$$

where x_1 and x_2 correspond to the real and imaginary parts of the electric field, respectively [30, 57, 58]. This will be investigated in detail in Sect. 3.5. Note that the choice $\varphi = 0$ recovers the diagonal control.

After the discussion of different realizations of the time-delayed feedback scheme, I will add some remarks on further extensions of this control method in the next section.

2.4 Extensions

In this section, I will summarize some aspects concerning possible modifications of the original time-delayed feedback control scheme as introduced in Sect. 2.1 and given by (2.1.2). These extensions include, without claim of completeness latency time effects, filtering of the control signal, multiple time delays, and nonlocal coupling.

Previously, I have assumed that the generation of the feedback and its application to the system under control happens instantaneously. In an experimental setup, however, there is always a latency involved. This control loop latency is associated with finite propagation speed of the control signal, delays in the measurement of the system to generate the control signal, or processing times for the calculation of the feedback [10, 57, 58, 60, 61]. In optical or opto-electronic systems, for instance, the length of an optical fiber used to transmit the control signal can become of crucial importance [35]. The same holds for electronic systems such as a fast diode resonator, where also a propagation delay act as limiting factor [34]. In neuronal systems, these propagation delays, e.g., in dendrites, play also an important role both in models and experiments [62, 63].

All latency times of different origins can be summed up to a new control parameter δ which acts as an additional time delay in all arguments of the control loop. Therefore, the controlled system of (2.1.2) can be rewritten as

$$\frac{d}{dt}\mathbf{x}(t) = \mathbf{f}(\mathbf{x}(t)) - \mathbf{h}[K\mathbf{F}(\mathbf{s}(t-\delta))] \tag{2.4.1a}$$

$$= \mathbf{f}(\mathbf{x}(t)) - \mathbf{h}[K(\mathbf{s}(t-\delta) - \mathbf{s}(t-\delta-\tau))] \tag{2.4.1b}$$

$$= \mathbf{f}(\mathbf{x}(t)) - \mathbf{h}[K\{\mathbf{g}(\mathbf{x}(t-\delta)) - \mathbf{g}(\mathbf{x}(t-\delta-\tau))\}]. \tag{2.4.1c}$$

Note that the problem of latency cannot be solved by the introduction of a rescaled time delays $\tilde{\tau}$, i.e., $\tilde{\tau} = \delta + \tau$, because the quantity δ occurs also in the argument of both $\mathbf{s}(t - \delta)$ and $\mathbf{s}(t - \delta - \tau)$. This issue will be elaborated in Sect. 3.4 in the context of stabilization of steady states. I will derive an analytical expression of upper bound for δ, i.e., a maximum latency time, such that control is still possible [10, 57, 58].

Another issue that needs to be taken into account is additional filter in the control loop. On the one hand, the reason can be a limited bandwidth of the experimental equipment such that filtering is unavoidable. On the other hand, filters can be built into the control loop on purpose. In the latter case, the aim is to reduce the influence of unwanted high frequencies in the control force which eventually lead to the stabilization of the wrong timescales. If, for instance, these high frequencies are present in the system and yield generation of a feedback with the same fast timescale, a low-pass filter can help to overcome this limitation [13, 14, 49]. To adjust the time-delayed feedback scheme to this obstacle, the control signal $\mathbf{s}$ in (2.1.2) needs to be replaced by a low-pass filtered version $\bar{\mathbf{s}}$ and the controlled system reads:

$$\frac{d}{dt}\mathbf{x}(t) = \mathbf{f}(\mathbf{x}(t)) - \mathbf{h}[K\mathbf{F}(\bar{\mathbf{s}}(t))] \quad (2.4.2a)$$

$$= \mathbf{f}(\mathbf{x}(t)) - \mathbf{h}[K(\bar{\mathbf{s}}(t) - \bar{\mathbf{s}}(t-\tau))], \quad (2.4.2b)$$

where the low-pass filter can be described in different ways. On one hand, it can be realized by an additional differential equation as follows:

$$\frac{d}{dt}\bar{\mathbf{s}}(t) = -\alpha\bar{\mathbf{s}}(t) + \alpha\mathbf{s}(t), \quad (2.4.3)$$

where α denotes the cutoff frequency of the filter. On the other hand, there exists an equivalent integral formula for $\bar{\mathbf{s}}$:

$$\bar{\mathbf{s}}(t) = \alpha \int_{-\infty}^{t} \mathbf{s}(t')e^{-\alpha(t-t')}\,dt'. \quad (2.4.4)$$

The equivalence can be seen by straight forward calculation inserting (2.4.4) into (2.4.3)

$$\frac{d}{dt}\bar{\mathbf{s}}(t) = \frac{d}{dt}\alpha \int_{-\infty}^{t} \mathbf{s}(t')e^{-\alpha(t-t')}\,dt' \quad (2.4.5a)$$

$$= \left[\frac{d}{dt}e^{-\alpha t}\right]\alpha \int_{-\infty}^{t} \mathbf{s}(t')e^{\alpha t'}\,dt' + e^{-\alpha t}\alpha\frac{d}{dt}\left[\int_{-\infty}^{t} \mathbf{s}(t')e^{\alpha t'}\,dt'\right]$$

$$= \underbrace{-\alpha e^{-\alpha t}\alpha \int_{-\infty}^{t} \mathbf{s}(t')e^{\alpha t'}\,dt'}_{\bar{\mathbf{s}}(t)} + \underbrace{e^{-\alpha t}\alpha\mathbf{s}(t)e^{\alpha t}}_{\alpha\mathbf{s}(t)} \quad (2.4.5b)$$

$$= -\alpha\bar{\mathbf{s}}(t) + \alpha\mathbf{s}(t). \quad (2.4.5c)$$

Alternatively, the solution of the linear, inhomogeneous, differential equation (2.4.3) can be written as

$$\bar{\mathbf{s}}(t) = \int_{-\infty}^{\infty} G(t-t')\mathbf{s}(t')\,dt' \tag{2.4.6}$$

with the Green's function

$$G(t) = \begin{cases} \alpha e^{-\alpha t}, & t \geq 0 \\ 0, & t < 0 \end{cases}, \tag{2.4.7}$$

which yields the integral formula (2.4.4). This issue will be become important again in Sect. 2.6, where a transfer function of the low-pass filter will be derived.

Similar to a low-pass filter, one can think of the introduction of a high-pass filter or a bandpass filter into the control loop [64, 65]. In optical systems, filtering of the feedback signal was also investigated for laser models of Lang–Kobayashi type [66, 67]. Furthermore, the introduction of a filtered feedback has also been proven to be important in the investigation of a Hopf bifurcation [33].

If there are multiple timescales in the system that need to be controlled, the introduction of additional control loops with different time delays τ_1, τ_2, ... can be a proper extension. In this case, there will be a complex interplay between the different timescales [68–71].

A second time delay needs also to be taken into account, if it is already part of the dynamics of the uncontrolled system:

$$\frac{d}{dt}\mathbf{x}(t) = \mathbf{f}(\mathbf{x}(t), \mathbf{x}(t-\tau_1)). \tag{2.4.8}$$

In this case, the delay τ_2 of the time-delayed feedback scheme can be chosen independently:

$$\frac{d}{dt}\mathbf{x}(t) = \mathbf{f}(\mathbf{x}(t), \mathbf{x}(t-\tau_1)) - \mathbf{h}[K(\mathbf{s}(t) - \mathbf{s}(t-\tau_2))], \tag{2.4.9a}$$

An example, where the intrinsic time delay occurs in the highest derivative, will be considered in Chap. 5. See also Refs [72, 73].

It is also of importance to consider the limits of small and large time delays. In the limit of vanishing τ and assuming the feedback gain is of the order τ^{-1}, i.e., $K = O(1/\tau)$, the time-delayed feedback force becomes

$$\lim_{\tau\to 0} K\mathbf{F}(\mathbf{s}(t)) = \lim_{\tau\to 0} K[\mathbf{s}(t) - \mathbf{s}(t-\tau)] \tag{2.4.10a}$$

$$= \tilde{K} \lim_{\tau\to 0} \frac{\mathbf{s}(t) - \mathbf{s}(t-\tau)}{\tau} \tag{2.4.10b}$$

$$= \tilde{K}\frac{d\mathbf{s}(t)}{dt} \tag{2.4.10c}$$

$$= \tilde{K}\frac{d\mathbf{g}(\mathbf{x}(t))}{dt}. \tag{2.4.10d}$$

Thus, the control force is proportional to the derivative of the control signal and the time-delayed feedback methods is equal to derivative control [74–76]. Derivative control has a disadvantage that it is sensitive to high frequencies. This becomes clear from the Fourier transform of (2.4.10c):

$$\mathcal{F}\left[\frac{d\mathbf{s}(t)}{dt}\right] = \int_{-\infty}^{\infty} \frac{d\mathbf{s}(t)}{dt} e^{-i\omega t}\, dt \tag{2.4.11a}$$

$$= \mathbf{s}(t)e^{-i\omega t}\big|_{-\infty}^{\infty} - \int_{-\infty}^{\infty} \mathbf{s}(t)\frac{d}{dt}e^{-i\omega t}\, dt \tag{2.4.11b}$$

$$= i\omega \int_{-\infty}^{\infty} \mathbf{s}(t)e^{-i\omega t}\, dt \tag{2.4.11c}$$

$$= i\omega\mathcal{F}[\mathbf{s}(t)] \tag{2.4.11d}$$

$$= i\omega\hat{\mathbf{s}}(\omega), \tag{2.4.11e}$$

where $\hat{\mathbf{s}}$ denotes the Fourier transform of the control signal $\mathbf{s}$. For instance in the presence of noise including high frequencies ω, this could lead to an arbitrarily large control force.

The second limit of large time delays, i.e., $\tau \to \infty$, will be discussed in Sect. 3.6 in the context of stabilization of steady states, when I will investigate the asymptotic behavior of the eigenvalues in the presence of time-delayed feedback.

I want to mention at last an extension to systems that have also spatial degrees of freedom. In these spatio-temporal systems, the feedback scheme can still be used in the previously discussed time-delayed form. However, it is also possible to use the spatial coordinate and a space delay Δx to generate a nonlocal control force. Restricting the discussion for notational convenience to systems $\mathbf{x}(x, t)$ with one spatial dimension, the control force $\mathbf{F}(x, t)$ can be written in analogy to the time-delayed case of (2.1.2c)

$$\mathbf{F}(x,t) = \mathbf{s}(x,t) - \mathbf{s}(x - \Delta x, t) \tag{2.4.12a}$$

$$= \mathbf{g}(\mathbf{x}(x,t)) - \mathbf{g}(\mathbf{x}(x - \Delta x, t)), \tag{2.4.12b}$$

where Δx denotes the spatial delay, i.e., the distance of the nonlocal coupling. This method was used for stabilization of unstable periodic pattern in spatio-temporal chaos in optical systems [77] as well as in neuronal systems [38, 78–80].

Opposed to the purely time-dependent case, where only information of the past is accessible, one can introduce in spatio-temporal systems a feedback which is symmetric in space

$$\mathbf{F}(x,t) = -\mathbf{s}(x + \Delta x, t) + 2\mathbf{s}(x,t) - \mathbf{s}(x - \Delta x, t) \tag{2.4.13a}$$

$$= -\mathbf{g}(\mathbf{x}(x - \Delta x, t)) + 2\mathbf{g}(\mathbf{x}(x, t)) - \mathbf{g}(\mathbf{x}(x - \Delta x, t)). \qquad (2.4.13b)$$

Further variations of the space-delayed feedback are possible. In general form, all different realization can be summarized using an integral kernel $\kappa(x')$ which specifies the coupling. Then, the control force can be written in a one-dimensional system as

$$\mathbf{F}(x,t) = \int_0^\infty \kappa(x')[\mathbf{s}(x,t) - \mathbf{s}(x + x', t)]\, dx' \qquad (2.4.14a)$$

$$= \int_0^\infty \kappa(x')[\mathbf{g}(\mathbf{x}(x,t)) - \mathbf{g}(\mathbf{x}(x + x', t))]\, dx'. \qquad (2.4.14b)$$

As examples, consider the control forces given by (2.4.12a) and (2.4.13a). In the first case, the kernel is a δ-function which is shifted by Δx, i.e., $\kappa(x') = \delta(x' + \Delta x)$. In the latter case, the kernel consists similarly of two δ-functions: $\kappa(x') = \delta(x' + \Delta x) + \delta(x' - \Delta x)$.

The integral kernel provides information of the structure, e.g., the symmetry, of the spatial coupling. If, for instance, a rotational symmetry is present in the coupling of a two-dimensional system, the control force is given by

$$\mathbf{F}(x,y,t) = \int_0^{2\pi}\int_0^\infty \kappa(r)[\mathbf{s}(x,y,t) - \mathbf{s}(x + r\cos\varphi, y + r\sin\varphi, t)]\, dr d\varphi$$
$$= \int_0^{2\pi}\int_0^\infty \kappa(r)[\mathbf{g}(\mathbf{x}(x,y,t)) - \mathbf{g}(\mathbf{x}(x + r\cos\varphi, y + r\sin\varphi, t))]\, dr d\varphi, \qquad (2.4.15)$$

where r and φ denote the polar coordinates and the kernel $\kappa(r)$ depends only on the radius $r = \sqrt{x^2 + y^2}$. If the kernel exhibits a minimum at $r_0 > 0$, the coupling is known as Mexican-hat potential which occurs, for instance, in the context of neural models realized as a two-dimensional reaction-diffusion system [78, 80]. In this context, the terminology Mexican-hat refers to the shape of the local connectivity network in the cortical tissue.

It is of course possible to combine all above mentioned extensions. For instance, additional filter in the feedback scheme can result in a non-zero control loop latency which arise from the time needed to constructs the filtered signal. In the framework of spatio-temporal dynamics, space delays and time delays can be mixed since nonlocal signals can be time-delayed [78].

To conclude this section, I will discuss an extension to an adaptive feedback controller to find a value of the feedback gain which yields successful stabilization [5]. The basic idea is to regard the feedback gain K as an additional dynamic variable. Therefore, one more differential equation must be added to the system to account for the temporal dynamics of K. In order to derive this additional equation, one can consider a state-dependent goal or error function Q given by [81]

$$Q(x(t), x(t-\tau)) = \frac{1}{2}[x(t) - x(t-\tau)]^2, \tag{2.4.16}$$

which vanishes if the system is stabilized. Substituting the right-hand side of the system's equation with control into the time derivative of Q yields the desired equation by the following relation

$$\frac{d}{dt}K(t) = -\gamma\frac{\partial}{\partial K}\left(\frac{dQ}{dt}\right), \tag{2.4.17}$$

where the parameter $\gamma > 0$ denotes the adaptation gain. For details see Ref. [81]. It has been shown that unstable steady states of focus type can be stabilized by this method which automatically chooses a suitable value for the feedback gain [81]. The adaptive control scheme can also be used to track periodic orbits in chaotic systems [24, 82].

The proposed adaptive algorithm is also known as gradient method [83–85]. A similar algorithm was used for adaptive synchronization of chaotic systems [86] and the control of a steady state in the Lorenz system by proportional control [87].

2.5 Linear Stability Analysis

In this section, I will elaborate the concept of linear stability analysis for systems subject to time-delayed feedback which will lead to a so-called characteristic equations. This technique will be used several times in the subsequent chapters.

Consider a small deviation $\delta\mathbf{x}(t)$ from a target state $\mathbf{x}^*(t)$, e.g., a steady state or a periodic orbit: $\delta\mathbf{x}(t) = \mathbf{x}(t) - \mathbf{x}^*(t)$. A linearization of the system equation including time-delayed feedback as given in (2.1.2) yields up to linear order in $\delta\mathbf{x}(t)$

$$\frac{d}{dt}\mathbf{x}(t) = \mathbf{f}(\mathbf{x}(t)) - \mathbf{h}[K\{\mathbf{g}(\mathbf{x}(t)) - \mathbf{g}(\mathbf{x}(t-\tau))\}] \tag{2.5.1a}$$

$$\begin{aligned}
&= \mathbf{f}(\mathbf{x}^*(t)) - \mathbf{h}[K\{\mathbf{g}(\mathbf{x}^*(t)) - \mathbf{g}(\mathbf{x}^*(t-\tau))\}] \\
&\quad + \underbrace{\left.\frac{\partial\mathbf{f}(\mathbf{x}(t))}{\partial\mathbf{x}(t)}\right|_{\mathbf{x}(t)=\mathbf{x}^*(t)}}_{\mathrm{I}}[\mathbf{x}(t) - \mathbf{x}^*(t)] \\
&\quad - \underbrace{\left.\frac{\partial\mathbf{h}[K\mathbf{g}(\mathbf{x}(t))]}{\partial K\mathbf{g}(\mathbf{x}(t))}\frac{\partial K\mathbf{g}(\mathbf{x}(t))}{\partial\mathbf{x}(t)}\right|_{\mathbf{x}(t)=\mathbf{x}^*(t)}}_{\mathrm{II}}[\mathbf{x}(t) - \mathbf{x}^*(t)] \\
&\quad + \overbrace{\left.\frac{\partial\mathbf{h}[K\mathbf{g}(\mathbf{x}(t))]}{\partial K\mathbf{g}(\mathbf{x}(t))}\frac{\partial K\mathbf{g}(\mathbf{x}(t))}{\partial\mathbf{x}(t)}\right|_{\mathbf{x}(t)=\mathbf{x}^*(t)}}[\mathbf{x}(t-\tau) - \mathbf{x}^*(t-\tau)].
\end{aligned} \tag{2.5.1b}$$

The first term $\mathbf{f}(\mathbf{x}^*(t))$ gives the derivative of the invariant solution $\mathbf{x}^*(t)$. The second term vanishes due to the noninvasive property of the control scheme. The term labelled as I represents the Jacobian matrix of $\mathbf{f}$ evaluated at $\mathbf{x}^*(t)$ and the matrix II can be abbreviated as $\mathbf{B}(t)$ summarizing the linearized control terms. This leads to an equivalent expression

$$\frac{d}{dt}\delta\mathbf{x}(t) = \mathbf{A}(t)\delta\mathbf{x}(t) - \mathbf{B}(t)[\delta\mathbf{x}(t) - \delta\mathbf{x}(t-\tau)], \tag{2.5.2}$$

where $\mathbf{A}(t)$ denotes the Jacobian matrix of the uncontrolled system. If the target state is a fixed point, i.e., $\mathbf{x}(t) \equiv \mathbf{x}^*$, the matrices $\mathbf{A}$ and $\mathbf{B}$ are time independent. Thus, an exponential ansatz for $\delta\mathbf{x}$ leads to a characteristic equation whose roots determine the stability.

If the linear stability analysis is performed at a periodic orbit with period τ, i.e., $\mathbf{x}(t) = \mathbf{x}^*(t) = \mathbf{x}^*(t-\tau)$, both matrices are periodic with that same period and Floquet theory will guarantees that solutions $\delta\mathbf{x}(t)$ of (2.5.2) can be decomposed into Floquet modes

$$\delta\mathbf{x}(t) = \sum_{j=0}^{\infty} c_j e^{\Lambda_j t}\mathbf{u}_j(t), \tag{2.5.3}$$

where $\mathbf{u}_j(t)$ and Λ_j denote the jth Floquet mode and corresponding complex Floquet exponent, respectively [12, 28, 45, 46, 61, 88–91]. Note that the Floquet modes are periodic with period τ, i.e., $\mathbf{u}_j(t) = \mathbf{u}_j(t+\tau)$. Inserting this decomposition into (2.5.2) yields

$$\Lambda\mathbf{u}(t) + \frac{d}{dt}\mathbf{u}(t) = \mathbf{A}(t)\mathbf{u}(t) - \mathbf{B}(t)\left[\mathbf{u}(t) - e^{-\Lambda\tau}\mathbf{u}(t-\tau)\right] \tag{2.5.4a}$$

$$= \mathbf{A}(t)\mathbf{u}(t) - \mathbf{B}(t)\left(1 - e^{-\Lambda\tau}\right)\mathbf{u}(t), \tag{2.5.4b}$$

where the subscript j is omitted for notational convenience. Finally, one arrives at a differential equation for $\mathbf{u}(t)$

$$\frac{d}{dt}\mathbf{u}(t) = \left[\mathbf{A}(t) - \mathbf{B}(t)\left(1 - e^{-\Lambda\tau}\right) - \Lambda\mathbf{Id}\right]\mathbf{u}(t) \tag{2.5.5}$$

with the $n \times n$ identity matrix $\mathbf{Id}$.

Using the fundamental matrix $\Phi(t)$ determined by the differential equation

$$\frac{d}{dt}\Phi(t) = [\mathbf{A}(t) - \mathbf{B}(t)]\Phi(t) \tag{2.5.6}$$

with $\Phi(0) = \mathbf{Id}$, one can write the solutions of (2.5.2) as $\delta\mathbf{x}(t) = \Phi(t)\mathbf{x}_{\text{init}}$ with some initial condition $\mathbf{x}_{\text{init}}$.

The spectrum of the Floquet exponents is given by the roots of the characteristic equation

$$\det\left[\mathbf{A}(t) - \mathbf{B}(t)\left(1 - e^{-\Lambda\tau}\right) - \Lambda\mathbf{Id}\right] = 0. \tag{2.5.7}$$

Note that a similar equation can be derived in the context of steady states $\mathbf{x}^*$ where an ansatz for $\delta\mathbf{x}(t)$ is given by $\delta\mathbf{x}(t) = e^{\Lambda t}\mathbf{x}_{\text{init}}$. Then, (2.5.7) describes a characteristic equation for the eigenvalues Λ of the controlled system.

The control scheme is successful if the real parts of all Floquet exponents or, in case of steady states, all eigenvalues are negative. An equivalent statement is that all multipliers μ defined as $\mu = \exp(\Lambda\tau)$ are located inside the unit circle in the complex plane. This Floquet multiplier can be understood as the rate how the distance from the invariant solution increases in an interval $[t, t + \tau]$.

In the case of extended time-delayed feedback as given by (2.2.1), the infinite series in the control force collapses as a geometric series

$$\mathbf{F}(t) = \sum_{n=0}^{\infty} R^n[\mathbf{s}(t - n\tau) - \mathbf{s}(t - (n+1)\tau)] \tag{2.5.8a}$$

$$= \sum_{n=0}^{\infty} R^n\left[e^{\Lambda(t-n\tau)}\mathbf{u}(t - n\tau) - e^{\Lambda[t-(n+1)\tau]}\mathbf{u}(t - (n+1)\tau)\right] \tag{2.5.8b}$$

$$= e^{\Lambda t}\left(1 - e^{-\Lambda\tau}\right)\sum_{n=0}^{\infty} R^n e^{-n\Lambda\tau}\mathbf{u}(t) \tag{2.5.8c}$$

$$= e^{\Lambda t}\frac{1 - e^{-\Lambda\tau}}{1 - Re^{-\Lambda\tau}}\mathbf{u}(t). \tag{2.5.8d}$$

Therefore, the characteristic equation in the case of extended time-delayed feedback control reads

$$\det\left[\mathbf{A}(t) - \mathbf{B}(t)\frac{1 - e^{-\Lambda\tau}}{1 - Re^{-\Lambda\tau}} - \Lambda\mathbf{Id}\right] = 0. \tag{2.5.9}$$

The following chapters will specify the matrices **A** and **B** such that (2.5.7) and (2.5.9) become a powerful tool for the stability analysis of the target state. In the case of phase-dependent coupling or additional latency as mentioned in Sect. 2.4, a characteristic equation can be derived as well. Details of the corresponding derivations will be discussed in Sects. 3.5 and 3.4.

2.6 Transfer Function

In this section, I will discuss effects of time-delayed feedback control in the frequency domain and I will show how the different frequencies in the control signal contribute to the control force. The transform into Fourier space is especially helpful if additional filters are present in the feedback loop. I will discuss both low-pass and bandpass filters and show how these filters influence the control signal in the formalism of transfer functions.

The transfer function of the extended time-delayed control scheme is derived in the following [34]. Starting from the definition of the control force $\mathbf{F}(t)$ given by (2.2.1a), the transfer function can be calculated by a Fourier transform

$$\mathbf{F}(t) = \sum_{n=0}^{\infty} R^n[\mathbf{s}(t - n\tau) - \mathbf{s}(t - (n+1)\tau)] \tag{2.6.1a}$$

$$\Rightarrow \hat{\mathbf{F}}(\omega) = \int_{-\infty}^{\infty} dt\, e^{-i\omega t} \sum_{n=0}^{\infty} R^n[\mathbf{s}(t - n\tau) - \mathbf{s}(t - (n+1)\tau)] \tag{2.6.1b}$$

$$= \sum_{n=0}^{\infty} R^n \left[e^{-in\omega\tau} \hat{\mathbf{s}}(\omega) - e^{-i(n+1)\omega\tau} \hat{\mathbf{s}}(\omega) \right] \tag{2.6.1c}$$

$$= \sum_{n=0}^{\infty} R^n e^{-in\omega\tau} (1 - e^{-i\omega\tau}) \hat{\mathbf{s}}(\omega). \tag{2.6.1d}$$

Using the geometric series

$$\sum_{n=0}^{\infty} R^n \left(e^{-i\omega\tau}\right)^n = \frac{1}{1 - Re^{-i\omega\tau}} \tag{2.6.2}$$

yields the Fourier transform $\hat{\mathbf{F}}(\omega)$ of the extended time-delayed control force

$$\hat{\mathbf{F}}(\omega) = T_1(\omega) \hat{s}(\omega), \tag{2.6.3}$$

where the transfer function $T_1(\omega)$ is given by

$$T_1(\omega) = \frac{1 - e^{-i\omega\tau}}{1 - Re^{-i\omega\tau}}. \tag{2.6.4}$$

Figure 2.2 depicts the absolute value of this transfer function $|T_1(\omega)|$ for different values of the memory parameter $R = 0, 0.3, 0.6$, and 0.9 as black (solid), (dashed), green (dotted), and blue (dash-dotted) curves, respectively. As discussed in Ref. [34] in the context of stabilization of periodic orbits, the transfer function drops to zero at multiples of the basic frequency τ^{-1}. In the context of stabilization of periodic orbits, this frequency belongs to the periodic orbit under control to guarantee the noninvasiveness of time-delayed feedback. One can see that the notches at these frequencies become narrower as R increases. Due to the notches, the frequency of the periodic orbit does not contribute to the control signal, so that the control force vanishes if stabilization is successful. The steeper notches for larger R indicate that the extended time-delayed feedback is more sensitive to frequencies different from the one to be controlled, so that more feedback is produced for these unwanted frequencies, which makes the control scheme more efficient.

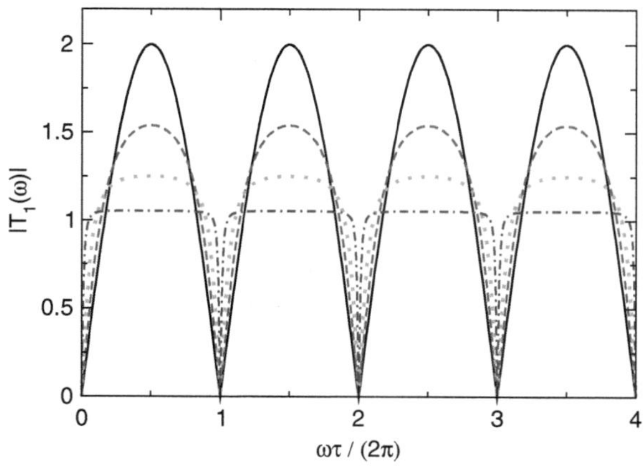

Fig. 2.2 Absolute value of the transfer function of the extended time-delayed feedback scheme according to (2.6.4). The *black (solid)*, *red (dashed)*, *green (dotted)*, and *blue (dash-dotted)* curves correspond to different memory parameters $R = 0$, 0.3, 0.6, and 0.9, respectively

The maximum value of the transfer function $|T_1(\omega)|$ approaches 1 for R closer to 1 and the plateaus become flatter. Therefore, intermediate frequencies generate a smaller response for larger R and thus are less likely to destabilize the system. The frequency comb shown in Fig. 2.2 can be realized experimentally for the stabilization of *cw* emission and intensity pulsations of a semiconductor laser via an all-optical feedback [30, 31, 92]. The feedback is implemented by a Fabry–Perrot interferometer attached to the laser.

As mentioned in Sect. 2.4, there is a variety of extensions of the original time-delayed feedback scheme. For example, it is desirable to include filter in the control loop in order to avoid unwanted frequencies. This filtered feedback has been successfully implemented in optical experiments [64, 65, 67, 93–99] as well as in nonlinear electronic circuits [33, 100], charge transport in semiconductor devices [13, 14, 49], and for the control of unstable steady states of focus type [10, 57, 58].

In the following, I will derive the transfer function of a low-pass filter. Then I will show that a combination of a low-pass filter and time-delayed feedback can be treated in the Fourier space by product of the respective transfer functions.

Similar to the derivation of (2.6.4), the transfer function of the low-pass filter is given by the Fourier transform of (2.4.4)

$$\bar{\mathbf{s}}(t) = \alpha \int_{-\infty}^{t} \mathbf{s}(t')e^{-\alpha(t-t')}\,dt' \tag{2.6.5a}$$

$$\Rightarrow \hat{\bar{\mathbf{s}}}(\omega) = \int_{-\infty}^{\infty} e^{-i\omega t}\alpha \int_{-\infty}^{t} \mathbf{s}(t')e^{-\alpha(t-t')}\,dt'dt. \tag{2.6.5b}$$

The second integral of the right-hand side of (2.6.5b) is a convolution of the form

$$f_1(t) * f_2(t) = \int_{-\infty}^{\infty} f_1(\tau) f_2(t-\tau)\, d\tau, \tag{2.6.6}$$

where the functions $f_1(t)$ and $f_2(t)$ are given in the present case by

$$f_1(t) = \mathbf{s}(t) \qquad \text{and} \qquad f_2(t) = \begin{cases} \alpha e^{-\alpha t}, & t \geq 0 \\ 0, & t < 0 \end{cases}. \tag{2.6.7}$$

With these definitions, (2.6.6) becomes

$$f_1(t) * f_2(t) = \alpha \int_{-\infty}^{t} \mathbf{s}(t') e^{-\alpha(t-t')}\, dt'. \tag{2.6.8}$$

Fourier theory yields that the Fourier transform of a convolution of two functions is the product of the single transformed functions

$$\mathcal{F}[f_1(t) * f_2(t)] = \mathcal{F}[f_1(t)] \mathcal{F}[f_2(t)]. \tag{2.6.9}$$

Therefore the Fourier transform of the low-pass filter is given by

$$\hat{\bar{\mathbf{s}}}(\omega) = \mathcal{F}[\mathbf{s}(t)] \mathcal{F}[f_2(t)] \tag{2.6.10a}$$

$$= \hat{\mathbf{s}}(\omega) \underbrace{\int_{-\infty}^{\infty} f_2(t) e^{-i\omega t}\, dt}_{\alpha \int_0^{\infty} e^{-\alpha t} e^{-i\omega t}\, dt} \tag{2.6.10b}$$

$$= \alpha \hat{\mathbf{s}}(\omega) \left[\frac{1}{-\alpha - i\omega} e^{-(\alpha + i\omega)t} \right]_0^{\infty} \tag{2.6.10c}$$

$$= \hat{\mathbf{s}}(\omega) \frac{1}{1 + i\frac{\omega}{\alpha}}. \tag{2.6.10d}$$

The last equation can be rewritten introducing the transfer function of the low-pass filter $T_2(\omega)$

$$\bar{\mathbf{s}}(\omega) = \frac{1}{1 + i\frac{\omega}{\alpha}} \mathbf{s}(\omega) = T_2(\omega) \mathbf{s}(\omega). \tag{2.6.11}$$

Figure 2.3 shows the frequency dependence of the absolute value of the transfer function $|T_2(\omega)|$ of the low-pass filter for different values of the cutoff frequency α. The black (solid), red (dashed), green (dotted), and blue (dash-dotted) curves refer to different values of $\alpha\tau = 0.25$, 0.5, 0.75, and 1, respectively. It can be seen that the absolute value of the transfer function is

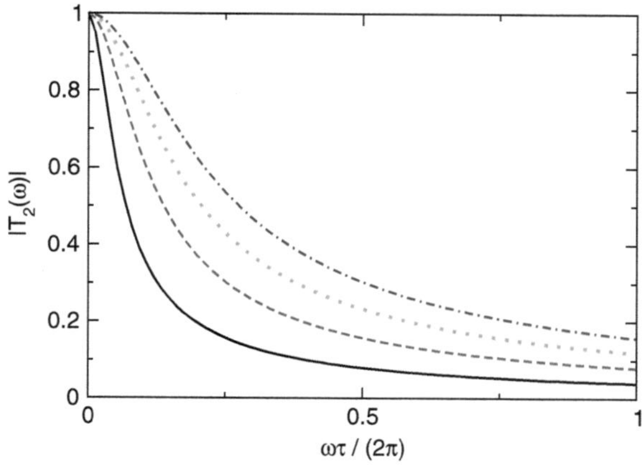

Fig. 2.3 Absolute value of the transfer function of the low-pass filter according to (2.6.11). The *black* (*solid*), *red* (*dashed*), *green* (*dotted*), and *blue* (*dash-dotted*) curves correspond to different values of the cutoff frequency $\alpha\tau = 0.25, 0.5, 0.75$, and 1, respectively

strictly monotonic decreasing. Hence, high frequencies are suppressed. For instance, the amplitude of the frequency at $\omega = \alpha$ is reduced by a factor of $1/\sqrt{2}$. Therefore α is called a cutoff frequency. One can see that larger values of α lead to a slower decrease of the transfer function. Thus higher frequencies can pass the filter to a larger extent.

Consider now the combination of the extended time-delayed control scheme and the low-pass filter. In this case the feedback of (2.2.1a) is given by

$$\mathbf{F}(t) = \sum_{n=0}^{\infty} R^n [\bar{\mathbf{s}}(t - n\tau) - \bar{\mathbf{s}}(t - (n+1)\tau)], \tag{2.6.12}$$

where $\bar{\mathbf{s}}(t)$ denotes again the low-pass filtered version of the control signal $\mathbf{s}(t)$ according to (2.4.4). A Fourier transform of (2.6.12) yields similar to the derivation of (2.6.3)

$$\hat{\mathbf{F}}(\omega) = \sum_{n=0}^{\infty} R^n e^{-in\omega\tau}(1 - e^{-i\omega\tau})\hat{\bar{\mathbf{s}}}(\omega) \tag{2.6.13a}$$

$$= \frac{1 - e^{-i\omega\tau}}{1 - Re^{-i\omega\tau}} \hat{\bar{\mathbf{s}}}(\omega). \tag{2.6.13b}$$

Using the notation of transfer functions, the last equation can be rewritten using (2.6.11)

$$\hat{\mathbf{F}}(\omega) = T_1(\omega)\hat{\bar{\mathbf{s}}}(\omega) \tag{2.6.14a}$$

$$= T_1(\omega)T_2(\omega)\hat{\mathbf{s}}(\omega) \tag{2.6.14b}$$

$$= T(\omega)\hat{\mathbf{s}}(\omega), \tag{2.6.14c}$$

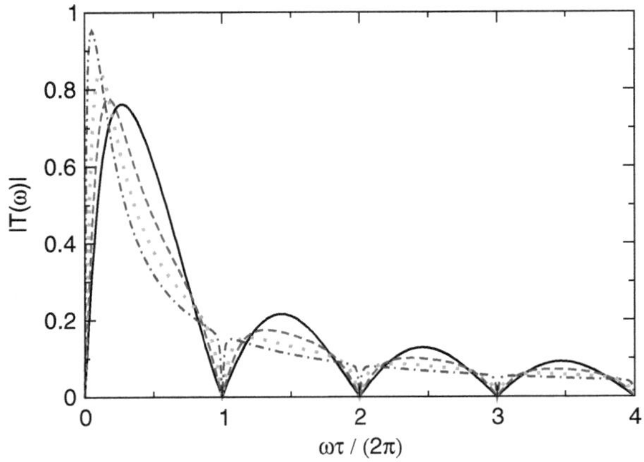

Fig. 2.4 Absolute value of the transfer function of the extended time-delayed feedback method combined with low-pass filtering according to (2.6.15). The *black* (*solid*), *red* (*dashed*), *green* (*dotted*), and *blue* (*dash-dotted*) curves correspond to different memory parameters $R = 0, 0.3, 0.6$, and 0.9, respectively. The cutoff frequency is fixed at $\alpha\tau = 1$

where the combined transfer function $T(\omega)$ is given by

$$T(\omega) = \frac{1 - e^{-i\omega\tau}}{1 - Re^{-i\omega\tau}} \frac{1}{1 + i\frac{\omega}{\alpha}}. \tag{2.6.15}$$

Note that the combined transfer function $T(\omega)$ is the product of the single functions for the extended time-delayed feedback $T_1(\omega)$ and the low-pass filter $T_2(\omega)$.

The absolute value of $|T(\omega)|$ is displayed in Fig. 2.4 for different memory parameters R and fixed cutoff frequency $\alpha\tau = 1$, where the black (solid), red (dashed), green (dotted), and blue (dash-dotted) curves refer to $R = 0, 0.3, 0.6$, and 0.9, respectively. As in Fig. 2.2, there are notches at multiples of the basic frequency τ^{-1} because the roots of $|T_1(\omega)|$ persist. The amplitudes of frequencies larger than the cutoff frequency α are reduced and thus are only minor contributions to the feedback response. This is important to notice in order to understand how the low-pass filter improves the controllability of the system as will be discussed in the following.

In the context of time-delayed feedback applied to the nonlinear electron transport in semiconductor devices like a superlattice [29, 101, 102], additional low-pass filtering has been successfully used to suppress chaotic current oscillations and to stabilize a periodic orbit [13, 48, 49]. This is not possible without the filter, because the unfiltered feedback scheme includes unwanted high frequencies which arise from well-to-well hopping of the electrons. As a consequence, the control scheme can stabilize the unwanted frequencies instead of the frequency of the target orbit. A low-pass filter given by (2.4.4) and (2.6.11) with an appropriate cutoff frequency eliminates these frequency components in the control signal.

Similar to a low-pass filter it is also possible to suppress low frequencies, which can be realized ba a high-pass filter, or include only an intermediate range of frequencies to enter the control force. The latter can be implemented by a bandpass

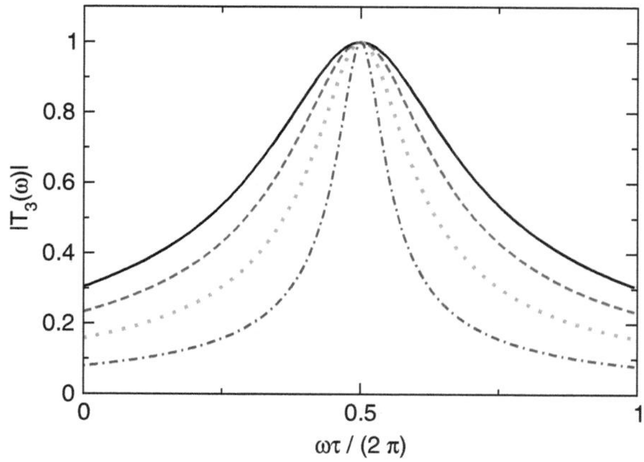

Fig. 2.5 Absolute value of the transfer function of the bandpass filter according to (2.6.16). The *black* (*solid*), *red* (*dashed*), *green* (*dotted*), and *blue* (*dash-dotted*) curves correspond to different values of $\alpha\tau = 0.25, 0.5, 0.75$, and 1, respectively. The frequency shift is fixed at $\omega_0\tau = \pi$

filter. The transfer function $T_3(\omega)$ is this device is given by a frequency shift ω_0 added to the low-pass filter $T_2(\omega)$. This yields

$$T_3(\omega) = \frac{1}{1 + i\frac{\omega-\omega_0}{\alpha}} \tag{2.6.16}$$

and the bandpass filter signal becomes

$$\hat{\bar{\mathbf{s}}}(\omega) = T_3(\omega)\hat{\mathbf{s}}(\omega). \tag{2.6.17}$$

Note that vanishing ω_0 recovers the low-pass filter. The equivalent differential equation reads

$$\frac{d}{dt}\bar{\mathbf{s}}(t) = -\alpha\bar{\mathbf{s}}(t) + \alpha\mathbf{s}(t) + i\omega_0\mathbf{s}(t). \tag{2.6.18}$$

Compare with the equation for the low-pass filter given by (2.4.3).

Figure 2.5 depicts the absolute value of the transfer function of the bandpass filter $|T_3(\omega)|$ for different values of $\alpha\tau = 0.25, 0.5, 0.75$, and 1 as black (solid), red (dashed), green (dotted), and blue (dash-dotted) curves, respectively. The frequency shift is fixed at $\omega_0\tau = \pi$. In analogy to the combination of low-pass filtering and time-delayed feedback, one could include a bandpass filter in the delay line. In Fourier space, this leads again the product of the respective transfer functions.

The extension of bandpass filtering in the control loop has been used to investigate a Hopf bifurcation in an experiment with a nonlinear electronic circuit including band-limited feedback [33, 100]. A bandpass filter has also a large relevance in optical systems. Filtered optical feedback has been used for both experimental and theoretical investigation of *cw* emission, frequency and relaxation oscillations in semiconductor lasers [64, 65, 67, 94, 97, 98]. These devices provide an interest experimental system since they exhibit a rich bifurcation scenario [93, 96, 103–106]. In this context, the complex electric field can be chosen as control signal **s** which is altered by the filter [58].

2.7 Intermediate Conclusion

In this chapter, I have introduced time-delayed feedback control. This feedback scheme generates a control signal from the difference $s(t) - s(t - \tau)$ between the present and an earlier value of an appropriate system variable s. It is noninvasive since the control forces vanish if the target state which could be, for instance, a periodic orbit or a steady state is reached. Due to this property, the unstable states themselves of the original system are not changed by the control, but only their neighborhood is adjusted such that neighboring trajectories converge to it, i.e., the control forces act only if the system deviates from the state to be stabilized. Involving no numerically expensive computations and applicable in experimental setups, time-delayed feedback control is capable of controlling systems with very fast dynamics still in real-time mode [30, 32, 35]. Moreover, detailed knowledge of the target state is not required.

In addition, I have discussed various modifications of the original control scheme such as multiple time-delayed feedback, control loop latency, filtering of the control signal, or nonlocal coupling with spatial delays. Furthermore, I have considered different realizations in terms of coupling functions which measure the system's variables for construction of the control signal and which specify the application of the control force back to the system. Towards the end of this chapter, I applied the concept of transfer functions to the control scheme which added an additional perspective and opened connections with experimental realizations.

This chapter will serve as central node connecting all subsequent chapters where I will apply the time-delayed feedback method to a variety of different dynamic systems. The next chapter will be devoted to the control of steady states. I will also consider next to the original Pyragas scheme many of the above-mentioned modifications which are experimentally relevant. In Chaps. 4–6, I will then discuss other classes of dynamic systems such as periodic states, neutral delay-differential equations, and excitable systems.

References

1. Ott E, Grebogi C, Yorke JA (1990) Controlling chaos. Phys Rev Lett 64:1196
2. Nijmeijer H, Schaft AVD (1996) Nonlinear dynamical control systems, 3rd edn. Springer, New York
3. Ogata K (1997) Modern control engineering. Prentice-Hall, New York.
4. Fradkov AL, Pogromsky AY (1998) Introduction to control of oscillations and chaos. World Scientific, Singapore

5. Fradkov AL, Miroshnik IV, Nikiforov VO (1999). Nonlinear and adaptive control of complex systems. Kluwer, Dordrecht
6. Schuster HG (Editor) (1999) Handbook of chaos control. Wiley-VCH, Weinheim
7. Schöll E, Schuster HG (eds) (2008) Handbook of chaos control. Wiley-VCH, Weinheim. Second completely revised and enlarged edition
8. Pyragas K (1992) Continuous control of chaos by self-controlling feedback. Phys Lett A 170:421
9. Socolar JES, Sukow DW, Gauthier DJ (1994) Stabilizing unstable periodic orbits in fast dynamical systems. Phys Rev E 50:3245
10. Hövel P, Schöll E (2005) Control of unstable steady states by time-delayed feedback methods. Phys Rev E 72:046203
11. Sieber J, Krauskopf B (2007) Control based bifurcation analysis for experiments. Nonlinear Dyn 51:365
12. Schlesner J (2002) Nichtlineare Oszillationen in Halbleiterübergittern unter zeitverzögerter Rückkopplung. Master's thesis, Technische Universität Berlin
13. Schlesner J, Amann A, Janson NB, Just W, Schöll E (2003) Self-stabilization of high frequency oscillations in semiconductor superlattices by time-delay autosynchronization. Phys Rev E 68:066208
14. Schlesner J, Amann A, Janson NB, Just W, Schöll E (2004) Self-stabilization of chaotic domain oscillations in superlattices by time-delayed feedback control. Semicond Sci Technol 19:S34
15. Kehrt M (2008) Zeitverzögerte Rückkopplungskontrolle eines global gekoppelten Reaktions-Diffusions-Modells. Master's thesis, Technische Universität Berlin
16. Hunt BR, Ott E (1996) Optimal periodic orbits of chaotic systems. Phys Rev Lett 76:2254
17. Hunt BR, Ott E (1996) Optimal periodic orbits of chaotic systems occur at low period. Phys Rev E 54:328
18. Yang T-H, Hunt BR, Ott E (2000) Optimal periodic orbits of continuous time chaotic systems. Phys Rev E 62:1950
19. Zoldi SM, Greenside HS (1998) Comment on optimal periodic orbits of chaotic systems. Phys Rev Lett 80:1790
20. Hunt BR, Ott E (1998) Hunt and Ott reply. Phys Rev Lett 80:1791
21. Just W, Reckwerth D, Möckel J, Reibold E, Benner H (1998) Delayed feedback control of periodic orbits in autonomous systems. Phys Rev Lett 81:562
22. Franceschini G, Bose S, Schöll E (1999) Control of chaotic spatiotemporal spiking by time-delay autosynchronisation. Phys Rev E 60:5426
23. Zoldi SM, Franceschini G, Bose S, Schöll E (2000) Stabilizing unstable periodic orbits in reaction-diffusion systems by global time-delayed feedback control. In: Fiedler B, Gröger K, Sprekels J (eds) Proceedings of Equadiff 99. World Scientific Publishing, Singapore, p 1311
24. Yu X (1999) Tracking inherent periodic orbits in chaotic dynamic systems via adaptive variable structure time-delay self control. IEEE Trans Circuits Syst 46:1408
25. Ando H, Boccaletti S, Aihara K (2007) Automatic control and tracking of periodic orbits in chaotic systems. Phys Rev E 75:066211
26. Pyragas K, Pyragas V, Kiss IZ, Hudson JL (2002) Stabilizing and tracking unknown steady states of dynamical systems. Phys Rev Lett 89:244103
27. Parmananda P (2003) Tracking fixed-point dynamics in an electrochemical system using delayed-feedback control. Phys Rev E 67:045202(R)
28. Unkelbach J, Amann A, Just W, Schöll E (2003) Time–delay autosynchronization of the spatiotemporal dynamics in resonant tunneling diodes. Phys Rev E 68:026204
29. Schöll E (2004) Pattern formation in semiconductors: control of spatio-temporal dynamics. Ann Phys (Leipzig) 13:403. Special Topic Issue, edited by Friedrich R, Kuhn T, Linz S
30. Schikora S, Hövel P, Wünsche HJ, Schöll E, Henneberger F (2006) All-optical noninvasive control of unstable steady states in a semiconductor laser. Phys Rev Lett 97:213902

31. Schikora S, Wünsche HJ, Henneberger F (2008) All-optical noninvasive chaos control of a semiconductor laser. Phys Rev E 78:025202
32. Blakely JN, Illing L, Gauthier DJ (2004) Controlling fast chaos in delay dynamical systems. Phys Rev Lett 92:193901
33. Illing L, Gauthier DJ (2005) Hopf bifurcations in time-delay systems with band-limited feedback. Physica D 210:180
34. Sukow DW, Bleich ME, Gauthier DJ, Socolar JES (1997) Controlling chaos in a fast diode resonator using time-delay autosynchronisation: experimental observations and theoretical analysis. Chaos 7:560
35. Gauthier DJ, Sukow DW, Concannon HM, Socolar JES (1994) Stabilizing unstable periodic orbits in a fast diode resonator using continuous time-delay autosynchronization. Phys Rev E 50:2343
36. Bleich ME, Socolar JES (1996) Stability of periodic orbits controlled by time-delay feedback. Phys Lett A 210:87
37. Bleich ME, Socolar JES (1996) Controlling spatiotemporal dynamics with time-delay feedback. Phys Rev E 54:R17
38. Schneider FM, Schöll E, Dahlem MA (2009) Controlling the onset of traveling pulses in excitable media by nonlocal spatial coupling and time delayed feedback. Chaos 19:015110
39. Hövel P, Shah SA, Dahlem MA, Schöll E (2009) Feedback-dependent control of stochastic synchronization in coupled neural systems. In: Fortuna L, Frasca M (eds) Proceedings of 4th international scientific conference on physics and control (PhysCon 09). IPACS Open Access Library http://lib.physcon.ru (e-Library of the International Physics and Control Society), http://arxiv.org/abs/0911.2334v1
40. Yeung MKS, Strogatz SH. (1999) Time delay in the Kuramoto model of coupled oscillators. Phys Rev Lett 82:648
41. Lysyansky B, Maistrenko Y, Tass PA (2008) Coexistence of numerous synchronized and desynchronized states in a model of two phase oscillators coupled with delay. Int J Bifur Chaos 18:1791
42. Rosenblum MG, Pikovsky AS (2004) Controlling synchronization in an ensemble of globally coupled oscillators. Phys Rev Lett 92:114102
43. Popovych OV, Hauptmann C, Tass PA (2005) Effective desynchronization by nonlinear delayed feedback. Phys Rev Lett 94:164102
44. Popovych OV, Hauptmann C, Tass PA (2005) Demand-controlled desynchronization of brain rhythms by means of nonlinear delayed feedback. In: Proceedings of IEEE Engineering and Medicine Biology. 27th Annual conference
45. Beck O, Amann A, Schöll E, Socolar JES, Just W (2002) Comparison of time-delayed feedback schemes for spatio-temporal control of chaos in a reaction-diffusion system with global coupling. Phys Rev E 66:016213
46. Baba N, Amann A, Schöll E, Just W (2002) Giant improvement of time-delayed feedback control by spatio-temporal filtering. Phys Rev Lett 89:074101
47. Just W, Popovich S, Amann A, Baba N, Schöll E (2003) Improvement of time–delayed feedback control by periodic modulation: analytical theory of Floquet mode control scheme. Phys Rev E 67:026222
48. Hövel P (2004) Effects of chaos control and latency in time-delay feedback methods. Master's thesis, Technische Universität Berlin
49. Schöll E, Hizanidis J, Hövel P, Stegemann G (2007) Pattern formation in semiconductors under the influence of time-delayed feedback control and noise. In: Schimansky-Geier L, Fiedler B, Kurths J, Schöll E (eds) Analysis and control of complex nonlinear processes in physics, chemistry and biology. World Scientific, Singapore, pp 135–183
50. Stegemann G, Balanov AG, Schöll E (2006) Delayed feedback control of stochastic spatiotemporal dynamics in a resonant tunneling diode. Phys Rev E 73:016203
51. Stegemann G, Schöll E (2007) Two-dimensional spatiotemporal pattern formation in the double-barrier resonant tunneling diode. New J Phys 9:55

52. Hizanidis J, Balanov AG, Amann A, Schöll E (2006) Noise-induced oscillations and their control in semiconductor superlattices. Int J Bifur Chaos 16:1701
53. Hizanidis J, Balanov AG, Amann A, Schöll E (2005) Control of noise-induced oscillations in superlattices by delayed feedback. In: Gonzales T, Mateos J, Pardo D (eds) Proceedings of 18th international conference on noise and fluctuations (ICNF-2005), vol 780. American Institute of Physics, Melville, New York), pp 41–44. ISBN 0-7354-0267-1
54. Hizanidis J, Schöll E (2008) Control of noise-induced spatiotemporal patterns in superlattices. Phys Stat Sol (c) 5:207
55. Hizanidis J (2008) Control of noise-induced spatio-temporal dynamics in superlattices. Ph.D. thesis, Technische Universität Berlin
56. Kehrt M, Hövel P, Flunkert V, Dahlem MA, Rodin P, Schöll E (2009) Stabilization of complex spatio-temporal dynamics near a subcritical Hopf bifurcation by time-delayed feedback. Eur Phys J B 68:557
57. Dahms T, Hövel P, Schöll E (2007) Control of unstable steady states by extended time-delayed feedback. Phys Rev E 76:056201
58. Dahms T, Hövel P, Schöll E (2008) Stabilizing continuous-wave output in semiconductor lasers by time-delayed feedback. Phys Rev E 78:056213
59. Balanov AG, Janson NB, Schöll E (2005) Delayed feedback control of chaos: bifurcation analysis. Phys Rev E 71:016222
60. Just W, Reckwerth D, Reibold E, Benner H (1999) Influence of control loop latency on time-delayed feedback control. Phys Rev E 59:2826
61. Hövel P, Socolar JES (2003) Stability domains for time-delay feedback control with latency. Phys Rev E 68:036206
62. Manor Y, Koch C, Segev I (1991) Effect of geometrical irregularities on propagation delay in axonal trees. Biophys J 60: 1424
63. Schwark HD, Jones EG (1989) The distribution of intrinsic cortical axons in area 3b of cat primary somatosensory cortex. Exp Brain Res 78:501
64. Yousefi M, Lenstra D (1999) Dynamical behavior of a semiconductor laser with filtered external optical feedback. IEEE J Quantum Electron 35:970
65. Fischer A, Andersen O, Yousefi M, Stolte S, Lenstra D (2000) Experimental and theoretical study of filtered optical feedback in a semiconductor laser. IEEE J Quantum Electron 36:375
66. Lang R, Kobayashi K (1980) External optical feedback effects on semiconductor injection laser properties. IEEE J Quantum Electron 16:347
67. Erzgräber H, Krauskopf B, Lenstra D, Fischer APA, Vemuri G (2006) Frequency versus relaxation oscillations in a semiconductor laser with coherent filtered optical feedback. Phys Rev E 73:055201(R)
68. Ahlborn A, Parlitz U (2004) Stabilizing unstable steady states using multiple delay feedback control. Phys Rev Lett 93:264101
69. Ahlborn A, Parlitz U (2005) Controlling dynamical systems using multiple delay feedback control. Phys Rev E 72:016206
70. Ahlborn A, Parlitz U (2006) Laser stabilization with multiple-delay feedback control. Opt Lett 31:465
71. Ahlborn A, Parlitz U (2007) Controlling spatiotemporal chaos using multiple delays. Phys Rev E 75:65202
72. Kyrychko YN, Blyuss KB, Hövel P, Schöll E (2009) Asymptotic properties of the spectrum of neutral delay differential equations. Dyn Sys 24:361
73. Blyuss KB, Kyrychko YN, Hövel P, Schöll E (2008) Control of unstable steady states in neutral time-delayed systems. Eur Phys J B 65:571
74. Johnston GA, Hunt ER (1993) Derivative control of the steady state in Chua's circuit driven in the chaotic region. IEEE Trans Circuits Syst 40:833
75. Parmananda P, Rhode MA, Johnson GA, Rollins RW, Dewald HD, Markworth AJ (1994) Stabilization of unstable steady states in an electrochemical system using derivative control. Phys Rev E 49:5007

76. Parmananda P, Eiswirth M (1996) Stabilizing unstable fixed points using derivative control. J Phys Chem 100:16568
77. Lu W, Yu D, Harrison RG (1996) Control of patterns in spatiotemporal chaos in optics. Phys Rev Lett 76:3316
78. Dahlem MA, Schneider FM, Schöll E (2008) Failure of feedback as a putative common mechanism of spreading depolarizations in migraine and stroke. Chaos 18:026110
79. Dahlem MA, Hiller G, Panchuk A, Schöll E (2009) Dynamics of delay-coupled excitable neural systems. Int J Bifur Chaos 19:745
80. Schneider FM Beeinflussung von neuronalen (2008) Erregungswellen durch raum-zeitliche Rückkopplung., Master's thesis, Technische Universität Berlin
81. Guzenko PY, Hövel P, Flunkert V, Fradkov AL, Schöll E (2008) Adaptive tuning of feedback gain in time-delayed feedback control. In: Fradkov A, Andrievsky B (eds) Proceedings of 6th EUROMECH nonlinear dynamics conference (ENOC-2008). IPACS Open Access Library http://lib.physcon.ru (e-Library of the International Physics and Control Society)
82. Yu X (1997) Tracking inherent periodic orbits in chaotic system via adaptive time delayed self-control. In: Proceedings of the 36th IEEE conference on decision and control vol 1, pp 401
83. Guzenko PY, Fradkov AL (1997) Gradient control of Henon map dynamics. Int J Bifur Chaos 7:701
84. Fradkov AL, Guzenko PY, Pavlov A (2000) Adaptive control of recurrent trajectories based on linearization of Poincare map. Int J Bifur Chaos 10:621
85. Astrov YA, Fradkov AL, Guzenko PY (2005) Suppression of a noise-induced transition by feedback control. In: Physics and control, 2005. Proceedings 2005 International Conference. IEEE, Piscataway, NJ, US, pp 662–667
86. Kakmeni FMM, Bowong S, Tchawoua C (2006) Nonlinear adaptive synchronization of a class of chaotic systems. Phys Lett A 355:47
87. Liao T-L, Lin S-H (1999) Adaptive control and synchronization of Lorenz systems. J Frankl Inst 336:925
88. Hale JK (1971) Functional differential equations. Applied mathematical sciences, vol 3. Springer, New York
89. Just W, Bernard T, Ostheimer M, Reibold E, Benner H (1997) Mechanism of time-delayed feedback control. Phys Rev Lett 78:203
90. N Baba: Stabilisierung instabiler räumlicher Muster durch zeitverzögerte Rückkopplung mit räumlichen Filtern. Master's thesis, Technische Universität Berlin (2001)
91. Amann A, Schöll E, Just W (2007) Some basic remarks on eigenmode expansions of time-delay dynamics. Physica A 373:191
92. Wünsche HJ, Schikora S, Henneberger F (2008) Noninvasive control of semiconductor lasers by delayed optical feedback. In:Schöll E, Schuster HG (eds) Handbook of chaos control. Wiley, VCH, Weinheim, second completely revised and enlarged edition
93. Erzgräber H, Krauskopf B, Lenstra D (2007) Bifurcation analysis of a semiconductor laser with filtered optical feedback. SIAM J Appl Dyn Syst. 6:1
94. Erzgräber H, Lenstra D, Krauskopf B, Fischer APA, Vemuri G (2007) Feedback phase sensitivity of a semiconductor laser subject to filtered optical feedback: experiment and theory. Phys Rev E 76: 026212
95. Erzgräber H, Krauskopf B (2007) Dynamics of a filtered-feedback laser: influence of the filter width. Opt Lett 32:2441
96. Green K Krauskopf B (2006) Mode structure of a semiconductor laser subject to filtered optical feedback. Opt. Commun. 258:243
97. Fischer APA, Yousefi M, Lenstra D, Carter MW, Vemuri G (2004) Filtered optical feedback induced frequency dynamics in semiconductor lasers. Phys Rev Lett 92:023901
98. Fischer APA, Yousefi M, Lenstra D, Carter MW, Vemuri G (2004) Experimental and theoretical study of semiconductor laser dynamics due to filtered optical feedback. IEEE J Sel Top Quantum Electron 10:944

99. Yousefi M, Lenstra D, Vemuri G (2003) Nonlinear dynamics of a semiconductor laser with filtered optical feedback and the influence of noise. Phys Rev E 67:046213
100. Illing L, Gauthier DJ (2006) Ultra-high-frequency chaos in a time-delay electronic device with band-limited feedback. Chaos 16: 033119
101. Wacker A (2002) Semiconductor superlattices: a model system for nonlinear transport. Phys Rep 357:1
102. Amann A (2003) Nonlinear and chaotic front dynamics in semiconductor superlattices. Ph.D. thesis, Technische Unversität Berlin
103. Wieczorek S, Krauskopf B, Lenstra D (1999) Unifying view of bifurcations in a semiconductor laser subject to optical injection. Opt Commun 172:279
104. Krauskopf B, Lenstra D (eds) (2000) Fundamental issues of nonlinear laser dynamics. In: AIP conference proceedings, vol 548. American Institute of Physics, Melville, New York
105. Wieczorek S, Krauskopf B, Lenstra D (2002) Multipulse excitability in a semiconductor laser with optical injection. Phys Rev Lett 88:063901
106. Wieczorek S, Krauskopf B, Simpson T, Lenstra D (2005) The dynamical complexity of optically injected semiconductor lasers. Phys Rep 416:1

Chapter 3
Control of Steady States

Was ist Chaos?

Es ist jene Ordnung, die man bei der Erschaffung der Welt zerstört hat.[1]

Stanisław Jerzy Lec, Sämtliche unfrisierte Gedanken

Next to periodic orbits, steady states or equilibria[2] play an important role in a large variety of optical, electronic, chemical, biological, and other nonlinear systems [1–6]. Although the effects of time-delayed feedback schemes on the stability of periodic orbits are understood to large extend [7–11], much less is known in the case of a fixed point. There are some results discussing the application of the extend time-delayed feedback method [12] and numerical simulations of Chua's circuit [5]. For the stabilization of periodic orbits, the period of the desired orbit gives a good hint for the best choice of the time delay. For steady states, such an intrinsic timescale is not obvious. A detailed theoretical investigation was presented recently [13–16]. Even though I will consider only steady states in this chapter, I will occasionally point out some analogies and connections to the control of unstable periodic orbits.

The structure of this chapter is the following: In the next Sect. 3.1, I will start with a description of the model that will be under investigation throughout this chapter. Sections 3.2 and 3.3 will be devoted to the stabilization by Pyragas control and extended time-delayed feedback, respectively. The proceeding Sects. 3.4 and 3.5 will deal with the discussion of both control schemes involving experimentally important modifications of control loop latency and phase-dependent coupling, respectively.

---

[1] What is chaos? It is the order which was destroyed at the creation of the world.

[2] In the mathematical literature *equilibria* denote fixed points or steady states of a dynamics system, not to be confused with *equilibrium* in the physical literature. All systems considered here are far from thermodynamic equilibrium.

P. Hövel, *Control of Complex Nonlinear Systems with Delay*,
Springer Theses, DOI: 10.1007/978-3-642-14110-2_3,

Asymptotic properties of time-delayed feedback control will be the topic of Sect. 3.6. Finally, I will summarize this chapter in an intermediate conclusion in Sect. 3.7.

3.1 Model Equations

In this section, I will introduce the model that will be used throughout the whole chapter. It consists of a special two-dimensional linear system which has a fixed point of focus type. Before discussing this model in detail, however, let me briefly review all possible scenarios of two-dimensional linear systems [17, 18]. These can be written as follows

$$\begin{pmatrix} \frac{dx(t)}{dt} \\ \frac{dy(t)}{dt} \end{pmatrix} = \begin{pmatrix} a & b \\ c & d \end{pmatrix} \begin{pmatrix} x(t) \\ y(t) \end{pmatrix} \tag{3.1.1a}$$

$$\Leftrightarrow \quad \frac{d\mathbf{x}(t)}{dt} = \mathbf{A}\mathbf{x}(t) \tag{3.1.1b}$$

with real constants a, b, c, and d and the state vector $\mathbf{x}(t) = (x(t), y(t))$. The eigenvalues Λ_0 of the matrix $\mathbf{A}$ allow for a characterization of possible scenarios of the linear system. They are given by the roots of the characteristic equation

$$0 = \det\left(\mathbf{A} - \Lambda_0 \mathbf{Id}\right) \tag{3.1.2a}$$

$$= \det \begin{pmatrix} a - \Lambda_0 & b \\ c & d - \Lambda_0 \end{pmatrix} \tag{3.1.2b}$$

$$= \Lambda_0^2 - (a+d)\Lambda_0 + ad - bc \tag{3.1.2c}$$

$$= \Lambda_0^2 - \mathrm{tr}(\mathbf{A})\Lambda_0 + \det \mathbf{A}, \tag{3.1.2d}$$

where $\mathbf{Id}$ denotes the 2×2 identity matrix. Therefore, the solution of this equation can be written in dependence on the trace $\mathrm{tr}(\mathbf{A})$ and the determinant $\det \mathbf{A}$

$$\Lambda_0 = \frac{\mathrm{tr}(\mathbf{A}) \pm \sqrt{[\mathrm{tr}(\mathbf{A})]^2 - 4 \det \mathbf{A}}}{2}. \tag{3.1.3}$$

Figure 3.1 depicts schematically the possible scenarios of the two-dimensional system (3.1.1a, 3.1.1b). Depending on the value of the trace and determinant of the matrix $\mathbf{A}$, the system exhibits a saddle point, a stable or unstable node, and a stable or unstable focus. A saddle exists for a negative determinant. For positive determinant, the sign of the argument of the square root in Eq. 3.1.3, i.e., $[\mathrm{tr}(\mathbf{A})]^2 - 4 \det \mathbf{A}$, determines whether a node or a focus occurs. For $[\mathrm{tr}(\mathbf{A})]^2 - 4 \det \mathbf{A} < 0$, the eigenvalue Λ_0 becomes complex and a focus exists. For positive values, a node is present in the system. The stability of node or focus depends on the sign of the trace

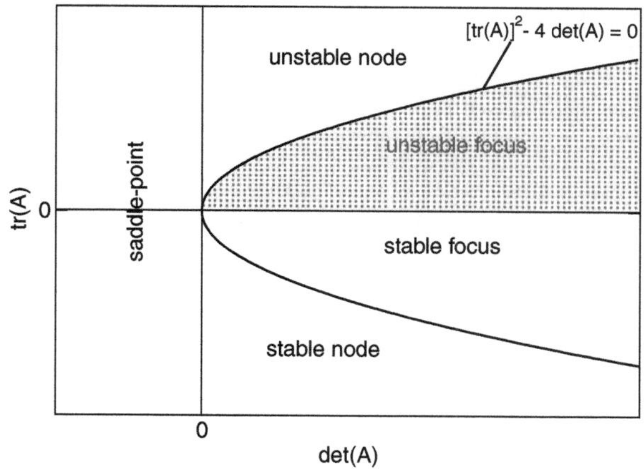

Fig. 3.1 Schematic diagram of possible scenarios in two-dimensional linear systems given by Eqs. 3.1.1a and 3.1.1b. The different scenarios are given in dependence on the trace and determinant of the matrix **A**. The *shaded red region* refers to the case of an unstable focus

such that unstable, diverging dynamics occur for positive traces and attractors for negative traces.

In the following, I will consider a fixed point of focus type as indicated by the shaded red region in Fig. 3.1. Without loss of generality, I choose coordinates $x(t)$ and $y(t)$ as center manifold coordinates for which the system has the following form

$$\frac{dx(t)}{dt} = \lambda x(t) + \omega y(t) \tag{3.1.4a}$$

$$\frac{dy(t)}{dt} = -\omega x(t) + \lambda y(t), \tag{3.1.4b}$$

where λ and ω are real constants. Note that the sign of ω determines the direction of rotation in the (x, y) plane. For $\omega > 0$, the trajectory revolves clockwise around the origin and it rotates counter-clockwise for negative ω. The eigenvalues of this system are $\Lambda_0 = \lambda \pm i\omega$. Introducing complex notation, i.e., $z(t) = x(t) + iy(t) = r(t)e^{i\theta(t)}$, yields equivalently to Eqs. 3.1.4a and 3.1.4b

$$\frac{dz(t)}{dt} = (\lambda \pm i\omega)z(t) \tag{3.1.5}$$

or written in amplitude $r(t)$ and phases $\theta(t)$

$$\frac{dr(t)}{dt} = \lambda r(t) \tag{3.1.6a}$$

$$\frac{d\theta(t)}{dt} = \pm\omega. \tag{3.1.6b}$$

One can see from the last system of equations that λ acts as a damping rate whereas ω influences the phase. For $\omega \neq 0$ and positive λ, the system describes an unstable focus. Note that the non-zero imaginary part of the eigenvalue, i.e.,

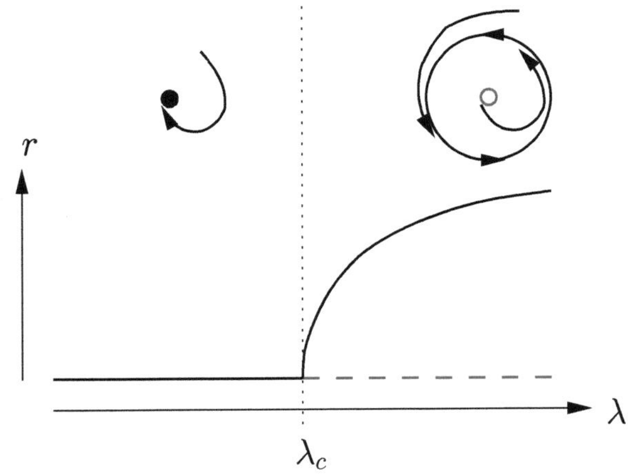

Fig. 3.2 *Lower part* Schematic diagram of the supercritical Hopf bifurcation, i.e., dependence of the amplitude r on the bifurcation parameter λ. At the critical value λ_c a bifurcation occurs which is in the present case at $\lambda_c = 0$. The *upper part* depicts schematically the periodic orbit and steady states of the system. *Solid lines* and *filled dots* denote stability, *dashed lines* and *empty dots* show instability. The *lines with arrow* heads show example trajectories

$\mathrm{Im}(\Lambda_0) = \pm\omega$ introduces a characteristic intrinsic timescale $T_0 = 2\pi/|\omega|$ which corresponds to the time it takes for the trajectory to revolve once around the fixed point. This timescale T_0 will serve as a natural reference value for the choice of time delays in the next sections. The case of vanishing ω decouples the system of Eqs. 3.1.4a and 3.1.4b and leads to an unstable node. I will briefly consider the impossibility of stabilizing a node or a saddle by time-delayed feedback in the following section.

The system under investigation can also be understood as a linearized version of the supercritical Hopf bifurcation

$$\frac{dr(t)}{dt} = \lambda r(t) - r(t)^3 \tag{3.1.7a}$$

$$\frac{d\theta(t)}{dt} = -\omega + br(t)^2 \tag{3.1.7b}$$

with a real constant b, where the linearization is performed at the fixed point located at the origin $(x^*, y^*) = (0, 0)$, i.e., $z^* = 0$. In complex notation using $z(t) = x(t) + iy(t) = r(t)e^{i\theta(t)}$ Eq. 3.1.7a can be rewritten as:

$$\frac{dz(t)}{dt} = (\lambda - i\omega)z(t) + (1 + ib)|z(t)|^2 z(t). \tag{3.1.8}$$

Omitting the cubic terms recovers the fixed point located at the origin and results in Eqs. 3.1.4a and 3.1.4b. Thus, the steady state of focus type can be seen as linearization of a Hopf bifurcation performed for the invariant solution at the origin.

Figure 3.2 displays a schematic diagram of the supercritical Hopf bifurcation where λ is used as a bifurcation parameter. The upper part shows possible periodic orbits and fixed points where the stability and instability is indicated by solid lines and filled dots or dashed lines and empty dots, respectively. The lines with arrow heads depict example trajectories. The lower part displays the characteristic square root dependence of the amplitude r on the bifurcation parameter λ. The case of a subcritical Hopf bifurcation can be seen in Fig. 4.2. For parameters below the critical value λ_c, e.g., $\lambda = 0$ in the present case, there exists only a stable fixed point. At the bifurcation at λ_c, the fixed point becomes unstable and a stable periodic orbit is born. Since the parameter λ is chosen positive, the system (3.1.4a, 3.1.4b) is a linearized supercritical Hopf normal form above the bifurcation.

After the introduction of the system under investigation, I will apply time-delayed feedback of various forms in the following sections. I will start with one of the simplest realizations of this control scheme and elaborate extensions and modifications in later sections.

3.2 Time-Delayed Feedback

In the previous Chap. 2, I have introduced time-delayed feedback in general form and especially in Sect. 2.3, I have shown the there is a large variety of coupling schemes to realize this feedback methods. In the following, I will apply time-delayed feedback control to two linear system. First, I will elaborate that unstable steady states of focus type can be stabilized by this control scheme in Sect. 3.2.1. If the feedback method is applied to linear systems without torsion, e.g., a repeller or a saddle point, stabilization fails as will be discussed in Sect. 3.2.2. The analysis and derivations of this section are mainly based on Ref. [13].

3.2.1 Unstable Focus

At this point, I will start with one of the simplest realizations where the function $\mathbf{g}$ which generates the control signal extracts single components of the system, and where the transducer function $\mathbf{h}$ applies the control force to the same components. Hence, the composition $\mathbf{h} \circ \mathbf{g}$ is the identity, i.e., $(\mathbf{h} \circ \mathbf{g})(\mathbf{x}) = \mathbf{x}$. Equivalently, the coupling matrix $\mathbf{A}$ introduced in Sect. 2.3 becomes the identity matrix. This specific configuration is called diagonal coupling and the system's equations (3.1.4a and 3.1.4b) of an unstable focus subject to time-delayed feedback become [13, 19]

$$\begin{pmatrix} \frac{dx(t)}{dt} \\ \frac{dy(t)}{dt} \end{pmatrix} = \begin{pmatrix} \lambda & \omega \\ -\omega & \lambda \end{pmatrix} \begin{pmatrix} x(t) \\ y(t) \end{pmatrix} - K \begin{pmatrix} x(t) - x(t-\tau) \\ y(t) - y(t-\tau) \end{pmatrix} \tag{3.2.1}$$

or equivalently

$$\frac{dx(t)}{dt} = \lambda x(t) + \omega y(t) - K[x(t) - x(t-\tau)] \tag{3.2.2a}$$

$$\frac{dy(t)}{dt} = -\omega x(t) + \lambda y(t) - K[y(t) - y(t-\tau)], \tag{3.2.2b}$$

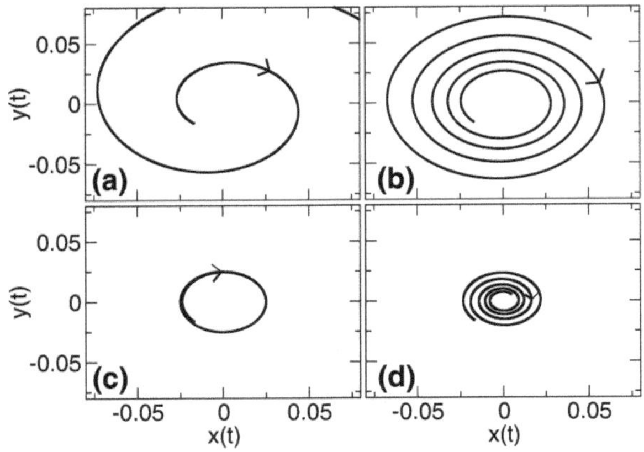

Fig. 3.3 Control of an unstable focus with $\lambda = 0.5$ and $\omega = \pi$ in the configuration space for different values of the feedback gain K. Panels **a**, **b**, **c**, and **d** correspond to $KT_0 = 0$, 0.4, 0.5, and 0.6, respectively. The time delay τ of the time-delayed feedback control scheme of Pyragas type is chosen as 1, corresponding to $\tau = T_0/2 = \pi/\omega$

where the feedback gain K and time delay τ are real control parameters. In a complex notation with $z(t) = x(t) + iy(t)$ the variables x and y correspond to the real and imaginary part, respectively. Thus, Eqs. 3.2.2a and 3.2.2b can be rewritten as

$$\frac{dz(t)}{dt} = (\lambda - i\omega)z(t) - K[z(t) - z(t-\tau)]. \tag{3.2.3}$$

Equivalently, one can separate the complex variable z in an amplitude $r(t)$ and a phase variable $\theta(t)$, i.e., $z(t) = r(t)e^{i\theta(t)}$. In this notation, the time-delayed system becomes

$$\frac{dr(t)}{dt} + i\frac{d\theta(t)}{dt}r(t) = (\lambda - i\omega)r(t) - K\left[r(t) - r(t-\tau)e^{i(\theta(t-\tau)-\theta(t))}\right], \tag{3.2.4}$$

which leads to two differential equations for the real and imaginary part, respectively

$$\frac{dr(t)}{dt} = \lambda r(t) - K[r(t) - r(t-\tau)\cos(\theta(t-\tau) - \theta(t))] \tag{3.2.5a}$$

$$\frac{d\theta(t)}{dt} = -\omega + K\frac{r(t-\tau)}{r(t)}\sin(\theta(t-\tau) - \theta(t)). \tag{3.2.5b}$$

Figure 3.3 depicts the dynamics of the controlled unstable focus with the parameters $\lambda = 0.5$ and $\omega = \pi$ in the (x, y) plane for different values of the feedback gain K. This value of ω yields an intrinsic timescale of $T_0 = 2\pi/\omega = 2$ which is the time needed for the trajectory to rotate around fixed point in the absence of control. Panels (a)–(d) correspond to increasing K, i.e., $KT_0 = 0$, 0.4, 0.5, and 0.6, respectively. The time delay of the Pyragas control scheme is chosen in all panels as $\tau = 1$ which corresponds to half the intrinsic timescale: $\tau = T_0/2$. Panel (a) displays the case of the absence of control, i.e., $K = 0$, and shows that the system is an unstable focus exhibiting undamped oscillations on a timescale T_0. It can be seen from panel (b) that increasing K reduces the instability. The system diverges more slowly to infinity indicated by the tighter spiral. Further increase of K stops the unstable behavior completely and produces periodic motion, i.e., a center as depicted in panel (c). The amplitude of the orbit depends on the initial conditions which are chosen as $x = 0.01$ and $y = 0.01$. For even larger feedback gains, the trajectory becomes an inward spiral and thus approaches the fixed point, i.e., the focus. Hence the time-delayed control scheme is successful.

In order to change the stability of the fixed point, the goal of the control force is to influence the sign of the real part of the eigenvalues. If all eigenvalues Λ of the controlled system (3.2.2a, 3.2.2b) have negative real part, the system will be stable. An exponential ansatz for $x(t)$ and $y(t)$ in Eqs. 3.2.2a and 3.2.2b, i.e., $x(t) \sim \exp(\Lambda t)$ and $y(t) \sim \exp(\Lambda t)$, reveals how the control force modifies the eigenvalues of the system. The characteristic equation becomes

$$0 = \det\left[\begin{pmatrix} \lambda - \Lambda & \omega \\ -\omega & \lambda - \Lambda \end{pmatrix} - K\begin{pmatrix} 1 - e^{-\Lambda\tau} & 0 \\ 0 & 1 - e^{-\Lambda\tau} \end{pmatrix}\right] \tag{3.2.6a}$$

$$= \left[\Lambda + K\left(1 - e^{-\Lambda\tau}\right) - \lambda\right]^2 + \omega^2 \tag{3.2.6b}$$

so that the complex eigenvalues Λ are given in the presence of a control force by the implicit equation

$$\lambda \pm i\omega = \Lambda + K\left(1 - e^{-\Lambda\tau}\right). \tag{3.2.7}$$

This formula will serve as a reference case when I will discuss effects on the characteristic equation due to modifications and generalizations of the control scheme in later sections. The factorization of Eq. 3.2.6b is possible because to control scheme is realized in diagonal form with the identity matrix as coupling matrix. Otherwise, the characteristic equation would involve, for instance, mixed terms of Λ and the exponential $e^{-\Lambda\tau}$. In the present case, however, the characteristic equation can be solved analytically using the Lambert function which is defined as the inverse function of $f(z) = ze^z$ for complex variable z [20–25]. The eigenvalues Λ are given by

$$\Lambda\tau = W\left(K\tau e^{-(\lambda \pm i\omega)\tau + K\tau}\right) + (\lambda \pm i\omega)\tau - K\tau. \tag{3.2.8}$$

Let me stress that there are infinitely many solutions in terms of $\Lambda \in \mathbb{C}$ of this characteristic equation. This reflects the infinite dimensions of delay differential equations and can be seen in the characteristic equation by the multivalued property of the complex exponential.

The Lambert function is also known as product logarithm and has infinitely many branches. I will need to consider only the branches with positive real part because the corresponding modes determine the stability of the system. For details about the Lambert function see Appendix A of Refs. [25] and [26].

Panel (a) of Fig. 3.4 shows the dependence of the largest real part of the complex eigenvalues Λ upon the time delay τ according to Eqs. 3.2.7 and 3.2.8 for

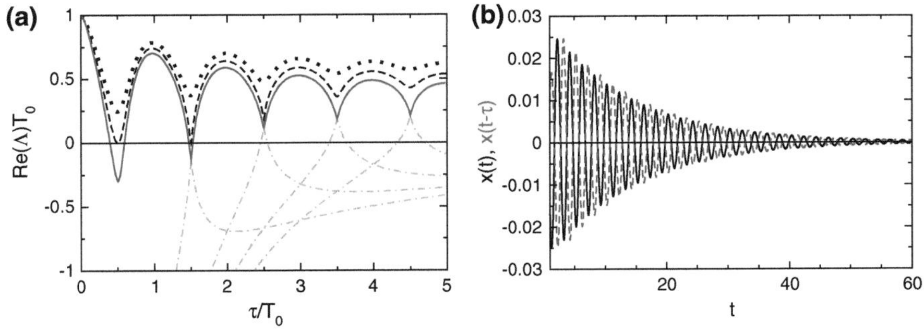

Fig. 3.4 Panel **a** Largest real part of the complex eigenvalues Λ versus τ for $\lambda = 0.5$ and $\omega = \pi$, i.e., $T_0 = 2$, for different feedback gains $KT_0 = 0.4$, 0.5, and 0.6 as *dotted, dashed black*, and *red (solid) curves*, respectively. Some lower eigenvalues are also displayed for $KT_0 = 0.6$ as *green (dash-dotted) curves*. Panel **b** Time series of the x component of the unstable focus: The *black (solid) curve* corresponds to $x(t)$, the *red (dashed) curve* to the delayed x component $x(t-\tau)$ with $\tau = 1$. The parameters of the unstable focus and the control scheme are as in panel **d** of Fig. 3.3

$\lambda = 0.5$ and $\omega = \pi$, which yields an intrinsic timescale $T_0 = 2$. The dotted, dashed black, and red (solid) curves correspond to a feedback gain of $KT_0 = 0.4$, 0.5, and 0.6, respectively. All curves start at $\mathrm{Re}(\Lambda) = \mathrm{Re}(\Lambda_0) = \lambda$ for $\tau = 0$, i.e., when no control is applied to system. For increasing time delay, the real part $\mathrm{Re}(\Lambda)$ decreases. It can be seen in the case of $KT_0 = 0.6$ that there exist values of the time delay for which $\mathrm{Re}(\Lambda)$ becomes negative. Thus, the control is successful. The curve for $KT_0 = 0.5$ shows the threshold case where $\mathrm{Re}(\Lambda)$ becomes zero for $\tau = 1 = T_0/2$, but does not change sign.

As mentioned above, the time-delayed control scheme generates an infinite number of additional eigenmodes and the corresponding eigenvalues are the solutions of the transcendental Eq. 3.2.7 [27, 28]. The real parts of the eigenvalues all originate from $-\infty$ for $\tau = 0$ [29, 30]. Some of these lower eigenvalues are displayed for $KT_0 = 0.6$ in Fig. 3.4a as green (dash-dotted) curves. The different branches of the eigenvalue spectrum originate from the multiple-leaf structure of the complex Lambert function. The real part of each eigenvalue branch exhibits a typical nonmonotonic dependence upon τ which leads to crossover of different branches resulting in an oscillatory modulation of the largest real part as a function of τ. Such behavior of the eigenvalue spectrum appears to be quite general, and has been found for various delayed feedback coupling schemes, including the Floquet spectrum of unstable periodic orbits [31, 32] and applications to noise-induced motion where the fixed point is stable [33].

The notch at $\tau/T_0 = 1$ corresponds to the time delay used in Fig. 3.3, so that at this value of τ the red, solid back, and dashed curves correspond to panels (d), (c), and (b) of Fig. 3.3, respectively. The notches at larger τ become less pronounced leading to less effective realization of the time-delayed control scheme, i.e., a smaller or no τ-interval with negative $\mathrm{Re}(\Lambda)$.

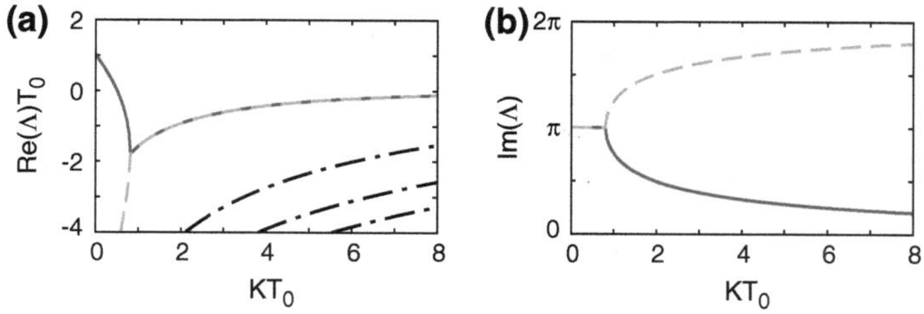

Fig. 3.5 Panels **a** and **b** depict the real and imaginary part of the complex eigenvalues Λ in dependence on the feedback gain K for fixed time delay $\tau = T_0/2 = 1$. The parameters of the unstable focus and the control scheme are as in panel **d** of Fig. 3.3

In the case of an unstable periodic orbit the optimal time delay is equal to the period T of the orbit to be stabilized. This yields the noninvasive property of time-delayed feedback control which is discussed in detail in Chap. 2. Note that in the case of an unstable steady state, however, the time delay is not so obviously related to a parameter of the system. I will investigate later which combinations of the feedback gain K and the time delay τ lead to successful control.

Panel (b) of Fig. 3.4 displays the time series of $x(t)$ and its time delayed counterpart $x(t - \tau)$ as black (solid) and red (dashed) curves, respectively. The control parameters are fixed at $KT_0 = 0.6$ and $\tau = T_0/2 = 1$ which leads to successful control as in panel (d) of Fig. 3.3. The x component of the control force can be calculated from the difference of the two curves and subsequent multiplication by K. Since $x(t)$ tends to zero in the limit of large t as the system reaches the focus located at the origin, the control force vanishes if the system is stabilized. Thus the control scheme is noninvasive. Note that the current signal in back and its delayed counterpart in red color are in anti-phase. This observation will become important in Sect. 3.4.

Figure 3.5 shows the real and imaginary part of the eigenvalues Λ in panels (a) and (b), respectively. In this plot, the feedback gain K is varied and the time delay is fixed at half the intrinsic timescale, i.e., $\tau = T_0/2 = 1$. The largest real part starts at the real part of the system's eigenvalue λ for $K = 0$. As the feedback gain increases, the largest real part becomes smaller and eventually changes sign at the flip threshold $K_{flip} = \lambda/2$. The real part of all eigenvalues are negative at this point and the system is stabilized. Panel (a) shows next to the red (solid) branch starting at $\mathrm{Re}(\Lambda) = \lambda$ four control-induced branches which rise from negative infinity. One can see in panel (b) that the largest delay-induced eigenvalue shown in green (dashed) color collides with the eigenvalue originating from the uncontrolled system and forms a complex pair. The other control-induced branches shown as black (dash-dotted) curves in panel (a) have larger imaginary parts. For larger feedback gain K, the fixed point can lose its stability again in a Hopf bifurcation. A derivation of the corresponding maximum feedback gain can be found in Ref. [26].

After having demonstrated successful stabilization of an unstable steady state, I will investigate details of the shape of the domain of control in the following. In order to discuss the domain of control in the (K, τ) plane, it is helpful to consider the real and imaginary part of the characteristic equation (3.2.7) separately. Introducing p and q as real and imaginary parts of the eigenvalue Λ, i.e., $\Lambda = p + iq$, this equation becomes

$$\lambda = p + K[1 - e^{-p\tau}\cos(q\tau)] \tag{3.2.9a}$$

$$\omega = q + Ke^{-p\tau}\sin(q\tau). \tag{3.2.9b}$$

If one is interested in analytical results about the domain of control, the calculation can be done for special points and conditions. For instance, the condition of vanishing real part, i.e., $\mathrm{Re}(\Lambda) = p = 0$, corresponds to the boundary of the domain of control. It is the threshold case between stability and instability as show in Fig. 3.3c. At the threshold of control the sign of the real part p of the eigenvalue Λ changes. Therefore setting p to zero in the real and imaginary parts of Eqs. 3.2.9a and 3.2.9b, respectively, yields

$$\lambda = K[1 - \cos(q\tau)] \tag{3.2.10a}$$

$$\omega = q + K\sin(q\tau). \tag{3.2.10b}$$

A first analytical result can be inferred from the range of the cosine function. Since it is bounded between -1 and 1, a lower bound for the feedback gain in case of successful stabilization follows from Eq. 3.2.10a of the real part, i.e.,

$$\frac{\lambda}{2} \leq K. \tag{3.2.11}$$

Thus, a minimum feedback gain $K_{\min}$ is given by $K_{\min} = \lambda/2$. It should be noted that a similar characteristic equation as Eq. 3.2.7 holds for the Floquet exponents of an unstable periodic orbit, where the lower bound, $K_{\min} = \lambda/2$, of the feedback gain has been shown to correspond to the flip threshold of control [7, 8].

The conditions on the cosine function in Eq. 3.2.10a lead to a vanishing sine term in Eq. 3.2.10b of the imaginary part. This yields $q\tau = (2n + 1)\pi$ with $n = 0, 1, 2, \ldots$.

In order to express the values of the time delay τ that correspond to the minimum K in terms of the parameters of the uncontrolled system, it is useful to consider even and odd multiples of π for $q\tau$, i.e., $q\tau = 2n\pi$ and $q\tau = (2n + 1)\pi$ for $n = 0, 1, 2, \ldots$. In both cases, the imaginary part of Eqs. 3.2.10a and 3.2.10b leads to $q = \omega$. Hence, in the latter case, the time delay τ for $K_{\min} = \lambda/2$ becomes

$$\tau = \frac{\pi}{\omega}(2n+1). \tag{3.2.12}$$

The last expression can be rewritten using the uncontrolled eigenperiod $T_0 = 2\pi/\omega$

$$\tau = T_0 \frac{2n+1}{2}. \tag{3.2.13}$$

This discussion has shown that the combination $K = \lambda/2$ and $\tau = T_0\ (2n + 1)/2$ with $n = 0, 1, 2, \ldots$ correspond to points of successful control in the (K, τ) plane with minimum feedback gain. See, for instance, the solid black curve in Fig. 3.4a as well as Fig. 3.5.

For even multiples, i.e., $q\tau = 2n\pi$ for $n = 0, 1, 2, \ldots$, it follows from Eq. 3.2.10a that control is impossible for finite values of K, since

$$\frac{K-\lambda}{K} = \cos(q\tau)\big|_{q\tau=2n\pi} \tag{3.2.14a}$$

$$\Leftrightarrow \quad 1 - \frac{\lambda}{K} = 1 \tag{3.2.14b}$$

which cannot be satisfied for $\lambda \neq 0$ and finite K. Furthermore, Eq. 3.2.10b yields that for time delays which are integer multiples of the eigenperiod, i.e., $\tau = T_0\, n = 2\pi n/\omega$ with $n = 0, 1, 2, \ldots$, the control scheme fails for any feedback gain. These time delays correspond to the local extrema in Fig. 3.4a.

Another result that can be derived from Eq. 3.2.7 is a shift of q for increasing K. For this, taking the square of the real and imaginary part of Eq. 3.2.7 and using trigonometrical identities leads to

$$\left(\frac{K-\lambda}{K}\right)^2 + \left(\frac{\omega-q}{K}\right)^2 = \cos^2(q\tau) + \sin^2(q\tau) \tag{3.2.15a}$$

$$\Leftrightarrow \left(\frac{K-\lambda}{K}\right)^2 + \left(\frac{\omega-q}{K}\right)^2 = 1. \tag{3.2.15b}$$

The last equation can be solved for the imaginary part $\mathrm{Im}(\Lambda) = q$

$$\omega - q = \pm K\sqrt{1 - \left(\frac{K-\lambda}{K}\right)^2} \tag{3.2.16a}$$

$$\Leftrightarrow \quad q = \omega \mp \sqrt{(2K-\lambda)\lambda}. \tag{3.2.16b}$$

Inserting Eq. 3.2.16a into the real part of Eq. 3.2.7 leads to an explicit expression for the dependence of time delay τ on the feedback gain K at the threshold of stability, i.e., the boundary of the control domain $p = 0$

$$\frac{K-\lambda}{K} = \cos(q\tau) \tag{3.2.17a}$$

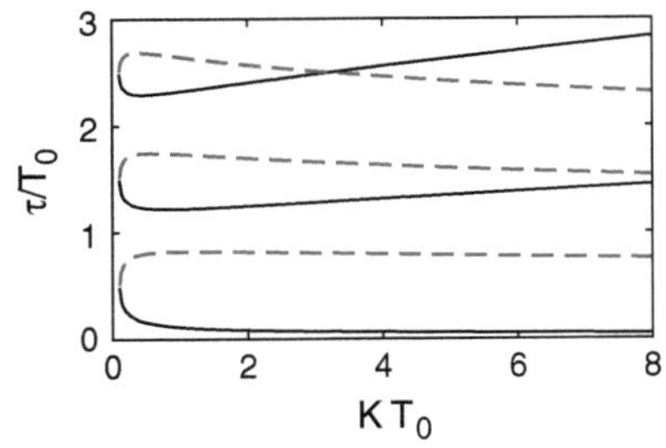

Fig. 3.6 Boundary of the domain of control in the (K, τ) plane according to Eqs. 3.2.18a, 3.2.18b and 3.2.18c. The *black* (*solid*) and *red* (*dashed*) *curves* correspond to the formula $\tau_1(K, n)$ and $\tau_2(K, n)$, respectively. The first three branches are plotted, i.e., $n = 0, 1, 2$. System's parameter: $\lambda = 0.1$ and $\omega = \pi$

$$\Leftrightarrow \quad \tau(K) = \frac{\arccos\left(\frac{K-\lambda}{K}\right)}{\omega \mp \sqrt{(2K-\lambda)\lambda}}. \tag{3.2.17b}$$

Keeping in mind the multivalued behavior of the arccos function and the different choices of the sign, there are three families of branches of solutions, where the non-negative integer n takes care of the different leaves of the involved multi-valued functions:

$$\tau_1(K,n) = \frac{2n\pi + \arccos\frac{K-\lambda}{K}}{\omega - \sqrt{(2K-\lambda)\lambda}}, \quad \frac{\lambda}{2} \leq K < \frac{\omega^2+\lambda^2}{2\lambda} \tag{3.2.18a}$$

$$\tau_2(K,n) = \frac{2(n+1)\pi - \arccos\frac{K-\lambda}{K}}{\omega + \sqrt{(2K-\lambda)\lambda}}, \quad \frac{\lambda}{2} \leq K \tag{3.2.18b}$$

$$\tau_3(K,n) = \frac{2(n+1)\pi - \arccos\frac{K-\lambda}{K}}{-\omega + \sqrt{(2K-\lambda)\lambda}}, \quad \frac{\omega^2+\lambda^2}{2\lambda} < K \tag{3.2.18c}$$

with non-negative integer n. The corresponding eigenvalues $\Lambda = iq$ are given by

$$q_{1,3} = \pm\left(\omega - \sqrt{(2K-\lambda)\lambda}\right) \tag{3.2.19a}$$

$$q_2 = \omega + \sqrt{(2K-\lambda)\lambda}. \tag{3.2.19b}$$

For the boundaries of the stability islands only the branches τ_1 and τ_2 are relevant.

Figure 3.6 depicts the boundary of the domain of control as it is analytically derived in Eqs. 3.2.18a, 3.2.18b and 3.2.18c. The solid and red (dashed) curves display the formula $\tau_1(K, n)$ and $\tau_2(K, n)$, respectively, and only the first three branches are plotted, i.e., $n = 0$, 1, and 2. Note that the control parameters are given in units of the intrinsic timescale T_0. The domain of control lies above the black (solid) curves and below the red (dashed) one.

The analytical expressions (3.2.18a, 3.2.18b, 3.2.18c) of the domain of control will be discussed in Sect. 3.6 in the context of large time delays. Note that at the points of minimum feedback gain and corresponding time delays

$$K_{\min} = \frac{\lambda}{2}, \quad \tau = \tau_{\min}(n) = \frac{(2n+1)\pi}{\omega} \tag{3.2.20}$$

the branch $\tau_1(K, n)$ ends, but is continued by $\tau_2(K, n)$.

An example of the combination of minimum feedback gain $K_{\min} = \lambda/2$ and corresponding time delay $\tau = T_0(2n+1)/2, n = 0, 1, 2, \ldots$ is shown in panel (c) of Fig. 3.3, where $K = \lambda/2 = 0.25$ and $\tau = T_0/2 = 1$. It describes the threshold case between stable and unstable fixed point.

Reconsidering the characteristic equations (3.2.9a) and (3.2.9b), another analytical result can be derived. This concerns a maximum divergence rate of a system such that time-delayed feedback control is still able to render it stable. The divergence rate is given by the real part of the uncontrolled eigenvalues λ. The imaginary part of the characteristic equation evaluated at the threshold of control as given by Eq. 3.2.9b leads to an expression for the feedback gain K

$$K = \frac{\omega - q}{\sin(q\tau)} \tag{3.2.21}$$

which can be inserted into the real part (3.2.9a) and thus yields

$$\lambda = \frac{(\omega - q)[1 - \cos(q\tau)]}{\sin(q\tau)}. \tag{3.2.22}$$

Choosing the time delay as $\tau = T_0/2 = \pi/\omega$ leads to

$$\lambda = (\omega - q)\frac{1 + \cos((\omega - q)\tau)}{\sin((\omega - q)\tau)} \tag{3.2.23a}$$

$$= \frac{(\omega - q)}{\tan\left(\frac{\omega - q}{2}\tau\right)}. \tag{3.2.23b}$$

The maximum value of λ is given for $\omega - q = 0$. This condition is clear from Fig. 3.5 where the largest real part in panel (a) is smallest at the collision of the red and green curves and the corresponding identity of the imaginary parts can be inferred from Fig. 3.5b. At this point, a complex pair of eigenvalues is born and their imaginary parts start a π which is the imaginary part ω of the uncontrolled eigenvalue. Thus, one obtains the following restriction on λ

$$\lambda = \left.\frac{\omega - q}{\tan\left(\frac{\omega - q}{2}\tau\right)}\right|_{\omega - q = 0} \tag{3.2.24a}$$

$$= \left.\frac{\frac{d}{d(\omega - q)}(\omega - q)}{\frac{d}{d(\omega - q)}\tan\left(\frac{\omega - q}{2}\tau\right)}\right|_{\omega - q = 0} \tag{3.2.24b}$$

$$= \frac{\tau}{2} \tag{3.2.24c}$$

$$\Leftrightarrow \lambda\tau = 2. \tag{3.2.24d}$$

If the collision of the red and green branches in Fig. 3.5 happens in the non-negative range of Re(Λ), the system cannot be stabilized by time-delayed feedback. Thus, the rate of divergence of the fixed point, i.e., λ, is restricted to $\lambda\,\tau \leq 2$.

In order to visualize the shape of the domain of control I will show how small deviations $\epsilon > 0$ from K_{min}, i.e, $K = \lambda/2+\epsilon$, influence the corresponding values of the time delay τ. This will yield a formula for the shape of the domain of control in the (k, τ) plane. For this, let $\eta > 0$ be small and $\tau = \frac{\pi}{\omega}(2n+1) \pm \eta$ a small deviation from τ at K_{min}. Inserting the expression for K and τ into Eq. 3.2.10a yields after some Taylor's expansions

$$-1 + \frac{4}{\lambda}\epsilon = -1 + \frac{1}{2}\left[\omega\eta \mp \frac{\pi}{\omega}(2n+1)\sqrt{2\lambda}\sqrt{\epsilon}\right]^2 \tag{3.2.25a}$$

$$\Leftrightarrow \eta = \left[\pm\frac{2\sqrt{2}}{\omega\sqrt{\lambda}} + \frac{\sqrt{2}\pi}{\omega^2}(2n+1)\sqrt{\lambda}\right]\sqrt{\epsilon}. \tag{3.2.25b}$$

A detailed derivation can be found in Chap. 2 of Ref. [26]. This equation describes the shape of the domain of control at the threshold of stabilization, i.e., $p = 0$, near the minimum feedback gain K at $\tau = T_0\ (2n + 1)/2$ in the (K, τ) control plane. Small deviations from K_{min} are connected to small deviations from $\tau = T_0\ (2n + 1)/2$ via a square root dependence.

Figure 3.7 displays the largest real part of the eigenvalues Λ in dependence on both the feedback gain K and the time delay τ for $\omega = \pi$ and two different values of λ, i.e., $\lambda = 0.5$ and 0.1 in panels (a) and (b), respectively. A discussion of the domain of control in the (λ, ω) plane can be found in Ref. [19].

Figure 3.7 summarizes the results of this section. The values of Λ are calculated using the analytic solution (3.2.8) of the characteristic equation (3.2.7). The two-dimensional projections at the bottom of each plot extract combinations of K and τ with negative Re(Λ), i.e., successful control of the system. Thus, the shaded areas indicate the domain of control whose boundary is given by the analytical expression (3.2.18a, 3.2.18b, 3.2.18c) and displayed in Fig. 3.6.

In the absence of a control force, i.e., $K = 0$, the real part of Λ starts at λ. Increasing the feedback gain decreases Re(Λ). For $K = K_{min} = \lambda/2$, the real part of the eigenvalue reaches 0 for certain time delays, i.e., $\tau = T_0\ (2n + 1)/2$ with $n = 0, 1, 2, \ldots$, and then changes sign. Thus, the system is stabilized. For values of the feedback gain slightly above the minimum value K_{min}, the domain of control shows a square root shape as predicted by Eq. 3.2.25b. It can be seen that for time delays of $\tau = T_0 n$ the largest real part of the eigenvalues remains positive for any feedback gain. For a smaller value of λ as in Fig. 3.7b, i.e., closer to the stability threshold of the fixed point, the domains of control become larger.

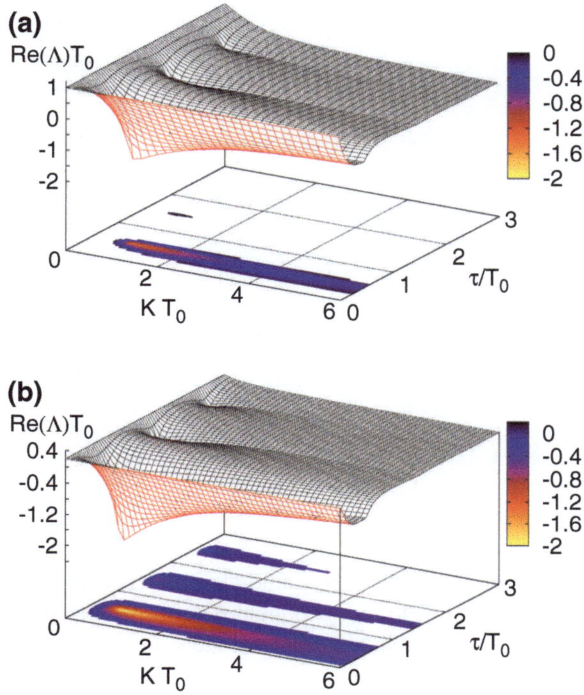

Fig. 3.7 Domain of control in the (K, τ) plane and largest real part of the complex eigenvalues Λ as a function of K and τ according to Eq. 3.2.8. The two-dimensional projection at the bottom shows combinations of τ and K, for which $\mathrm{Re}(\Lambda)$ is negative and thus the control successful. Panel **a**: $\lambda = 0.5$ and $\omega = \pi (T_0 = 2)$, panel **b** $\lambda = 0.1$ and $\omega = \pi (T_0 = 2)$

The real part of the complex eigenvalue Λ corresponds to a damping rate. For $\mathrm{Re}(\Lambda)$ it is a rate of convergence with which the trajectory approaches the fixed point. Therefore, one can define a transient time T_{trans} as the time needed to enter a disc of radius ϵ around the fixed point starting from an initial distance r_0. An analytical formula can be derived in the absence of control from the amplitude equation of the system (3.1.6a). Separation of variables leads to

$$T_{\mathrm{trans}}(r_0, \epsilon) = \int_{r_0}^{\epsilon} \frac{1}{\lambda r} dr \tag{3.2.26a}$$

$$= \frac{1}{\lambda} \log(r) \Big|_{r_0}^{\epsilon} \tag{3.2.26b}$$

$$= \frac{1}{\lambda} \log\left(\frac{\epsilon}{r_0}\right), \tag{3.2.26c}$$

which relates the real part of the uncontrolled eigenvalue $\Lambda_0 = \lambda \pm i\omega$ to the transient time T_{trans}. In the presence of control, the control force changes the eigenvalues of the system. Thus, one can substitute λ in the formula (3.2.26c) by the largest real part $\mathrm{Re}(\Lambda)$ calculated from the characteristic equation (3.2.7) if the different branches are well separated. Then, the transient time can be rewritten as

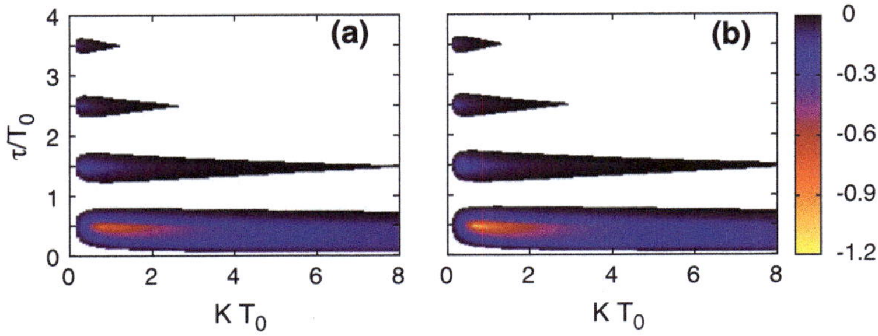

Fig. 3.8 Largest real part of the eigenvalue Λ in dependence on the feedback gain K and time delay $\tau = 1$. Panel **a** shows the value of Re(Λ) according to Eq. 3.2.27 and Panel **b** displays Re(Λ) as calculated from the characteristic equation (3.2.7). The *color code* shows only negative values of Re(Λ). *Parameters* $\lambda = 0.1$ and $\omega = \pi(T_0 = 2)$

$$T_{\text{trans}}(\Lambda; r_0, \epsilon) = \frac{1}{\text{Re}(\Lambda)} \log\left(\frac{\epsilon}{r_0}\right). \tag{3.2.27}$$

A similar relation can be found for systems with periodic orbits where the real part of the leading Floquet exponent determines the stability. See Ref. [34] for details concerning various classes of models including chaotic dynamics.

Figure 3.8 depicts the largest of the complex eigenvalue Λ for $\lambda = 0.1$ and $\omega = \pi(T_0 = 2)$. Panel (b) shows the value of Re(Λ) as calculated from the characteristic equation (3.2.7) and is identical to the bottom projection of Fig. 3.7b. Panel (a) depicts Re(Λ) calculated from simulations of the system's dynamics (3.2.1) according to Eq. 3.2.7. The color code shows only negative values of Re(Λ).

The radius which determines the size of the neighborhood around the fixed point is chosen as $\epsilon = 0.01$ and the trajectory is calculated for 10^7 timesteps with a time step of $\Delta t = 0.001$, i.e., $t_{\max} = 10{,}000$. The simulation starts at $(x_0, y_0) = (0.1, 0.1)$ without control. The control is switched on at $t = 10$. This yields an initial distance for the approach of the fixed point of $r_0 = \sqrt{x_0^2 + y_0^2}\exp(10\lambda)$. One can see that panels (a) and (b) coincide very well. Thus, the real part of the eigenvalue can be interpreted as a damping rate with which the trajectory approaches the fixed point or moves away from it. Small discrepancies at the boundaries, for instance, for large time delays are due to a very slow convergence/divergence of the system and a maximum transient time corresponding to the simulation time, i.e., $T_{\text{trans}} \leq t_{\max}$. In the limit $t_{\max} \longrightarrow \infty$ and vanishing ϵ, the domains of control from the two panels should coincide.

3.2.2 Saddle Point

Before discussing an extension of the Pyragas control in the next section, let me discuss effects of time-delayed feedback of Pyragas type on fixed points with one unstable dimension, but without torsion, i.e., saddle points. At first, I will consider a one dimensional model and show that this system cannot be stabilized by time-delayed feedback. In addition, I will also investigate a two dimensional model which has one stable and one unstable manifold.

In one dimension, an unstable fixed point is a repeller which can be written including time-delayed feedback control as

$$\frac{dx(t)}{dt} = \alpha x(t) - K[x(t) - x(t-\tau)], \tag{3.2.28}$$

where α is a positive constant denoting the divergence rate. As in the previous discussion, the control force is determined by the feedback gain K and the time delay τ and using an exponential ansatz, i.e., $x(t) \sim e^{\Lambda t}$, yields a characteristic equation

$$\Lambda = \alpha - K\left(1 - e^{-\Lambda\tau}\right) \tag{3.2.29}$$

which can be solved in terms of the Lambert function W such that the root of Eq. 3.2.29 are given by

$$\Lambda\tau = W\left(K\tau e^{-\alpha\tau + K\tau}\right) + (\alpha - K)\tau. \tag{3.2.30}$$

In the following, I will show analytically that time-delayed feedback cannot stabilize the repeller. Consider the characteristic equation (3.2.29), which can be separated in real and imaginary parts using $\Lambda = \lambda + i\omega$

$$\lambda = \alpha - K\left[1 - e^{-\lambda\tau}\cos(\omega\tau)\right] \tag{3.2.31a}$$

$$\omega = -Ke^{-\lambda\tau}\sin(\omega\tau). \tag{3.2.31b}$$

If the saddle point became stable, there would be a combination of K and τ such that the real part of the complex eigenvalue Λ became zero, i.e., $\lambda = 0$. Then, Eqs. 3.2.31a and 3.2.31b become

$$\alpha = K[1 - \cos(\omega\tau)] \tag{3.2.32a}$$

$$\omega = -K\sin(\omega\tau). \tag{3.2.32b}$$

The latter equation has always the solution $\omega = 0$ which reflects that the eigenvalue of the one-dimensional saddle point is real. This solution, however, violates Eq. 3.2.32b for the real part since the value of α is a positive.

Equation 3.2.31b provides the following expression for the feedback gain K at the threshold of control

$$-K = \frac{\omega}{\sin(\omega\tau)} \tag{3.2.33}$$

which yields together with the Eq. 3.2.31a for the real part

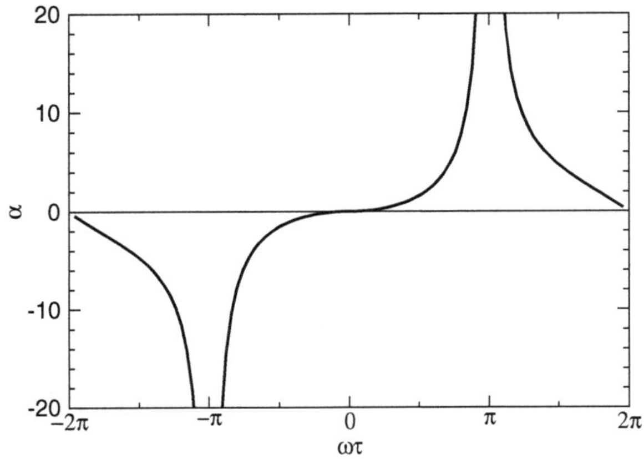

Fig. 3.9 Functional dependence of the parameter α on ω according to Eq. 3.2.34c

$$\alpha = K[1 - \cos(\omega\tau)] \tag{3.2.34a}$$

$$= -\omega \frac{1 - \cos(\omega\tau)}{\sin(\omega\tau)} \tag{3.2.34b}$$

$$= \pm\omega \sqrt{\frac{1 - \cos(\omega\tau)}{1 + \cos(\omega\tau)}}. \tag{3.2.34c}$$

The last equation is a functional dependence which must be satisfied in case of stabilization. This dependence is depicted in Fig. 3.9 which show a divergence for $\omega\tau = \pm\pi$.

Since the eigenvalue of the uncontrolled system $\Lambda_0 = \alpha$ is real and a solution of Eqs. 3.2.32a and 3.2.32b with $\omega = 0$ has been ruled out, a mode generated by the control method can collide with the unstable mode of the system for $\omega\tau = \pm\pi$, i.e., the corresponding eigenvalue must be real. As depicted by Fig. 3.9, the corresponding value of α would be infinity according to Eq. 3.2.34c.

Therefore it is clear that time-delayed feedback control fails for a one-dimensional repeller. Solving the characteristic equation numerically supports this result as will be discussed in the following.

Figure 3.10 displays solutions of Eq. 3.2.29 according to Eq. 3.2.30. Note that the Lambert function is multivalued. Thus, Fig. 3.10 depicts only six branches with largest real part. The system's parameter is chosen as $\alpha = 0.1$ and the time delay τ is fixed at $\tau = 1$. It can be seen that the largest curve shown in black (solid) starting at $\Lambda = \alpha$ remains positive for all values of the feedback gain K. Thus, the one-dimensional system (3.2.28) cannot be stabilized by time-delayed feedback. All other modes starting from $-\infty$ remain negative. See red (dashed) curves.

A similar results hold for a two-dimensional saddle point which can be written in the presence of a diagonal time-delayed feedback scheme as follows

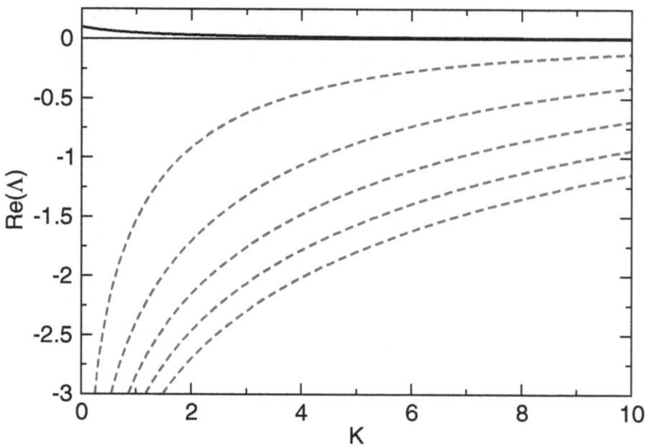

Fig. 3.10 Real part of the eigenvalue Λ in dependence on the feedback gain K for fixed time delay $\tau = 1$ according to Eq. 3.2.30. *Parameter* $\alpha = 0.1$

$$\begin{pmatrix} \frac{dx(t)}{dt} \\ \frac{dy(t)}{dt} \end{pmatrix} = \begin{pmatrix} \alpha_1 & 0 \\ 0 & -\alpha_2 \end{pmatrix} \begin{pmatrix} x(t) \\ y(t) \end{pmatrix} + K \begin{pmatrix} x(t-\tau) - x(t) \\ y(t-\tau) - y(t) \end{pmatrix}, \tag{3.2.35}$$

where α_1 and α_2 are real, positive constants. Without loss of generality, one can consider the case $\alpha_1 = 1$ by rescaling the time. Since the eigenvalues of the systems are given by $\lambda_1 = \alpha_1 > 0$ and $\lambda_2 = -\alpha_2 < 0$, the system has one stable and one unstable manifold. Figure 3.11 displays the vector field of a saddle point and a landscape of a corresponding potential $U(x, y) = -\alpha_1 x^2/2 + \alpha_2 y^2/2$ in the absence of a control scheme.

It was demonstrated above that the one-dimensional repeller cannot be stabilized by a diagonal control scheme. Since the systems of a saddle point decouples into two independent one-dimensional systems, time-delayed feedback control fails for the two-dimensional saddle, as well. It is tempting to introduce some

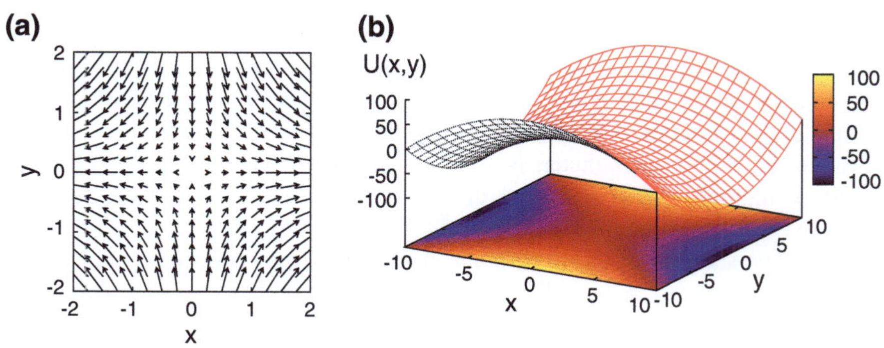

Fig. 3.11 Panel **a**: Vector field of a saddle point. The *arrows* show example trajectories. Panel **b**: Corresponding potential $U(x, y) = -\alpha_1 x^2/2 + \alpha_2 y^2/2$ corresponding to Eq. 3.2.35 for parameters $\alpha_1 = \alpha_2 = 1$. The *figure at the bottom* shows a projection which gives the value of the potential in *color code*

torsion by nondiagonal coupling of the control force. I will show in Sect. 3.5.2 that this does not yield a stabilization of the saddle point.

To summarize this section, I have demonstrated that unstable steady states of focus type can be stabilized by time-delayed feedback of Pyragas type where the control force included states of the system that are delayed by a time unit τ. The control scheme fails for saddle points.

In the next section, I will consider an extension of this control scheme which takes more information of the system's history into account. To be precise, the control will include all states which are delayed by an integer multiple of a basic time delay.

3.3 Extended Time-Delayed Feedback

Shortly after the rise of the idea of time-delayed feedback by Pyragas [9], Socolar et al. extended the original feedback scheme such that the control force did not only include the difference of the current state of the system to its delayed version, but also differences between states which were delayed by an integer multiple of a basic time delay [35]. I discussed the general form of this extended time-delayed feedback already in Sect. 2.2. Following Ref. [15], the system of an unstable focus as given by Eqs. 3.1.4a and 3.1.4b becomes in the presence of this control scheme

$$\frac{dx(t)}{dt} = \lambda x(t) + \omega y(t) - K\sum_{n=0}^{\infty} R^n[x(t-n\tau) - x(t-(n+1)\tau)] \tag{3.3.1a}$$

$$\frac{dy(t)}{dt} = -\omega x(t) + \lambda y(t) - K\sum_{n=0}^{\infty} R^n[y(t-n\tau) - y(t-(n+1)\tau)], \tag{3.3.1b}$$

where I consider the special case of diagonal coupling of the control force realized by an identity matrix as coupling matrix. Since I prepare the uncontrolled system as an unstable focus, I choose positive and non-zero values for the parameters λ and ω, respectively. The additional real control parameter R whose absolute value is smaller than 1 acts as a weight of information that are further in the past. Thus, it is called the memory parameter. Note that vanishing R recovers the time-delayed feedback method of Pyragas type discussed in the previous section. See, for instance, Eqs. 3.2.2a and 3.2.2b.

While the stability of the fixed point in the absence of control is given by the eigenvalues of matrix **A** as in Eq. 3.1.1b, i.e., $\Lambda_0 = \lambda \pm i\omega$, one has to solve the following characteristic equation in the case of an extended time-delayed control force:

$$\Lambda + K\frac{1 - e^{-\Lambda\tau}}{1 - Re^{-\Lambda\tau}} = \lambda \pm i\omega. \tag{3.3.2}$$

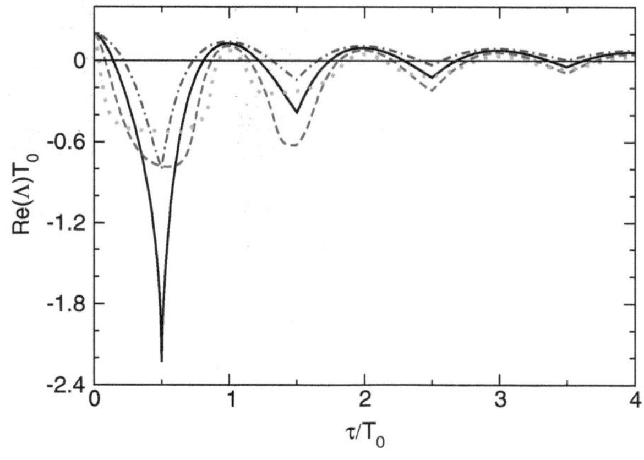

Fig. 3.12 Largest real part of the complex eigenvalues Λ as a function of τ for different values of R. The *blue (dash-dotted)*, *black (solid)*, *red (dashed)*, and *green (dotted) curves* correspond to $R = -0.35$, 0, 0.35, and 0.7, respectively. The parameters of the unstable focus are chosen as $\lambda = 0.1$ and $\omega = \pi$ which yields an intrinsic timescale $T_0 = 2\pi/\omega = 2$. The feedback gain K is fixed at $KT_0 = 0.6$

Due to the exponential, this characteristic equation becomes transcendental and possesses an infinite but countable set of complex solutions Λ [27, 28]. Note that vanishing memory parameter, i.e., $R = 0$, leads to the characteristic equation of Pyragas control as given in Eq. 3.2.7 in the previous Sect. 3.2. In this case of simple time-delayed feedback, the characteristic equation can be solved analytically in terms of the Lambert function [13, 20, 23, 25] as shown in Eq. 3.2.8. Let me stress that for non-zero memory parameter R, however, such a compact analytic expression is not possible. Thus, one has to solve Eq. 3.3.2 numerically.

Figure 3.12 depicts the dependence of the largest real parts of the eigenvalue Λ upon the time delay τ according to Eq. 3.3.2 for different memory parameters R and fixed feedback gain $KT_0 = 0.6$. The blue (dash-dotted), black (solid), red (dashed), and green (dotted) curves of Re(Λ) correspond to $R = -0.35$, 0, 0.35, and 0.7, respectively. The parameters of the unstable focus are chosen as $\lambda = 0.1$ and $\omega = \pi$. Note that the time delay τ is given in units of the intrinsic timescale $T_0 = 2\pi/\omega$. When no control is applied to the system, i.e., $\tau = 0$, all curves start at λ which corresponds to the real part of the uncontrolled eigenvalue. For increasing time delay, the real part of Λ decreases and eventually changes sign. Thus, the fixed point becomes stable. Note that there is a minimum of Re(Λ) indicating strongest stability if the time delay τ is equal to half the intrinsic period. For larger values of τ, the real part increases and becomes positive again. Hence, the system loses its stability. Above $\tau = T_0$, the cycle is repeated but the minimum of Re(Λ) is not so deep. The control method is less effective because the system has already evolved further away from the fixed point. For Pyragas control, i.e., $R = 0$, as discussed in Sect. 3.2, the minimum is deepest, however, the control interval, i.e., values of τ with negative real parts of Λ, increases for larger R. Therefore, the extended time-delayed control method is superior in comparison to the Pyragas scheme.

The control parameter space is three-dimensional and spanned by the time delay τ, the feedback gain K, and the memory parameter R. Thus, I will discuss in the following several one-dimensional projections as depicted, for instance, in

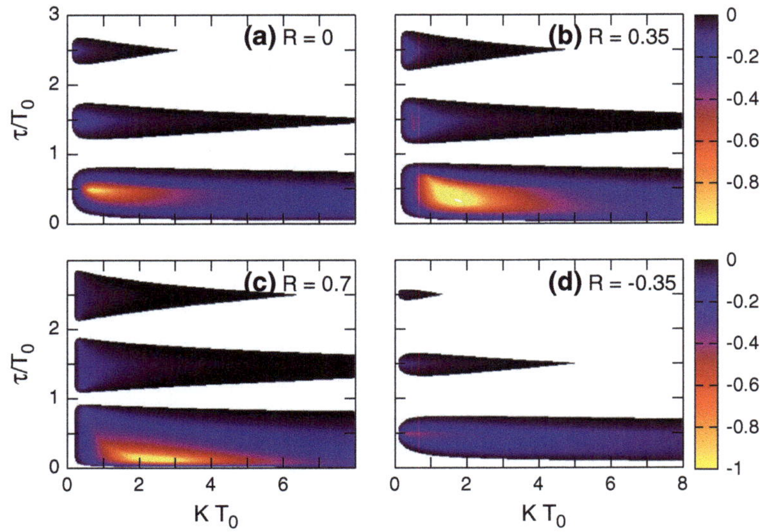

Fig. 3.13 Domain of control in the (K, τ) plane for different values of $R = 0$, 0.35, 0.7, and -0.35 in panels **a**, **b**, **c**, and **d**, respectively. The *color code* shows only negative values of the largest real part of the complex eigenvalues Λ according to Eq. 3.3.2. The parameters of the system are as in Fig. 3.12

Fig. 3.12 as well as two-dimensional projections where the third control parameter remains fixed.

Figure 3.13 shows the domain of control in the plane parameterized by the feedback gain K and time delay τ for different values of $R = 0$, 0.35, 0.7, and -0.35 in panels (a), (b), (c), and (d), respectively. The color code indicates only negative values of the largest real parts of the complex eigenvalue Λ. Therefore, Fig. 3.12 can be understood as a vertical cut through Fig. 3.13 for a fixed value of $KT_0 = 0.6$. Each panel displays several islands of stability which shrink for larger time delays τ. Note that no stabilization is possible if τ is equal to an integer multiple of the intrinsic period T_0. The domains of control become larger if the memory parameter R is closer to 1.

In order to obtain some analytic information of the domain of control, it is helpful to separate the characteristic equation (3.3.2) into real and imaginary parts. This yields using $\Lambda = p + iq$

$$K(1 - e^{-p\tau}\cos q\tau) = \lambda - p - Re^{-p\tau}[(\lambda - p)\cos q\tau \pm (\omega - q)\sin q\tau] \quad (3.3.3a)$$

$$Ke^{-p\tau}\sin q\tau = \pm(\omega - q) + Re^{-p\tau}[(\lambda - p)\sin q\tau \pm (\omega - q)\cos q\tau]. \quad (3.3.3b)$$

The boundary of the domain of controls is determined by a vanishing real part of Λ, i.e., $\mathrm{Re}(\Lambda) = p = 0$. With this constraint, Eqs. 3.3.3a and 3.3.3b can be rewritten as

$$K(1 - \cos q\tau) = \lambda - R[\lambda \cos q\tau \pm (\omega - q) \sin q\tau] \tag{3.3.4a}$$

$$K \sin q\tau = \pm(\omega - q) + R[\lambda \sin q\tau \pm (\omega - q) \cos q\tau]. \tag{3.3.4b}$$

At the threshold of control determined by a purely imaginary eigenvalue $\Lambda = i\omega$, there is—as in the case of Pyragas control in Sect. 3.2—a certain value of the time delay which will serve as a reference in the following given by

$$\tau = \frac{(2n+1)\pi}{\omega} = \left(n + \frac{1}{2}\right) T_0, \tag{3.3.5}$$

where n is any non-negative integer. For this special choice of the time delay, the range of possible feedback gains K in the domain of control becomes largest as can be seen in Fig. 3.13. Hence, this τ value will be referred to as optimal time delay in the following.

The value of a minimum feedback gain in dependence on the memory parameter R can be derived from Eq. 3.3.4a considering the extremal value of the trigonometrical function. Replacing cos $q\tau$ by its smallest values -1 yields

$$2K_{\text{min}} = \lambda + R\lambda \tag{3.3.6a}$$

$$\Rightarrow \quad K_{\text{min}}(R) = \frac{\lambda(1+R)}{2}. \tag{3.3.6b}$$

A detailed derivation of the maximum feedback gain at which the system loses its stability again via a Hopf bifurcation can be found in Ref. [7, 26, 36].

Extracting an expression for $\sin(q\tau)$ from Eq. 3.3.4a and inserting it into the equation for the imaginary part (3.3.4a) leads after some algebraic manipulation to a general dependence of K on the imaginary part q of Λ

$$K(q) = \frac{(1+R)[\lambda^2 + (\omega - q)^2]}{2\lambda}. \tag{3.3.7}$$

Taking into account the multivalued properties of the arcsine function, Eqs. 3.3.4a and 3.3.4b yield an analytical expressions of the time delay in dependence on q

$$\tau_1(q) = \frac{\arcsin\left(\frac{2\lambda(1-R^2)(\omega-q)}{\lambda^2(1-R^2)^2+(\omega-q)^2(1+R)^2}\right) + 2n\pi}{q}, \tag{3.3.8a}$$

$$\tau_2(q) = \frac{-\arcsin\left(\frac{2\lambda(1-R^2)(\omega-q)}{\lambda^2(1-R^2)^2+(\omega-q)^2(1+R)^2}\right) + (2n+1)\pi}{q}, \tag{3.3.8b}$$

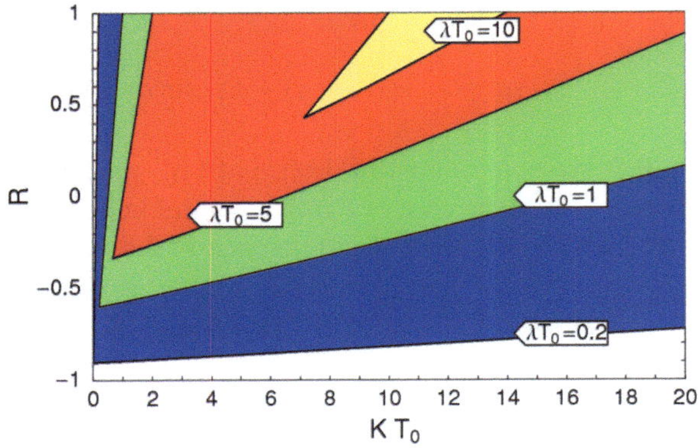

Fig. 3.14 Domain of control in the (K, R) plane for different values of λ. The *blue, green, red, and yellow (black, medium gray, dark gray, and light gray) domains* correspond to $\lambda T_0 = 0.2$, 1, 5, and 10, respectively, as indicated. The time delay is chosen as $\tau = T_0/2$ and $\omega = \pi$

where n is a non-negative integer. Together with Eq. 3.3.7, these formulas describe the boundary of the domain of control in Fig. 3.13. Note that two expressions τ_1 and τ_2 are necessary to capture the complete boundary. The case of simple time-delayed control is included as special choice of $R = 0$ and will be analyzed for large delays in Sect. 3.6.

For a better understanding of effects due to the memory parameter R, it is instructive to consider the domain of control in the plane parameterized by R and the feedback gain K. The results can be seen in Fig. 3.14, where the blue, green, red, and yellow areas correspond to the domain of control for $\lambda T_0 = 0.2$, 1, 5, and 10, respectively. The other system parameter is chosen as $\omega = \pi$. I keep the time delay constant at $\tau = T_0/2$. Note that the K interval for successful control increases for larger values of R. In fact, while the original Pyragas scheme, i.e., $R = 0$, fails for $\lambda T_0 = 10$, the extended time-delayed feedback method is still able to stabilize the fixed point. The upper left boundary corresponds to Eq. 3.3.6b. The lower right boundary can be described by a parametric representation which can be derived from the characteristic equation (3.3.2)

$$R = \frac{\lambda\tau - \vartheta\tan(\vartheta/2)}{\lambda\tau + \vartheta\tan(\vartheta/2)}, \tag{3.3.9a}$$

$$K\tau = \frac{\vartheta^2 + (\lambda\tau)^2}{\lambda\tau + \vartheta\tan(\vartheta/2)}, \tag{3.3.9b}$$

where the abbreviation $\vartheta = (q - \omega)\tau$ is used for notational convenience. For a detailed derivation of theses formulas see Ref. [37]. The range of ϑ is given by $\vartheta \in [0, \pi)$. A linear approximation leads to an analytic dependence of R and the feedback gain K given by a function $R(K)$ instead of the parametric equations (3.3.9a, 3.3.9b). A Taylor expansion around $\vartheta = \pi$ yields the following maximum feedback gain [37]

$$K_{\max}(R) = \frac{\lambda^2 + \pi^2}{2\lambda}(R + 1) + 2(R - 1). \tag{3.3.10}$$

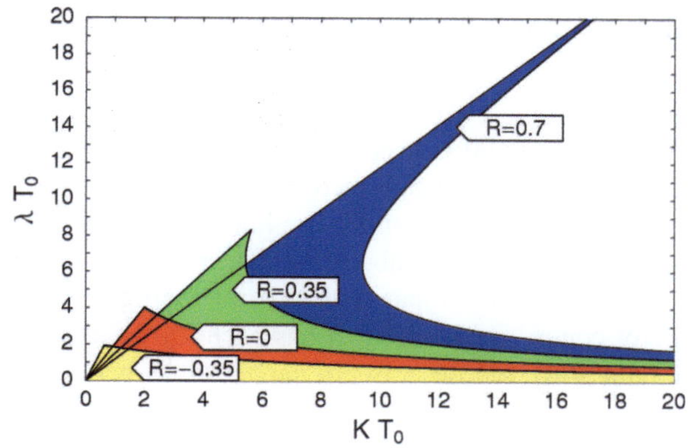

Fig. 3.15 Domain of control in the (K, λ) plane for different memory parameters R. The *yellow, red, green, and blue* (*light gray, dark gray, medium gray, and black*) *areas* correspond to $R = -0.35$, 0 (Pyragas control), 0.35, and 0.7, respectively. The time delay is fixed at $\tau = T_0/2$

Another representation of the superior control ability of extended time-delayed feedback is depicted in Fig. 3.15. The domain of control is given in the (K, λ) plane for different values of R. The yellow, red, green, and blue areas refer to $R = -0.35$, 0 (Pyragas control), 0.35, and 0.7, respectively. The time delay is chosen as $\tau = T_0/2$. One can see that for increasing R, the extended time-delayed feedback method can stabilize systems in a larger λ range [15, 26]. However, the corresponding K interval for successful control can become small. See, for instance, the blue area for $R = 0.7$ and large λ. A similar behavior was found in the case of stabilization of an unstable periodic orbit by extended time-delayed feedback control [8]. Let me stress that, as in the case of periodic orbits, the boundaries of the shaded areas can be calculated analytically from the following expression

$$K\tau = \frac{(1-R)\vartheta}{\tan(\vartheta/2)}\left[\left(\frac{1+R}{1-R}\right)^2 + \tan^2(\vartheta/2)\right] \tag{3.3.11a}$$

$$\lambda\tau = \frac{\vartheta}{\tan(\vartheta/2)}\left(\frac{1+R}{1-R}\right), \tag{3.3.11b}$$

where $\vartheta = (q-\omega)\tau$ with $\vartheta \in [0, \pi)$ is used as in Eqs. 3.3.9a and 3.3.9b. The maximum value for λ, which can be stabilized, is given by the special case $\vartheta = 0$ as

$$\lambda_{\max}\tau = 2\frac{1+R}{1-R}. \tag{3.3.12}$$

Note that the limit of vanishing memory parameter recovers the formula for the maximum divergence rate in the case of Pyragas control as derived in Eq. 3.2.24d.

Until now I have considered only a diagonal, instantaneous coupling of the control force. In the following sections, I will take latency effects as well as phase-dependent coupling into account. Both extensions are relevant in an experimental realization of time-delayed feedback control.

3.4 Latency Effects

In this section, I will consider a modifications of time-delayed feedback both in the case of the Pyragas scheme (Sect. 3.4.1) and the case of extended time-delayed feedback (Sect. 3.4.2) that involves non-zero latency times. These can be associated with the generation and injection of the feedback signal [38]. In optical realizations of the feedback scheme, for instance, the control loop latency enters due to the finite propagation time of the light between the laser and the Fabry–Perot control device [39]. It has been shown experimentally in a diode resonator circuit [40] for the case of an unstable periodic orbit that latency can have important effects on the controllability of the system and might limit the success of the time-delayed feedback method. Let me stress that in the case of unstable periodic orbits, Just has shown that longer latency times reduce the control abilities of the time-delayed feedback of Pyragas type [8, 41]. Similar results were found for extended time-delayed feedback [42]. Here I will discuss how latency times change the domain of control in the case of an unstable steady state. The discussions of this section follow the lines of reasoning of Refs. [13] and [15].

3.4.1 Time-Delayed Feedback

The latency time acts as an additional delay in all arguments of the Pyragas control force given by Eqs. 3.2.2a and 3.2.2b. Following Ref. [13], the system of an unstable focus subject to time-delayed feedback control including a latency δ can be written as

$$\frac{dx(t)}{dt} = \lambda x(t) + \omega y(t) - K[x(t-\delta) - x(t-\delta-\tau)] \tag{3.4.1a}$$

$$\frac{dy(t)}{dt} = -\omega x(t) + \lambda y(t) - K[y(t-\delta) - y(t-\delta-\tau)], \tag{3.4.1b}$$

where the control force is coupled to the system by an identity matrix, i.e., diagonal coupling. As in the case of vanishing latency times, the roots of a characteristic equation provide information about the stability of the system. This equation is given by

$$\lambda \pm i\omega = \Lambda + Ke^{-\Lambda\delta}\left(1 - e^{-\Lambda\tau}\right). \tag{3.4.2}$$

Note that in contrast to the case of instantaneous feedback as in Eq. 3.2.7 the latency time δ enters as an additional exponential. This has the consequence that solving Eq. 3.4.2 analytically by the Lambert function is no longer possible for non-zero latency times.

Figure 3.16 displays the dependence of the largest real part of the complex eigenvalues Λ on the time delay τ according to Eq. 3.4.2 for $\lambda = 0.5$ and $\omega = \pi$,

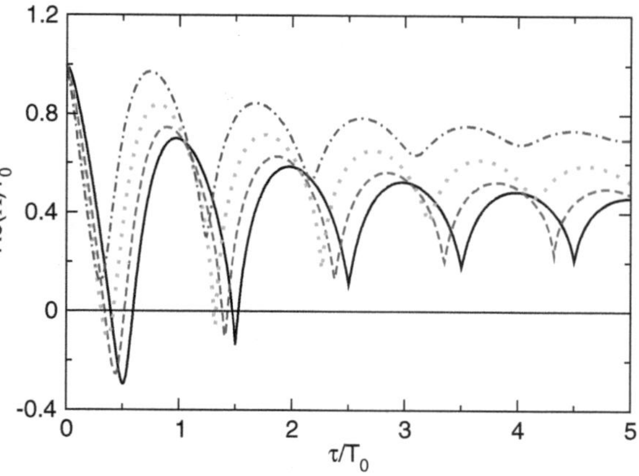

Fig. 3.16 Largest real part of the eigenvalues Λ versus τ for $\lambda = 0.5$, $\omega = \pi$ ($T_0 = 2$), and fixed feedback gain $KT_0 = 0.6$ as given by Eq. 3.4.2. The *black* (*solid*), *red* (*dashed*), *green* (*dotted*), and *blue* (*dash-dotted*) *curves* correspond to a latency time of $\delta = 0$, $0.05T_0$, $0.1T_0$, and $0.15T_0$, respectively

and different values of the latency time δ. The eigenvalues are calculated by solving Eq. 3.4.2 numerically. The black (solid), red (dashed), green (dotted), and blue (dash-dotted) curves correspond to $\delta = 0$, $0.05T_0$, $0.1T_0$, and $0.15T_0$, respectively. The feedback gain is fixed at part of the eigenvalues Λ versus τ for $\lambda = 0.5$, $\omega = \pi$ ($T_0 = 2$), and fixed feedback gain $KT_0 = 0.6$. The case of zero latency time is also displayed; it corresponds to the solid curve in Fig. 3.4a.

In order to understand the effects of control loop latency on the success of time-delayed feedback, it is helpful to separate the characteristic equation (3.4.2) into real and imaginary parts

$$p + K\left[e^{-p\delta}\cos(q\delta) - e^{-p(\tau+\delta)}\cos(q(\tau+\delta))\right] = \lambda \tag{3.4.3a}$$

$$q - K\left[e^{-p\delta}\sin(q\delta) - e^{-p(\tau+\delta)}\sin(q(\tau+\delta))\right] = \omega, \tag{3.4.3b}$$

where p and q denote again the real and imaginary part of Λ, respectively. Setting the real part to zero at threshold of control, i.e., $\mathrm{Re}(\Lambda) = p = 0$, results in a minimum feedback gain $K_{\min}$ if $\mathrm{Im}(\Lambda)\tau = q\tau = \pi(2n+1)$ for $n = 0, 1, 2, \ldots$ as derived on Sect. 3.2. This value of the imaginary part yields

$$K_{\min}(\delta) = \frac{\lambda}{2\cos\left[\pi(2n+1)\frac{\delta}{\tau}\right]} \geq \frac{\lambda}{2}. \tag{3.4.4}$$

Note that case of instantaneous coupling gives a lower bound in the minimum feedback gain. For non-zero latency times, the value of the feedback gain is larger. Compare also the case of vanishing δ that recovers Eq. 3.2.20.

The effects of the cosine function in Eq. 3.4.4 can be understood by considering Fig. 3.4b which depicts the time series of $x(t)$ and $x(t-\tau)$ and Fig. 3.17 which shows the dependence of the minimum feedback gain $K_{\min}$ on the latency time for $\lambda = 0.5$. Increasing latency time increases the value of $K_{\min}$. If δ becomes larger than half the time delay τ, $K_{\min}$ changes sign and control is possible only for

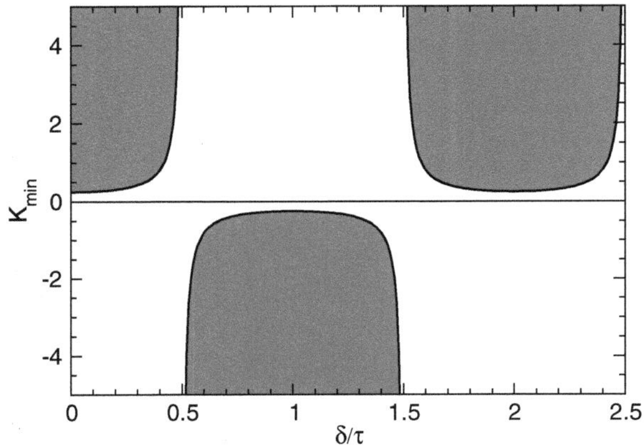

Fig. 3.17 Minimum feedback gain K_{min} versus relative latency δ/τ for $\lambda = 0.5$ and $\omega = \pi$ according to Eq. 3.4.4. The *shaded areas* show the domain of control for suitably chosen τ

negative K with $K < K_{\mathrm{min}}$ and suitably chosen τ. Note that in Fig. 3.4b the difference $x(t) - x(t - \tau)$ has to be positive for successful control. For $0.5\ \tau < \delta < 1.5\tau$ both $x(t - \delta)$ and $x(t - \delta - \tau)$ becomes closer to zero. Therefore in order to achieve control, the feedback gain becomes larger. In the limit $\delta/\tau \to 1/2$ the difference $x(t - \delta) - x(t - \tau - \delta)$ vanishes and thus the minimum feedback gain diverges. For even larger values of δ the above-mentioned difference changes its sign forcing K_{min} to do the same. Otherwise the control scheme would generate a force that pulls the system away from the target fixed point.

Figure 3.18 depicts the domain of control for system parameters $\lambda = 0.5$ and 0.1 in panels (a), (b) and (c), (d), respectively. The latency times are fixed at $\delta = 0.05T_0$ in panels (a) and (c) and $\delta = 0.15T_0$ in panels (b) and (d). The largest real part of the complex eigenvalues Λ is shown by color code in the (K, τ) plane, where only negative values corresponding to stability are plotted. It can be seen that increasing latency times reduce the domain of control. For instance, the small range at a time delay of $\tau = 1.5T_0$ in panel (a), where control is possible for $\delta = 0.05T_0$, vanishes for $\delta = 0.15T_0$ in (b). Note that non-zero latency times lead to a loss of the symmetry around $\tau = (2n + 1)T_0/2$ for $n = 1, 2, \ldots$ of the domain of control. See also the case of zero latency as displayed in Fig. 3.7.

3.4.2 Extended Time-Delayed Feedback

As discussed in Ref. [15] in the case of extended time-delayed feedback, the control force including a latency time is given for diagonal coupling by

$$\mathbf{F}(t) = K \begin{pmatrix} \sum\limits_{n=0}^{\infty} R^n[x(t - \delta - n\tau) - x(t - \delta - (n+1)\tau)] \\ \sum\limits_{n=0}^{\infty} R^n[y(t - \delta - n\tau) - y(t - \delta - (n+1)\tau)] \end{pmatrix} \tag{3.4.5a}$$

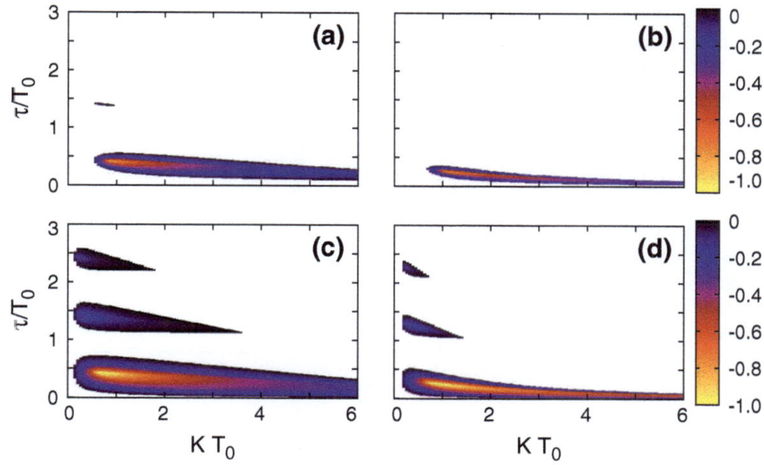

Fig. 3.18 Domain of control in the (K, τ) plane for different latency times: panels **a** and **c** show the result for $\delta = 0.05T_0$; panels **b** and **d** for $\delta = 0.15T_0$. The *shaded areas* indicate combinations of τ and K, for which the largest real part of the complex eigenvalues Λ is negative and thus control is successful. The value of $\mathrm{Re}(\Lambda)$ is indicated by the *color code*. The parameters of the unstable focus are chosen as $\omega = \pi(T_0 = 2)$ in all panels and $\lambda = 0.5$ in panels (**a**, **b**) and $\lambda = 0.1$ in (**c**, **d**)

$$= K \begin{pmatrix} x(t-\delta) - (1-R)\sum_{n=1}^{\infty} R^{n-1}x(t-\delta-n\tau) \\ y(t-\delta) - (1-R)\sum_{n=1}^{\infty} R^{n-1}y(t-\delta-n\tau) \end{pmatrix} \tag{3.4.5b}$$

$$= [\mathbf{x}(t-\delta) - \mathbf{x}(t-\delta-\tau)] + R\mathbf{F}(t-\tau). \tag{3.4.5c}$$

Similar to the case of Pyragas control discussed in the previous Sect. 3.4.1, the characteristic equation (3.3.2) needs to be modified by an additional exponential as follows

$$\lambda \pm i\omega = \Lambda + Ke^{-\Lambda\delta}\frac{1-e^{-\Lambda\tau}}{1-Re^{-\Lambda\tau}}. \tag{3.4.6}$$

Figure 3.19 depicts the dependence of the largest real part of the eigenvalues Λ on the time delay τ for fixed values of $R = 0.7$ and $KT_0 = 0.6$ but different latency times. Note that T_0 denotes again an intrinsic timescale given by $T_0 = 2\pi/\omega = 2$ as discussed in previous sections. The black (solid), red (dashed), green (dotted), and blue (dash-dotted) curves correspond to $\delta = 0$, $0.05T_0$, $0.15T_0$, and $0.25T_0$, respectively. It can be seen that the control scheme is less successful for longer latency times. The τ interval with negative real parts of Λ becomes smaller. In the case of $\delta = 0.25T_0$, for instance, control can only be achieved in a narrow range of small τ and the second minimum does not reach down to negative $\mathrm{Re}(\Lambda)$ anymore.

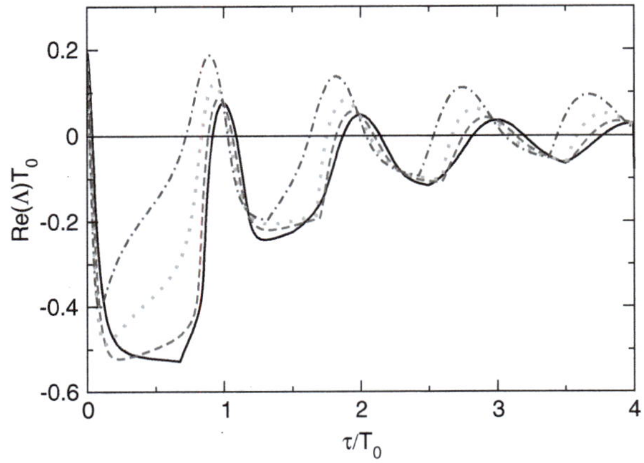

Fig. 3.19 Largest real part of the eigenvalues Λ as a function of τ for different latency times δ. The *black* (*solid*), *red* (*dashed*), *green* (*dotted*), and *blue* (*dash-dotted*) *curves* correspond to $\delta = 0$, $0.05T_0$, $0.15T_0$, and $0.25T_0$, respectively. The other control parameters are fixed as $R = 0.7$ and $KT_0 = 0.6$. *Parameters of the system* $\lambda = 0.1$ and $\omega = \pi (T_0 = 2)$

In addition, the minima of the real parts are distorted and shifted towards smaller time delays.

Taking also a varying feedback gain K into account, the domain of control can be seen in Fig. 3.20 in a projection on the (K, τ) plane. The remaining control parameter is fixed $R = 0.7$. Figures 3.20a–d correspond to values of $\delta = 0$, $0.05T_0$, $0.1T_0$, and $0.15T_0$, respectively. As in Fig. 3.13 of the Sect. 3.3, the color code corresponds to the largest real part of the complex eigenvalues which are

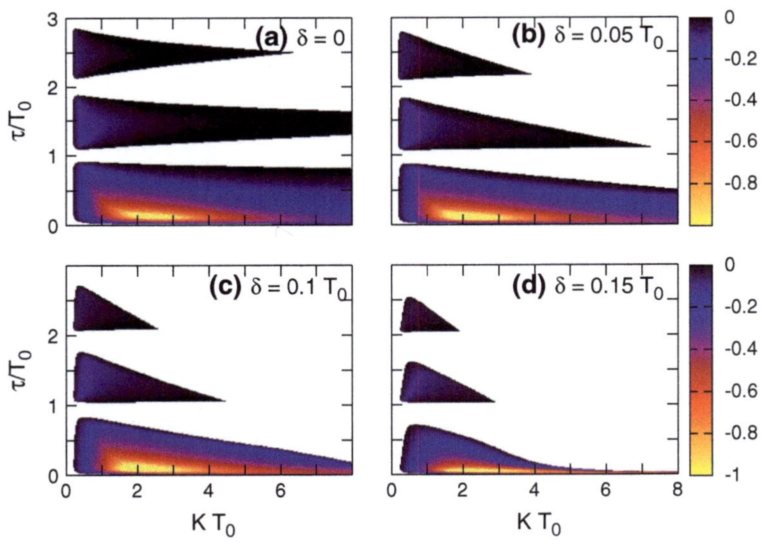

Fig. 3.20 Domain of control in the (K, τ) plane for different values of the latency time δ and fixed memory parameter $R = 0.7$. Panels **a**, **b**, **c**, and **d** correspond to values of $\delta = 0$, $0.05T_0$, $0.1T_0$, and $0.15T_0$, respectively. The *color code* shows the largest real part of the complex eigenvalues Λ as given by Eq. 3.4.6. Only negative values are displayed. The parameters of the system are as in Fig. 3.19

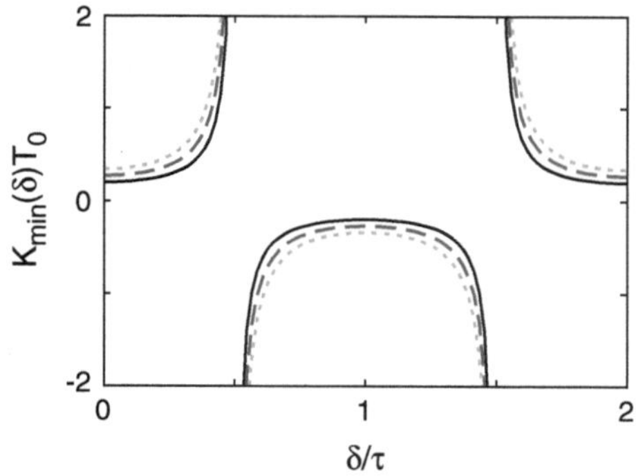

Fig. 3.21 Minimum feedback gain in dependence on the latency time δ according to Eq. 3.4.7 for different memory parameters R. The *black* (*solid*), *red* (*dashed*), and *green* (*dotted*) *curves* refer to values of $R = 0, 0.35$, and 0.7, respectively. The parameters of the system are as in Fig. 3.19

calculated from Eq. 3.4.6. Note that only negative values are depicted. For increasing latency time, the domains of control shrink. Similar to the discussion in the one-dimensional projection of Fig. 3.19, the islands are distorted towards smaller time delays.

Separating the characteristic equation (3.4.6) into real and imaginary part, one can derive, in analogy to Sect. 3.3, an expression for the minimum feedback gain

$$K_{\min}(\delta) = \frac{\lambda(1+R)}{\cos[(2n+1)\pi\delta/\tau]} \tag{3.4.7}$$

which is consistent with the case of Pyragas case discussed the previous Sect. 3.4.1. A detailed derivation can be found in Refs. [26, 37]. Figure 3.21 shows the minimum feedback gain $K_{\min}(\delta)$ for different memory parameters R which are fixed at $R = 0$, 0.35, and 0.7 in the black (solid), red (dashed), and green (dotted) curves, respectively. Note that $K_{\min}$ is smallest for the Pyragas case $R = 0$.

As another two-dimensional projection of the parameter space, Fig. 3.22 displays the domain of control in the (K, R) plane for different values of the latency time δ. The blue, green, red, and yellow areas refer to values of $\delta = 0, 0.05T_0, 0.1T_0$, and $0.15T_0$, respectively. Panel (a) shows the case of optimal choice of the time delay $\tau = T_0/2$ and panel (b) refers to $\tau = T_0/8$. In the first case, the domain of control shrinks considerably for increasing δ, whereas in the latter case, this change is less pronounced.

Similar to the previous sections, it is possible to derive a parametric expression for the boundary of the domain of control in terms of functions $R(\vartheta)$ and $K(\vartheta)$

$$R = \frac{-\vartheta\cos\left[(\omega\tau+\vartheta)\frac{\delta}{\tau}\right] + \vartheta\cos\left[(\omega\tau+\vartheta)\left(1+\frac{\delta}{\tau}\right)\right] + \lambda\tau\left(\sin\left[(\omega\tau+\vartheta)\frac{\delta}{\tau}\right] - \sin\left[(\omega\tau+\vartheta)\left(1+\frac{\delta}{\tau}\right)\right]\right)}{\vartheta\cos\left[(\omega\tau+\vartheta)\frac{\delta}{\tau}\right] - \vartheta\cos\left[(\omega\tau+\vartheta)\left(1-\frac{\delta}{\tau}\right)\right] - \lambda\tau\left(\sin\left[(\omega\tau+\vartheta)\frac{\delta}{\tau}\right] + \sin\left[(\omega\tau+\vartheta)\left(1-\frac{\delta}{\tau}\right)\right]\right)}, \tag{3.4.8}$$

$$K\tau = \frac{\left[\vartheta^2 + (\lambda\tau)^2\right]\cos\left[\frac{1}{2}(\omega\tau+\vartheta)\right]}{\lambda\tau\cos\left[\left(\frac{\delta}{\tau}-\frac{1}{2}\right)(\omega\tau+\vartheta)\right] + \vartheta\sin\left[\left(\frac{\delta}{\tau}-\frac{1}{2}\right)(\omega\tau+\vartheta)\right]}. \tag{3.4.9}$$

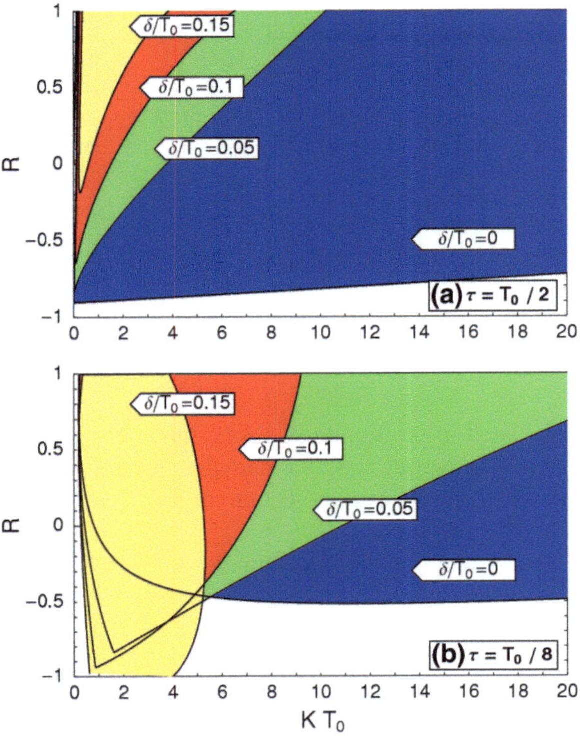

Fig. 3.22 Domain of control in the (K, R) plane for different values of the latency time δ. The *blue, green, red, and yellow* (*black, medium gray, dark gray, and light gray*) *areas* refer to values of $\delta = 0$, $0.05T_0$, $0.1T_0$, and $0.15T_0$, respectively. Panel **a** corresponds to an optimal time delay $\tau = T_0/2$, panel **b** to nonoptimal $\tau = T_0/8$. The parameters of the system are as in Fig. 3.19

The derivation of this parametric representation is lengthy and can be found in detail in Sect. 4.4 of Ref. [37]. Until this point, I have only considered time-delayed feedback with diagonal coupling. In the next section, I will extend the discussion to non-diagonal coupling matrices where an additional control parameter will introduced.

3.5 Phase-dependent Coupling

In the previous section, I have restricted the investigation to cases where the feedback is applied in diagonal form, i.e., the coupling matrix **A** which describes the coupling of the control force to the system is the identity matrix. See Sect. 2.3 for details concerning the notation. In this section, I will consider also nondiagonal coupling of time-delayed feedback. This will lead to an additional control parameter which can be interpreted as a feedback phase. Especially in optical systems, this modification of time-delayed feedback becomes important and can be related to the phase of the complex electric field, for instance, of a semiconductor laser [39, 43]. A schematic diagram of an all-optical realization of time-delayed feedback can be found in Fig. 3.27.

Following Ref. [39], I will investigate at first the effects of this new parameter in the context of simple time-delayed feedback applied to an unstable focus as well as to a saddle point in Sects. 3.5.1 and 3.5.2, respectively. Extended time-delayed feedback and phase-dependent coupling will then be the subject of Sect. 3.5.3 [15]. Section 3.5.4 concludes the discussion of phase-dependent coupling when a second phase parameter is introduced which affects only the delayed parts of the control force.

3.5.1 Unstable Focus

The generic model of an unstable steady state of focus type in the presence of phase-dependent coupling is given by the following equation in matrix form

$$\begin{pmatrix} \frac{dx(t)}{dt} \\ \frac{dy(t)}{dt} \end{pmatrix} = \begin{pmatrix} \lambda & \omega \\ -\omega & \lambda \end{pmatrix} \begin{pmatrix} x(t) \\ y(t) \end{pmatrix} - K \underbrace{\begin{pmatrix} \cos\varphi & -\sin\varphi \\ \sin\varphi & \cos\varphi \end{pmatrix}}_{\mathbf{A}} \begin{pmatrix} x(t) - x(t-\tau) \\ y(t) - y(t-\tau) \end{pmatrix}. \tag{3.5.1}$$

As in the previous sections, I choose positive parameter λ and non-zero ω which corresponds to an unstable fixed point of focus type in the absence of control. Note that the coupling matrix $\mathbf{A}$ becomes a rotational matrix in this realization of time-delayed feedback control. The new parameter φ acts as a feedback phase whose effects will be investigated in this section. The special choice of vanishing phase, i.e., $\varphi = 0$, recovers the diagonal feedback of Eq. 3.2.1 when the coupling matrix $\mathbf{A}$ becomes the identity matrix. In complex notation, i.e., $z(t) = x(t) + iy(t)$ the system can be rewritten as

$$\frac{dz(t)}{dt} = (\lambda \pm i\omega)z(t) - Ke^{i\varphi}[z(t) - z(t-\tau)]. \tag{3.5.2}$$

From this notation, it is clear that the new control parameter φ is indeed the phase of the complex feedback gain $Ke^{i\varphi}$. The control phase will be crucial in Chap. 4 where a similar phase factor will be used to overcome a topological limitation of time-delayed feedback control known as odd number limitation theorem which refers to the case of an unstable periodic orbit with an odd number of real Floquet multipliers larger than unity [44–48].

Similar to Sect. 3.2, one can derive a characteristic equation whose roots determine the stability of the system. The characteristic equation can be derived by the following determinant

$$\det\left[\begin{pmatrix} \lambda - \Lambda & \omega \\ -\omega & \lambda - \Lambda \end{pmatrix} - K \begin{pmatrix} \cos\varphi & -\sin\varphi \\ \sin\varphi & \cos\varphi \end{pmatrix} \begin{pmatrix} 1 - e^{-\Lambda\tau} & 0 \\ 0 & 1 - e^{-\Lambda\tau} \end{pmatrix}\right] = 0$$

which leads to

$$\Lambda + Ke^{\pm i\varphi}\left(1 - e^{-\Lambda\tau}\right) = \lambda \pm i\omega. \tag{3.5.3}$$

Note that, as in the case of diagonal control shown as Eqs. 3.2.1 in Sect. 3.2.1, the roots in terms of the eigenvalue Λ can be written by Lambert function

$$\Lambda\tau = W\left(K\tau e^{\pm i\varphi - (\lambda \pm i\omega)\tau + K\tau e^{\pm i\varphi}}\right) + (\lambda \pm i\omega)\tau - K\tau e^{\pm i\varphi}. \tag{3.5.4}$$

For details about the Lambert function see Refs. [20–25].

In order to obtain some analytical result, it is helpful to separate the characteristic equation (3.5.3) into real and imaginary parts with $\Lambda = p + iq$

$$p = \lambda - K\cos(\varphi)[1 - e^{-p\tau}\cos(q\tau)] - K\sin(\varphi)e^{-p\tau}\sin(q\tau) \tag{3.5.5a}$$

$$q = \omega + K\sin(\varphi)[1 - e^{-p\tau}\cos(q\tau)] - K\cos(\varphi)e^{-p\tau}\sin(q\tau). \tag{3.5.5b}$$

At the threshold of stability, the real part p vanishes. This simplification and trigonometrical identities lead to

$$\lambda = K[\cos(\varphi) - \cos(\varphi + q\tau)] \tag{3.5.6a}$$

$$\omega = q + K[\sin(\varphi + q\tau) - \sin(\varphi)]. \tag{3.5.6b}$$

The real part (3.5.6a) gives an expression for the minimum feedback gain in dependence on the control phase

$$K_{\min}(\varphi) = \frac{\lambda}{\cos(\varphi) - \cos(\varphi + q\tau)}. \tag{3.5.7}$$

Note that the case of vanishing φ leads to the smallest value of $K_{\min}(0) = \lambda/2$ which was already derived as Eq. 3.2.11 in Sect. 3.2.

Figure 3.23 shows the domain of control, i.e., $\mathrm{Re}(\Lambda) < 0$, in dependence on the parameters φ, K, and τ, where the feedback gain K and time delay τ are given in units of the intrinsic period $T_0 = 2\pi/\omega$, i.e., $T_0 = 2$ for the present choice of $\omega = \pi$. The parameter λ is chosen as $\lambda T_0 = 0.2$ in all plots. Panels (a) and (b) represent the (φ, K) plane for fixed values of the time delay $\tau/T_0 = 0.5$ and 0.9, respectively. Note that $\tau = T_0/2$ yields a symmetric domain of control with respect to $\varphi = 0$, which is the case of diagonal coupling as discussed in Sect. 3.2. For values other than this optimal time delay, the domain of control is distorted and shrinks. In the situation shown in Fig. 3.23b, control can no longer be achieved for $\varphi = 0$, but only for positive phase $\varphi > 0$. Panels (c) and (d) show the domain of control in the (φ, τ) plane for fixed feedback gain $KT_0 = 1$ and 2, respectively. It consists of isolated islands with a horizontal extension that becomes maximum and symmetric with respect to $\varphi = 0$ at delays of $\tau = (n + 1/2)T_0$ with $n = 0, 1, 2, \ldots$.

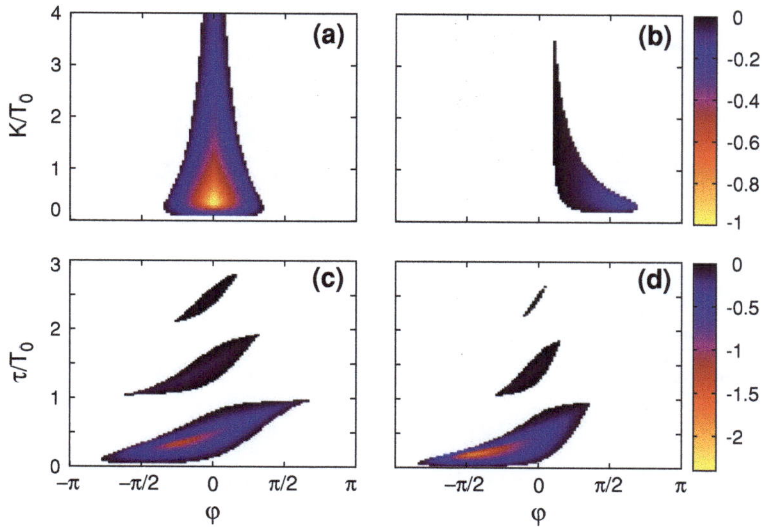

Fig. 3.23 Domain of control in dependence on φ,K, and τ in units of $T_0 = 2\pi/\omega = 2$. The largest real part of the complex eigenvalues Λ is shown in *color code*. Panels **a**, **b**: domain of control in the (φ, K) plane for fixed delay $\tau = T_0/2$ and $0.9T_0$, respectively. **c**, **d**: domain of control in the (φ, τ) plane for fixed feedback gain $KT_0 = 1$ and 2, respectively. *System's parameters* $\lambda T_0 = 0.2$ and $\omega = \pi$

Note that control is impossible for integer τ/T_0. For a range of τ-values in between, stabilization can be achieved by appropriately chosen φ. When crossing the islands at fixed φ, resonance-type behavior of the damping rate $-\text{Re}(\Lambda)$ occurs. With increasing n, the size of the islands decreases so that they eventually disappear at some critical value determined by the feedback strength K. After this first discussion of phase-dependent coupling, I will explore the domain of control in more detail in the following.

Including the feedback phase φ, there are now three control parameters where the other two are given by the feedback gain K and the time delay τ. For a complete picture, I will consider two-dimensional projections of this three.dimensional control parameter space in the following. These are projections in the (φ, τ), (φ, K), ad (K, τ) planes where the respective third parameter is fixed. The system's parameter will remain fixed as in Fig. 3.23 at $\lambda = 0.1$ and $\omega = \pi$. The latter yields an intrinsic timescale of $T_0 = 2$.

Figure 3.24 displays the domain of stability in the (φ, τ) plane. The color code shows the largest real part of the complex eigenvalue Λ according to Eq. 3.5.3 where it is negative. The feedback gain is fixed at $KT_0 = 0.2, 0.4, 0.6, 0.8, 1$, and 1.2 in panels (a)–(f), respectively. Note that panel (e) showing $KT_0 = 1$ is identical to Fig. 3.23c. In all panels the regions of stability are given by separated islands centered around $(\varphi, \tau) = (0, (n + 1/2)T_0)$ with $n = 0, 1, 2, \ldots$. These distinct islands become smaller as the time delay increases. While the areas of stability are elongated for small K as in panel (a), they gain size in the range of small τ and

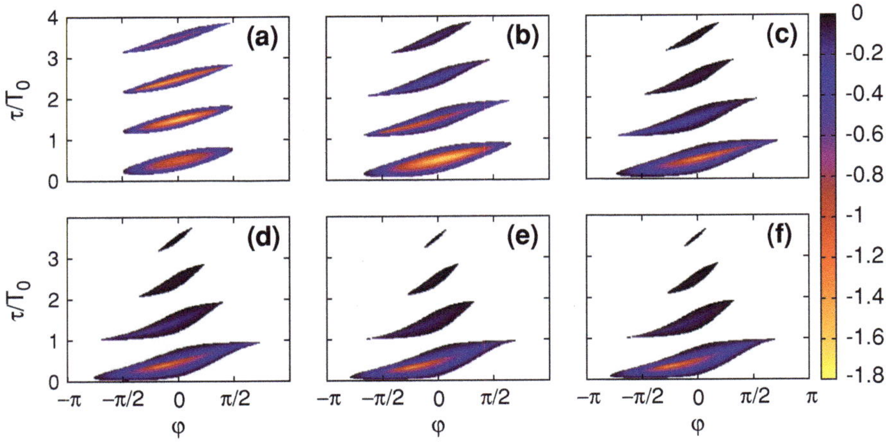

Fig. 3.24 Domain of control in dependence on φ and τ in units of $T_0 = 2\pi/\omega = 2$. The largest real part of the complex eigenvalues Λ is shown in *color code*. Panels **a–f** correspond to a feedback gain $KT_0 = 0.2$, 0.4, 0.6, 0.8, 1, and 1.2, respectively. System's parameters as in Fig. 3.23

their shape becomes distorted. Note that at the same time the islands for large delays shrink.

Figure 3.25 shows the domain of control in the projection spanned by the feedback phase φ and the feedback gain K while the time delay τ is fixed at

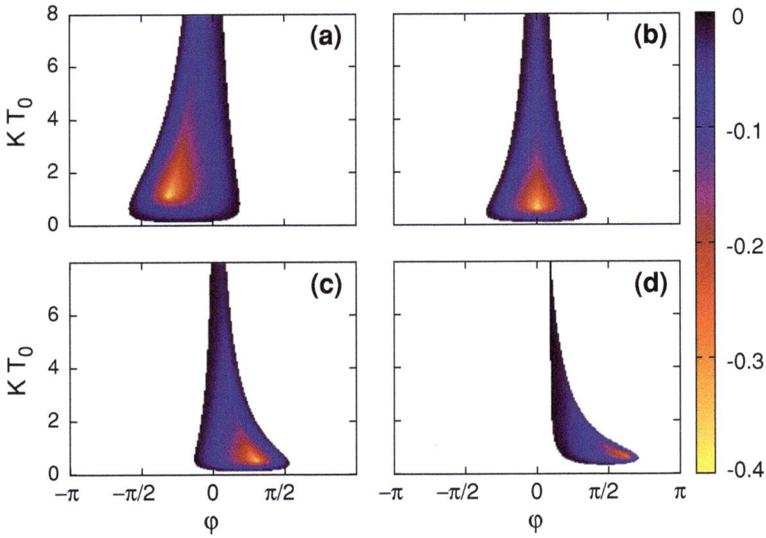

Fig. 3.25 Domain of control in dependence on φ and K in units of $T_0 = 2\pi/\omega = 2$. The largest real part of the complex eigenvalues Λ is shown in *color code*. Panels **a–d** correspond to a time-delay $\tau = 0.3T_0$, $0.5T_0$, $0.7T_0$, and $0.9T_0$, respectively. System's parameters as in Fig. 3.23

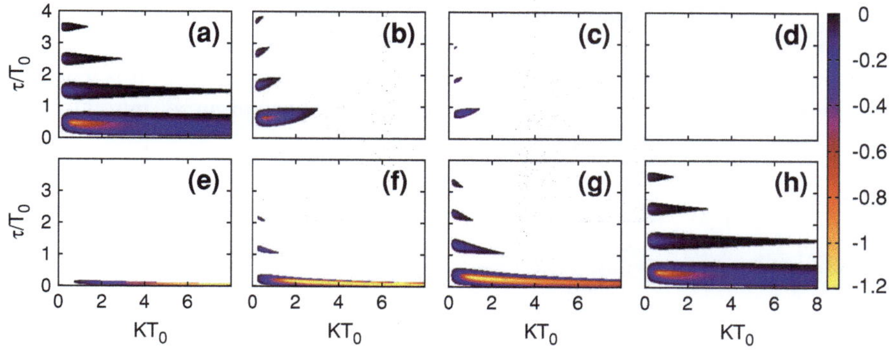

Fig. 3.26 Domain of control in dependence on K and τ for different feedback phases φ with normalization in units of $T_0 = 2\pi/\omega = 2$. Panels **a–h** correspond to a value of $\varphi = 0$, $\pi/4$, $\pi/2$, $3\pi/4$, $5\pi/4$, $3\pi/2$, $7\pi/4$, and 2π, respectively. The largest real part of the complex eigenvalues Λ is shown in *color code*. Only negative values are displayed. System's parameters as in Fig. 3.23

$\tau = 0.3T_0$, $0.5T_0$, $0.7T_0$, and $0.9T_0$ in panels (a)–(d), respectively. The color code indicates combinations of φ and K which yield stability and corresponds to $\text{Re}(\Lambda) < 0$ in Eq. 3.5.3. The domain of control is symmetric with respect to $\varphi = 0$ in panel (b) corresponding for optimal delay of $\tau = T_0/2$ as already discussed in Fig. 3.23. For smaller values of τ the domain is distorted towards negative phases as shown in panel (a). For $\tau > T_0/2$ the domain tends to positive values of φ until the diagonal case of $\varphi = 0$ can no longer yield stabilization with any feedback gain K as presented in panel (d).

Figure 3.26 depicts the domain of control for different control phases φ in the (K, τ) plane. Panels (a)–(h) correspond to a value of $\varphi = 0$, $\pi/4$, $\pi/2$, $3\pi/4$, $5\pi/4$, $3\pi/2$, $7\pi/4$, and 2π, respectively. The color code refers to the largest real part of the complex eigenvalues Λ and only negative values are shown. Panel (a) displays the case of vanishing control phase and is identical to Fig. 3.7b of Sect. 3.2. The domain of control is shifted towards larger time delays for $\varphi = 1$. Note that control is impossible for control phases around $\varphi = \pi$. See, for instance, panel (c) which shows only a very small domain of control or panel (d) corresponding to $\varphi = 3\pi/4$ when no control is possible for any combinations of K and τ. For φ values well beyond π and in panels (e)–(h), the domain reappears for slightly above integer multiples of T_0. Finally, the domain of control in panel (h) shown in the case of $\varphi = 2\pi$ is identical to panel (a) due to the 2π periodicity of the coupling matrix **A**.

3.5.2 Saddle Point

For further discussion of phase-dependent coupling of the Pyragas control method, let me consider steady states with real eigenvalues and one unstable dimension, i.e., saddle points. This type of steady states was already under investigation in

Sect. 3.2. There it was shown that control is impossible via diagonal coupling of the Pyragas feedback. See Eqs. 3.2.28–3.2.35. The reason was that the two equations remain decoupled for the diagonal coupling scheme. Thus, it is tempting to apply time-delayed feedback via phase-dependent coupling as introduced in Sect. 3.5.1. Then, the system's equation reads

$$\begin{pmatrix} \frac{dx(t)}{dt} \\ \frac{dy(t)}{dt} \end{pmatrix} = \begin{pmatrix} \alpha_1 & 0 \\ 0 & -\alpha_2 \end{pmatrix} \begin{pmatrix} x(t) \\ y(t) \end{pmatrix} + K \begin{pmatrix} \cos\varphi & -\sin\varphi \\ \sin\varphi & \cos\varphi \end{pmatrix} \begin{pmatrix} x(t-\tau) - x(t) \\ y(t-\tau) - y(t) \end{pmatrix}. \tag{3.5.8}$$

Without loss of generality, one can consider the case $\alpha_2 = 1$ by rescaling the time. As a reminder on the saddle point consider Fig. 3.11. In order to obtain information about the stability of the delayed system, one has to calculate the complex roots Λ of the characteristic equation, which is given by

$$\begin{aligned} 0 &= \begin{vmatrix} \alpha_1 - \Lambda + K\cos(\varphi)\left(e^{-\Lambda\tau} - 1\right) & -K\sin(\varphi)\left(e^{-\Lambda\tau} - 1\right) \\ K\sin(\varphi)\left(e^{-\Lambda\tau} - 1\right) & -1 - \Lambda + K\cos(\varphi)\left(e^{-\Lambda\tau} - 1\right) \end{vmatrix} \\ &= (\alpha_1 - \Lambda)(-1 - \Lambda) - 2\Lambda K\cos(\varphi)\left(e^{-\Lambda\tau} - 1\right) \\ &\quad + K\cos(\varphi)(\alpha - 1)\left(e^{-\Lambda\tau} - 1\right) + K^2\left(e^{-\Lambda\tau} - 1\right)^2. \end{aligned} \tag{3.5.9}$$

At the threshold of control, the eigenvalue Λ becomes purely imaginary, i.e., $\Lambda = iq$. Thus, Eq. 3.5.9 can be rewritten as

$$\begin{aligned} 0 &= -\alpha_1 - q^2 - iq(\alpha - 1) \\ &\quad - 2qK\cos(\varphi)\sin(q\tau) - i2qK\cos(\varphi)[\cos(q\tau) - 1] \\ &\quad + K\cos(\varphi)(\alpha - 1)[\cos(q\tau) - 1] - iK\cos(\varphi)(\alpha - 1)\sin(q\tau) \\ &\quad + 2K^2\cos(q\tau)[\cos(q\tau) - 1] - i2K^2\sin(q\tau)[\cos(q\tau) - 1]. \end{aligned} \tag{3.5.10}$$

In the following, I will split Eq. 3.5.10 into real and imaginary part

$$\begin{aligned} \mathrm{Re}:\ 0 = &-\alpha_1 - q^2 - 2qK\cos(\varphi)\sin(q\tau) + K\cos(\varphi)(\alpha_1 - 1)[\cos(q\tau) - 1] \\ &+ 2K^2\cos(q\tau)[\cos(q\tau) - 1] \end{aligned} \tag{3.5.11a}$$

$$\begin{aligned} \mathrm{Im}:\ 0 = &-q(\alpha_1 - 1) - 2qK\cos(\varphi)[\cos(q\tau) - 1] - K\cos(\varphi)(\alpha_1 - 1)\sin(q\tau) \\ &- 2K^2\sin(q\tau)[\cos(q\tau) - 1] \end{aligned} \tag{3.5.11b}$$

and show by analytic means that, assuming a change of stability, i.e., $\mathrm{Re}(\Lambda) = 0$, is possible, I will arrive at a contradiction.

Since the unstable eigenvalue of the uncontrolled system, i.e., $\Lambda_0 = \alpha_1$, is real, Λ has to be real for vanishing feedback gain, too. This leaves for the imaginary part two possibilities, i.e., $\mathrm{Im}(\Lambda) = q = 0$ and $\mathrm{Im}(\Lambda) = q = \pi/\tau$. Considering the real part of the characteristic equation yields in the first case

$$0 = -\alpha_1. \tag{3.5.12}$$

This is a contradiction because the parameter α_1 was initially assumed to be positive. In the second case, i.e., $\mathrm{Im}(\Lambda) = q = \pi/\tau$, the characteristic equation (3.5.11a, 3.5.11b) becomes

$$\mathrm{Re}: \; 0 = -\alpha_1 - \left(\frac{\pi}{\tau}\right)^2 - 2\mathrm{K}\cos(\varphi)(\alpha_1 - 1) + 4\mathrm{K}^2 \tag{3.5.13a}$$

$$\mathrm{Im}: \; 0 = -\frac{\pi}{\tau}(\alpha_1 - 1) + 4\frac{\pi}{\tau}K\cos(\varphi). \tag{3.5.13b}$$

The imaginary part leads to an expression for K:

$$K = \frac{\alpha_1 - 1}{4\cos(\varphi)} \Rightarrow 2K\cos(\varphi) = \frac{\alpha_1 - 1}{2}. \tag{3.5.14}$$

From the real part (3.5.13a), one obtains the following relation:

$$\left(\frac{\pi}{\tau}\right)^2 = -\alpha_1 - 2K\cos(\varphi)(\alpha_1 - 1) + 4K^2 > 0. \tag{3.5.15}$$

Inserting the formula for the feedback gain K of Eq. 3.5.14 into Eq. 3.5.15 yield

$$-\alpha_1 - 2K\cos(\varphi)(\alpha_1 - 1) + 4K^2 > 0 \tag{3.5.16a}$$

$$\Leftrightarrow \quad -\alpha_1 - 2\frac{\alpha_1 - 1}{4\cos(\varphi)}\cos(\varphi)(\alpha_1 - 1) + 4K^2 > 0 \tag{3.5.16b}$$

$$\Leftrightarrow \quad -\alpha_1^2 - 1 + 8K^2 > 0 \tag{3.5.16c}$$

$$\Leftrightarrow \quad 8K^2 - 1 > \alpha_1^2. \tag{3.5.16d}$$

The last relation yields a contradiction in the uncontrolled case, i.e., $K = 0$,

$$-1 > \alpha_1^2, \tag{3.5.17}$$

because α_1 is a real parameter. Thus, saddle points cannot be stabilized by time-delayed feedback of the form (3.5.8).

After the discussion of Pyragas control applied via phase-dependent coupling, I will investigate the effects of an additional memory parameter on this specific choice of nondiagonal coupling which leads to extended time-delayed feedback in the next section.

3.5.3 Extended Time-Delayed Feedback

In this section, I will investigate the effects of phase-dependent coupling on the extended time-delayed feedback scheme [15, 49]. As in the Sect. 3.5.1, the control force is applied to an unstable focus by a rotational matrix

$$\begin{pmatrix} \frac{dx(t)}{dt} \\ \frac{dy(t)}{dt} \end{pmatrix} = \begin{pmatrix} \lambda & \omega \\ -\omega & \lambda \end{pmatrix} \begin{pmatrix} x \\ y \end{pmatrix} - \begin{pmatrix} \cos\varphi & -\sin\varphi \\ \sin\varphi & \cos\varphi \end{pmatrix} \times \begin{pmatrix} \sum_{n=0}^{\infty} R^n [x(t-n\tau) - x(t-(n+1)\tau)] \\ \sum_{n=0}^{\infty} R^n [y(t-n\tau) - y(t-(n+1)\tau)] \end{pmatrix}. \quad (3.5.18)$$

In this case, the control-parameter space is four dimensional and consists of the time delay τ, the feedback gain K, the memory parameter R, and the feedback phase φ. In optical systems like semiconductor lasers with external optical feedback [39, 50], this feedback phase can be seen as the phase of the electric field. Experimentally, this phase of the feedback can be varied by tuning the distance between the laser and an external Fabry–Perot resonator. Figure 3.27 depicts schematically the experimental configuration of an all-optical time-delayed feedback controller. The memory parameter R accounts for multiple reflections in the resonator.

It has also been demonstrated that a feedback phase plays an important role in the suppression of collective synchrony in a globally coupled oscillator network [51].

Solving the characteristic equation which reads

$$\Lambda + Ke^{\mp i\varphi} \frac{1 - e^{-\Lambda\tau}}{1 - Re^{-\Lambda\tau}} = \lambda \pm i\omega \quad (3.5.19)$$

provides again useful information about the stability of the system (3.5.18). Note that, as in the case of simple time-delayed feedback considered in Sect. 3.5.1, the feedback phase φ enters the characteristic equation as an additional exponential. Compare also to the case of vanishing phase given by Eq. 3.3.2. In order to derive a minimum feedback gain $K_{\min}(\varphi)$, one can follow a similar strategy as in the case

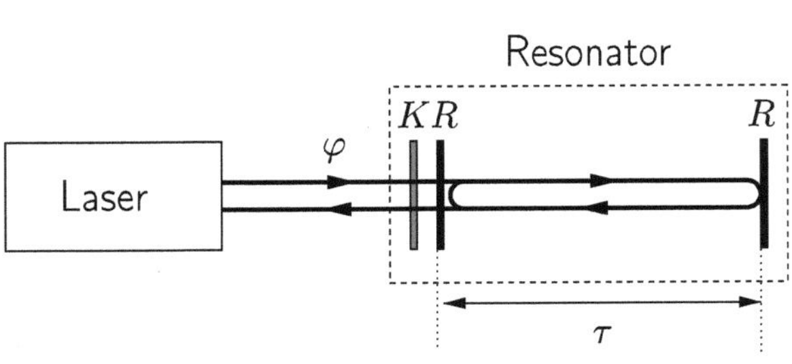

Fig. 3.27 Schematic diagram of a laser with a Fabry–Perot resonator

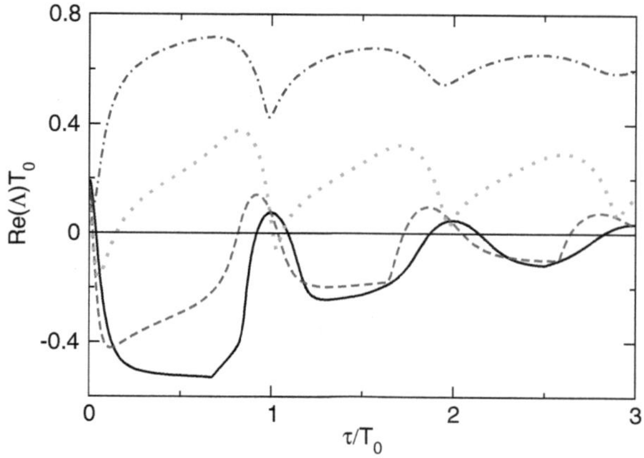

Fig. 3.28 Largest real part of the eigenvalues Λ as a function of τ for different phases φ. The *black (solid)*, *red (dashed)*, *green (dotted)*, and *blue (dash-dotted) curves* correspond to $\varphi = 0, \pi/4, \pi/2$, and $3\pi/4$, respectively. The other control parameters are fixed as $R = 0.7$ and $KT_0 = 0.6$. *System's parameters* $\lambda\ T_0 = 0.2$ and $\omega = \pi$

of $\varphi = 0$ as elaborated in Sect. 3.3. This involves a separation of the characteristic equation (3.5.19) into real and imaginary parts and considering the threshold case $\mathrm{Re}(\Lambda) = 0$. This will lead to a formula similar to Eq. 3.3.6b, i.e.,

$$K_{\min}(\varphi) = \frac{\lambda(1+R)}{2\cos\varphi}. \tag{3.5.20}$$

The behavior of $K_{\min}(\varphi)$ for different memory parameters resembles the minimum feedback gain $K_{\min}(\delta)$ in the presence of a latency time. For this, compare Eq. 3.4.7 and Fig. 3.21. Note that the time delay that corresponds to this value of $K_{\min}$ is no longer given by Eq. 3.2.12 and is not the optimal time delay in the general case of non-zero phase. Nevertheless, Eq. 3.5.20 can be used as a coarse estimate of the minimum feedback gain for the regime of small values of φ, if the time delay is chosen as Eq. 3.2.12.

Figure 3.28 depicts the dependence of the largest real part of the eigenvalues Λ on the time delay τ for fixed values of $R = 0.7$ and $KT_0 = 0.6$, but different values of the phase. The black (solid), red (dashed), green (dotted), and blue (dash-dotted) curves correspond to feedback phases $\varphi = 0$, $\pi/4$, $\pi/2$, and $3\pi/4$, respectively. It can be observed that the control is overall less effective for larger φ, as the curves are shifted up towards positive reals parts for increasing the phase. The range of possible values for the time delay shrinks. The optimal time delay is shifted towards smaller values for larger φ, which can be seen for the case of $\varphi = \pi/2$, where the optimal time delay is in the range of $\tau = 0.1T_0$ instead of $0.5T_0$, which was the optimal time delay for $\varphi = 0$ according to Eq. 3.2.12.

Since the space of control parameters is now spanned by four parameters, i.e., time delay τ, the feedback gain K, the memory parameter R, and the feedback phase φ, I will discuss in the following different two-dimensional projections onto planes of two of these parameters while the remaining two are fixed.

At first, I will consider the domain of control in the plane parameterized by K and φ. Hence, one keeps the other remaining control parameters R and φ fixed. Figures 3.29 and 3.30 show the domain of control for a time delay of $T_0/2$ and

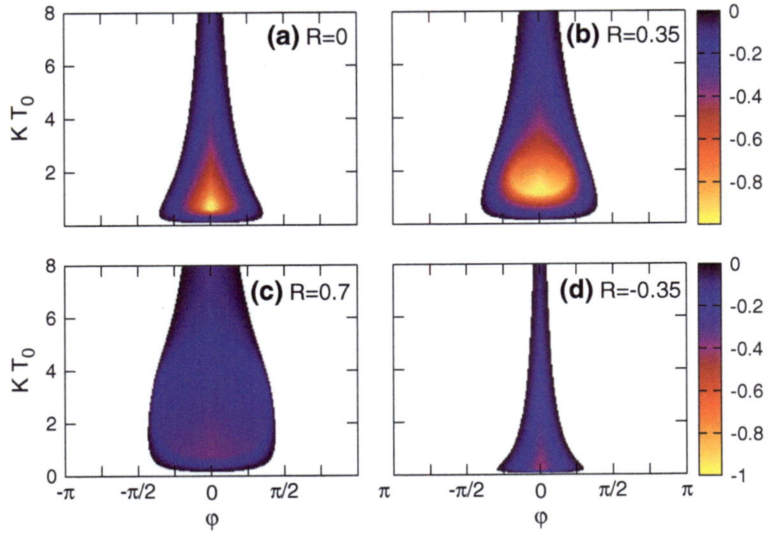

Fig. 3.29 Domain of control in the (φ, K) plane for optimal time delay $\tau = T_0/2$. Panels **a**, **b**, **c**, and **d** correspond to a memory parameter R of 0, 0.35, 0.7, and -0.35, respectively. The *color code* shows the largest real part of the complex eigenvalues Λ as given by Eq. 3.5.19. Only negative values are displayed. The parameters of the system are as in Fig. 3.28

$0.1T_0$, respectively. In each figure, the memory parameter R is chosen as $R = 0$, 0.35, 0.7, and -0.35 in panels (a), (b), (c), and (d), respectively. The color code corresponds to the largest real part of the complex eigenvalues as calculated from Eq. 3.5.19. Only negative values are depicted, i.e., those combinations of K and φ for which the control scheme is successful. Note that the case $R = 0$ corresponds to the simple time-delayed feedback control method discussed in the previous section. An increase of the memory parameter R leads to a larger domain of control. Even though the system can be stabilized for a larger range of K and φ, the system becomes over all less stable, since the real part of Λ is closer to zero. See the less pronounced yellow area in Figs. 3.29 and 3.30 for increasing R. For negative values of R, the domain of control shrinks. Note that also in the case of non-optimal time delay as in Fig. 3.30 the range of choice for possible feedback gain and phase is enlarged.

For a better understanding of the effects of the feedback phase, Fig. 3.31 depicts the domain of control in the (K, R) plane. The blue, green, red, and yellow areas correspond to successful control for $\varphi = 0$, $\pi/8$, $\pi/4$, and $3\pi/8$, respectively. Panel (a) shows the case of optimal time delay, i.e., $\tau = T_0/2$; panel (b) displays the case of $\tau = 0.1T_0$. Note that an increase of φ leads to a smaller domain of control in the case of $\tau = T_0/2$. This effect, however, is reversed for non-optimal choices of τ, where the phase φ compensates for the bad choice of the time delay. Thus, control is possible again, for instance, in the Pyragas case of vanishing memory parameter $R = 0$. Following the strategy introduced in Sect. 3.3, one can

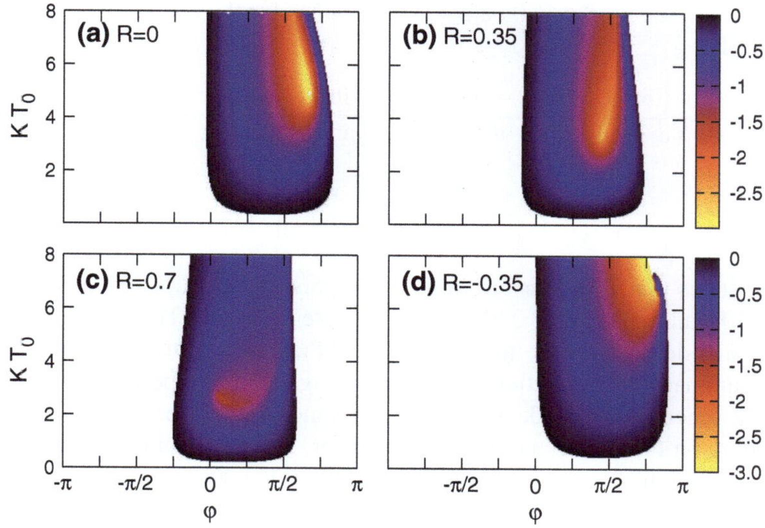

Fig. 3.30 Domain of control in the (φ, K) plane for time delay $\tau = 0.1T_0$. Panels **a**, **b**, **c**, and **d** correspond to a memory parameter R of 0, 0.35, 0.7, and -0.35, respectively. The *color code* shows the largest real part of the complex eigenvalues Λ as given by Eq. 3.5.19. Only negative values are depicted. The parameters of the system are as in Fig. 3.28

Fig. 3.31 Domain of control in the (K, R) plane for different values of the feedback phase φ. The *blue, green, red, and yellow* (*black, medium gray, dark gray, and light gray*) *areas* correspond to $\varphi = 0$, $\pi/8$, $\pi/4$, and $3\pi/8$, respectively. Panel **a** displays the domain of control for optimal $\tau = T_0/2$; panel **b** for $\tau = 0.1T_0$. The parameters of the system are as in Fig. 3.28

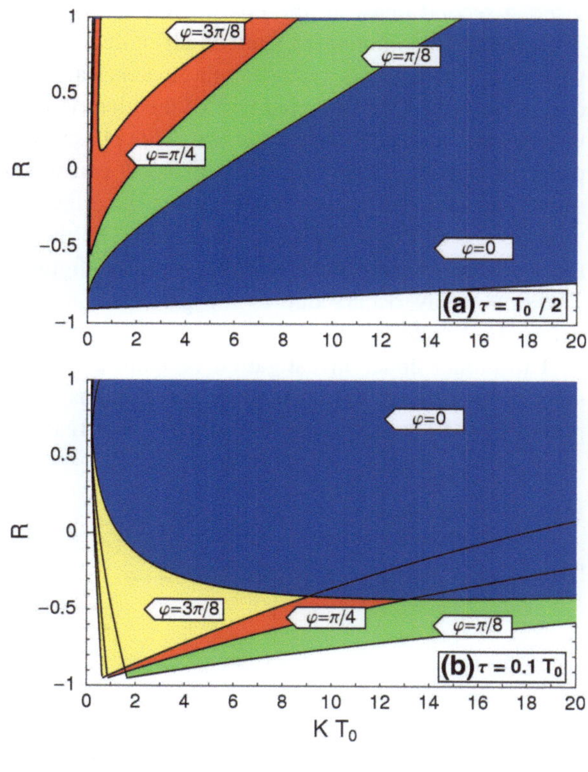

derive also in the case $\varphi \neq 0$ parametric formulas for the boundary of the domain of control:

$$R = \frac{\vartheta[\cos(\omega\tau + \vartheta + \varphi) - \cos\varphi] + \lambda\tau[\sin\varphi - \sin(\omega\tau + \vartheta + \varphi)]}{\vartheta[\cos\varphi - \cos(\omega\tau + \vartheta + \varphi)] - \lambda\tau[\sin\varphi + \sin(\omega\tau + \vartheta + \varphi)]} \tag{3.5.21a}$$

$$K\tau = \frac{(\vartheta^2 + \lambda^2\tau^2)\cos\left[\frac{1}{2}(\omega\tau + \vartheta)\right]}{\lambda\tau\cos\left[\frac{1}{2}(\omega\tau + \vartheta) - \varphi\right] - \vartheta\sin\left[\frac{1}{2}(\omega\tau + \vartheta) - \varphi\right]}. \tag{3.5.21b}$$

Let me stress that it is possible to derive expression of $K(q)$ and $\tau(q)$ similar to Eqs. 3.3.7 and 3.3.8a and 3.3.8b also in the case of $\varphi \neq 0$. These calculations are lengthy and do not produce more insight and thus are omitted here.

To summarize this section, I have investigated an experimentally relevant coupling scheme of time-delayed feedback which introduces a control phase φ. Depending on this additional control parameter, I have calculated the domains of control in various two-dimensional projections of the control parameter space.

3.5.4 Two Feedback Phases

As another extension of phase-dependent coupling of the Pyragas scheme, I will insert a second phase in the control force which acts only on the delayed parts. This extension is motivated from an all-optical experimental setup [39, 43, 52, 53] where the feedback is realized via an external resonator. Then, the two phases occur naturally as the delayed part of the electric field can be shifted by a phase ψ with respect to the instantaneous part. This phase shift is independent of a common phase φ of both components. The phase ψ can be modified by fine tuning of the mirror position in the external cavity, i.e., the length of the external cavity, whereas the common phase φ is accessible by the distance between the laser and the resonator. Figure 3.32 visualizes this configuration by a schematic diagram. The time delay occurs due to the propagation time of the electric field in the external resonator.

The model under investigation is again a fixed point of focus type which is subject to time-delayed feedback of Pyragas type [54] including two phase parameters. In complex notation, this system reads

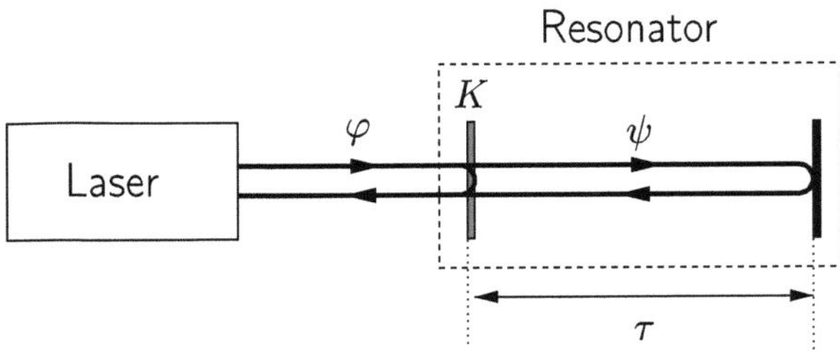

Fig. 3.32 Schematic diagram of a laser with external resonator

$$\frac{dz(t)}{dt} = (\lambda \pm i\omega)z(t) - Ke^{i\varphi}\left[z(t) - e^{i\psi}z(t-\tau)\right], \tag{3.5.22}$$

where τ and K denote the time delay and the feedback gain, respectively. As mentioned above there are two phase parameters: φ, which corresponds to a common phase shift in both the delayed and current signal, and ψ, which occurs in the delayed term only. Separated in real and imaginary parts $z(t) = x(t) + iy(t)$, one can rewrite the equations as follows

$$\begin{aligned}\begin{pmatrix} \frac{dx(t)}{dt} \\ \frac{dy(t)}{dt} \end{pmatrix} &= \begin{pmatrix} \lambda & \omega \\ -\omega & \lambda \end{pmatrix}\begin{pmatrix} x(t) \\ y(t) \end{pmatrix} \\ &- K\begin{pmatrix} \cos\varphi & -\sin\varphi \\ \sin\varphi & \cos\varphi \end{pmatrix}\begin{pmatrix} x(t) - \cos\psi x(t-\tau) \\ y(t) - \cos\psi y(t-\tau) \end{pmatrix} \\ &- K\begin{pmatrix} \sin\varphi & \cos\varphi \\ -\cos\varphi & \sin\varphi \end{pmatrix}\begin{pmatrix} \sin\psi x(t-\tau) \\ \sin\psi y(t-\tau \end{pmatrix}.\end{aligned} \tag{3.5.23}$$

Note that the case $\psi = 0$ recovers the system already investigated in Sect. 3.5.1 [39].

The characteristic equation of the system above reads

$$\Lambda = \lambda \pm i\omega - Ke^{i\varphi}\left(1 - e^{i\psi - \Lambda\tau}\right), \tag{3.5.24}$$

where the second phase enters in the last exponential function as a phase shift. Compare with the characteristic equation (3.5.3) of Sect. 3.5.1. Let me stress that Eq. 3.5.24 can be solved using the Lambert function $W(z)$ as discussed in Sects. 3.2.1 and 3.5.1

$$\begin{aligned}\Lambda\tau = {} & W\left[K\tau\exp\left(i(\psi + \varphi) - (\lambda \pm i\omega)\tau + K\tau e^{i\varphi}\right)\right] \\ & + \lambda \pm i\omega - K\tau e^{i\varphi}.\end{aligned} \tag{3.5.25}$$

As can be seen from the system equation with phase-dependent time-delayed feedback, i.e., Eq. 3.5.23, and from the corresponding characteristic equation (3.5.24), there are four control parameters: time delay τ, feedback gain K, and the feedback phases φ and ψ. This leads to a four-dimensional parameter space which will be investigated by two-dimensional projections onto planes parameterized by only two control parameters while the other two parameters are fixed. Next to the projection onto the (K, τ) plane for different phases ψ, I will focus on the projections where ψ is one of the two varying parameters.

The calculations are performed with the following set of system's parameters unless stated otherwise: $\lambda = 0.1$ and $\omega = \pi$, which yields an initially unstable focus with intrinsic period $T_0 = 2\pi/\omega = 2$. The color code in all figures of this section displays the domains of stability, i.e., the largest real part of the complex eigenvalue Λ according to Eq. 3.5.24 where it is negative. The parameters K and τ are normalized in units of the intrinsic timescale $T_0 = 2$.

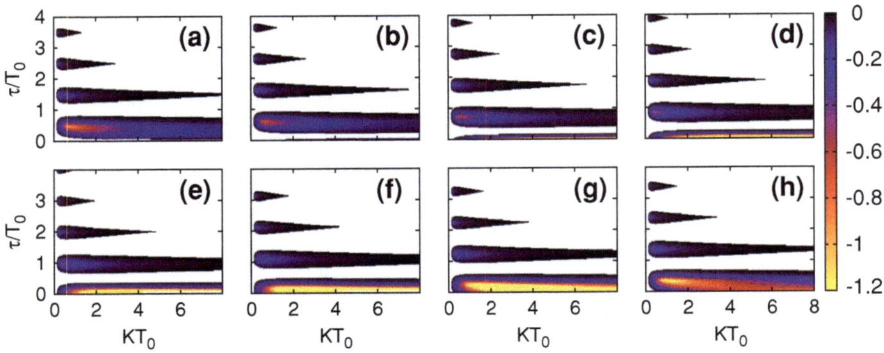

Fig. 3.33 Domain of control in dependence on K and τ units of $T_0 = 2\pi/\omega = 2$ for different control phases $\psi = 0, 0.25\pi, 0.5\pi, 0.75\pi, \pi, 1.25\pi, 1.5\pi$, and 1.75π in panels (**a**)–(**h**), respectively. The largest real part of the complex eigenvalues Λ is shown in *color code*. The feedback phase φ is fixed at $\varphi = 0$. *System's parameters* $\lambda T_0 = 0.2$ and $\omega = \pi$

Figure 3.33 depicts the domain of control in the (K, τ) plane for different control phases ψ, but fixed feedback phase $\varphi = 0$. Panels (a)–(h) correspond to $\psi = 0, 0.25\pi, 0.5\pi, 0.75\pi, \pi, 1.25\pi, 1.5\pi$, and 1.75π, respectively. Panel (a) for $\psi = 0$ was already shown in Sect. 3.5.1. For comparison see Fig. 3.26a.

The domains of control are given by separate islands of stability which shrink in size as the time delay τ increases. For increasing ψ in the range $[0, \pi]$, the domains are shifted towards smaller time delays and a reverse shift towards the original position can be seen as ψ approaches 2π. This is in contrast to the dependence on the feedback phase φ discussed in Fig. 3.26. There, the domains of stability became smaller for increasing φ and eventually vanished completely around $\varphi \approx \pi$. In the case of two phases, an adjustment of the time delay can compensate for nonzero control phases ψ.

After the discussion of the stability domains in the plane parameterized by the feedback gain K and time delay τ, I will consider in the following projections of the control parameter space which involve ψ as one of the two varying control parameters. Thus Figs. 3.34, 3.36, and 3.35 display the domains of control in the (ψ, K), (ψ, τ), and (φ, ψ) planes, respectively. In Figs. 3.34 and 3.36 the feedback phase φ is set to zero, and accordingly, the feedback gain K is fixed at $KT_0 = 1$ in Fig. 3.35.

Panels (a)–(d) of Fig. 3.34 correspond to different time delays of $\tau = 0.3T_0$, $0.5T_0$, $0.7T_0$, and $0.9T_0$, respectively. One can see that the shape of the stability regions is similar in all panels keeping in mind the 2π-periodicity of the controlled system with respect to ψ. Note that there is a shift of the domain depending on τ. The domain is centered around $\psi = 0$ if the time delay is one half of the intrinsic period T_0 as in panel (b). If the time delay is smaller than $T_0/2$, the region of smallest real part is found for negative control phases ψ depicted in panel (a) or equivalently, ψ between π and 2π. The largest real part of the eigenvalues is smallest for small time delays as shown by the large yellow area in panel (a).

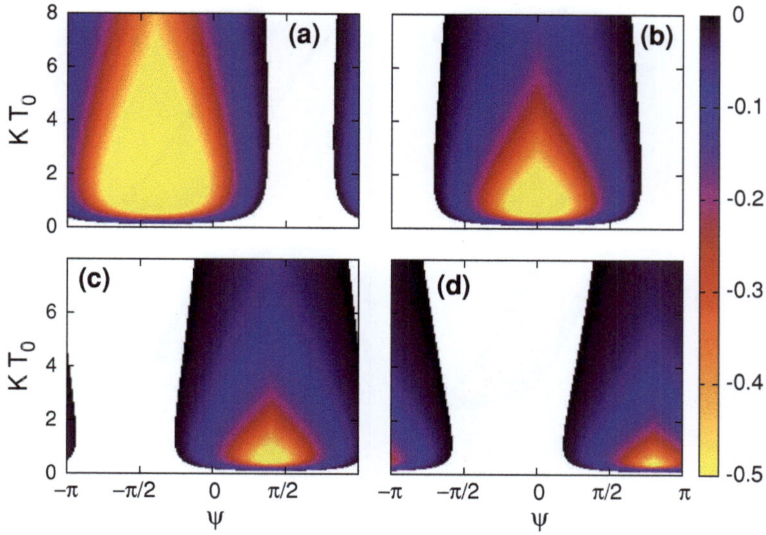

Fig. 3.34 Domain of control in dependence on ψ and K in units of $T_0 = 2$ for different time delays $\tau = 0.3T_0$, $0.5T_0$, $0.7T_0$, and $0.9T_0$ in panels (**a**)–(**d**), respectively. The largest real part of the complex eigenvalues Λ is shown in *color code*. The feedback phase is fixed at $\varphi = 0$. Other parameter as in Fig. 3.33

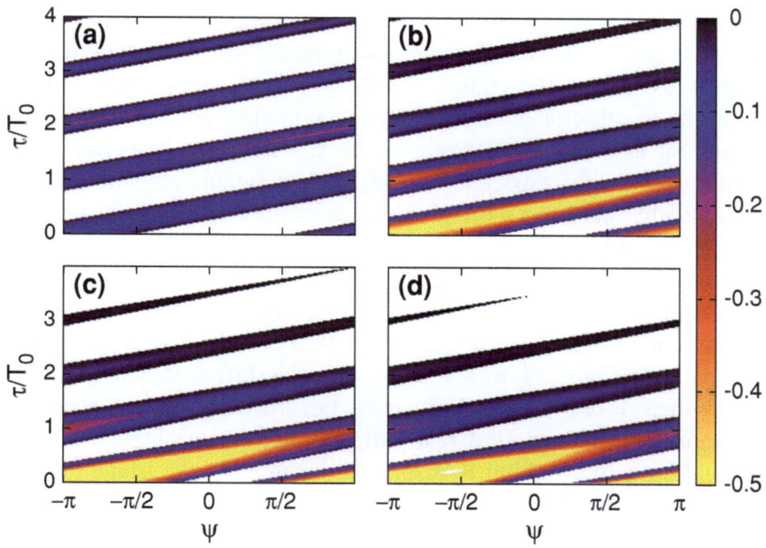

Fig. 3.35 Domain of control in dependence on ψ and τ in units of $T_0 = 2$ for different feedback gains $KT_0 = 0.2$, 0.6, 1, and 1.4 in panels (**a**)–(**d**), respectively. The largest real part of the complex eigenvalues Λ is shown in *color code*. The feedback phase is fixed at $\varphi = 0$. Other parameter as in Fig. 3.33

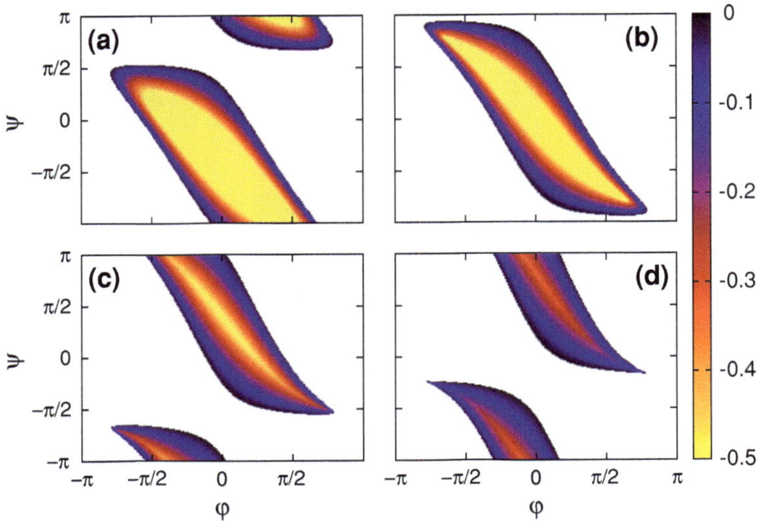

Fig. 3.36 Domain of control in dependence on φ and ψ for different time delays $\tau = 0.3T_0$, $0.5T_0$, $0.7T_0$, and $0.9T_0$ in panels **(a)**–**(d)**, respectively. The largest real part of the complex eigenvalues Λ is shown in *color code*. The feedback gain is fixed at $KT_0 = 1$. Other parameter as in Fig. 3.33

For $\tau > T_0/2$, the domain of control is shifted towards positive values of ψ as shown in panels (c) and (d). Recall that the dependence of the islands of stability on the feedback phase φ led to a distortion which was shown in Fig. 3.25.

Figure 3.35 depicts the stability regions in the (ψ, τ) plane where the feedback gains varies as $KT_0 = 0.2$, 0.6, 1, and 1.4 in panels (a)–(d), respectively. In general, the location of the maximum of the control domain is shifted by one intrinsic period as ψ increases from $-\pi$ to π, where the 2π-periodicity of the controlled system with respect to ψ can be seen again. The maximum of the domain of stability becomes more pronounced as the feedback gain increases, but the range of possible time delays for stabilization is smaller. See, for instance, panel (d), where the domain ends for $\tau \approx 3T_0$.

As a last projection which involves a ψ-axis, Fig. 3.36 shows the (φ, ψ) plane for different time delays $\tau = 0.3T_0$, $0.5T_0$, $0.7T_0$, and $0.9T_0$ in panels (a)–(d), respectively. As in Fig. 3.34, a change of τ yields a shift of the domain of control, whereas the shape is only slightly influenced. For small time delays, the system has the eigenvalue with the smallest real part indicated by the large yellow area in panel (a) and for τ approaching T_0 displayed in panel (d), the island of stability is more pointed at its ends.

To conclude this discussion of a second phase in the control scheme, the control phase ψ adds another degree of freedom which can compensate for a non-optimal choices of the other control parameters in order to stabilize the fixed point. The topic of the following section will still be the stabilization of a steady state where the focus is on the asymptotic behavior for large time delays.

3.6 Asymptotic Properties for Large Delays

The last section of this chapter is devoted to a deeper analytical insight into the time-delayed feedback control of steady states for large delays. Long delay times have been found of importance for frequency discretization in coupled chaotic oscillators [55]. In optical systems with feedback, large delays arise from the fast dynamics of the laser [39, 43]. In these devices the photon lifetime is much shorter than the round-trip time in an external cavity. Properties of stationary states in laser dynamics are investigated in Refs. [56–58] in the framework of the Lang-Kobayashi equation [59].

Following the chain of reasoning in Ref. [14], I will relate asymptotic properties of the eigenvalue spectrum with exact solutions and discuss the shape of the domain of control in the space spanned by the control parameters, i.e., feedback gain K and time delay τ.

Throughout this section, I will consider again the generic model of an unstable fixed point of focus type which is subject to diagonal Pyragas feedback as already introduced on Eq. 3.2.1 in Sect. 3.2. The system's equation is reproduced for convenience

$$\frac{dx(t)}{dt} = \lambda x(t) + \omega y(t) - K[x(t) - x(t-\tau)] \tag{3.6.1a}$$

$$\frac{dy(t)}{dt} = -\omega x(t) + \lambda y(t) - K[y(t) - y(t-\tau)], \tag{3.6.1b}$$

where I assume again positive parameter λ and non-zero ω yielding an unstable focus. Opposed to the previous Sects. 3.5.1–3.5.4, the feedback is applied to the system by diagonal coupling where the coupling matrix is the 2×2 identity matrix.

Note that three different timescales are of importance in the present control problem: (i) the inverse divergence rate of trajectories around the unstable fixed point $1/\lambda$, (ii) the period of undamped oscillations around the fixed point $T_0 = 2\pi/\omega$, where ω is the oscillation frequency, and (iii) the delay time τ used in the feedback control loop. Here, I consider the case $\tau \gg 1/\lambda$.

As discussed in the beginning of this chapter in Sect. 3.2, the stability of the fixed point is determined by the roots Λ of the characteristic equation

$$\lambda \pm i\omega = \Lambda + K\left(1 - e^{-\Lambda\tau}\right) \tag{3.6.2}$$

which can be solved analytically by the Lambert function

$$\Lambda\tau = W\left(K\tau e^{-(\lambda \pm i\omega)\tau + K\tau}\right) + (\lambda \pm i\omega)\tau - K\tau. \tag{3.6.3}$$

For details see the derivations and references in Sect. 3.2.

Figure 3.37 shows the real parts of the critical eigenvalues Λ as a function of τ for different values of $KT_0 = 0.5$, 1, 1.5, and 2 in panels (a), (b), (c), and (d),

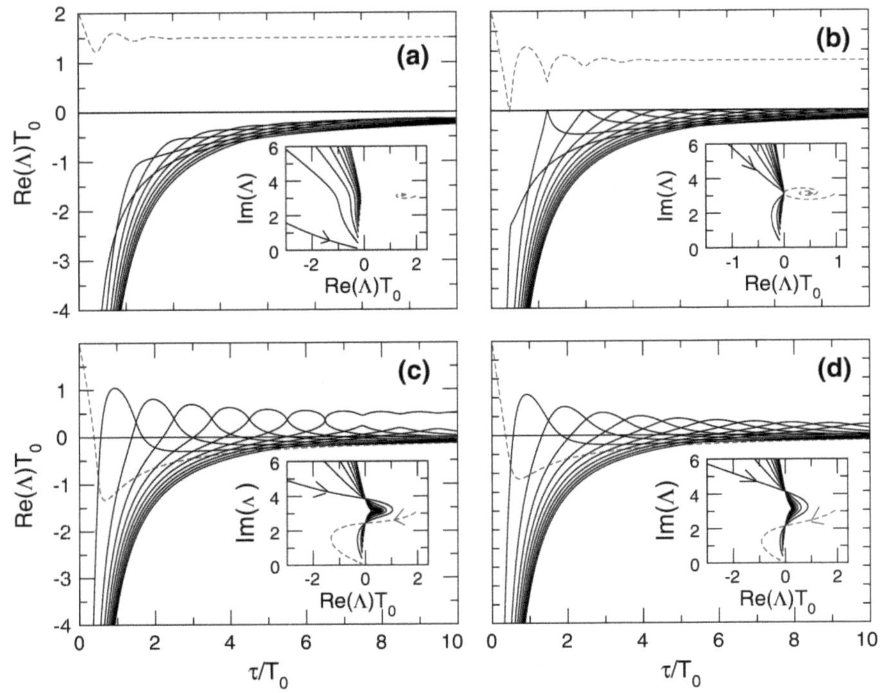

Fig. 3.37 Real parts of the complex eigenvalues Λ as a function of τ calculated from the characteristic Eq. 3.6.2 for 10 modes with the largest real parts. The feedback gain K is chosen as $KT_0 = 0.5$, 1, 1.5, and 2 in panels **a**, **b**, **c**, and **d**, respectively. The *inset* shows eigenmodes Λ in the complex plane for $\tau \in [0, 20]$ in units of the intrinsic timescale $T_0 = 2\pi/\omega = 2$ where the *red (dashed) curves* indicate the eigenvalue originating from the uncontrolled system and the *black (solid) curves* refer to eigenmodes created by the delay control. *Parameters* $\lambda = 1$ and $\omega = \pi(T_0 = 2)$

respectively. See also Fig. 3.4a. The panels show the 10 modes with the largest real part. The insets show the same eigenvalues as curves in the complex plane, parameterized by τ. Note that the eigenvalue originating from the uncontrolled system displayed in red (dashed) color is the most unstable one for sufficiently small K and does not couple to the eigenvalues generated by the delay as can be seen in Figs. 3.37a and b. The countable set of eigenvalues generated by the delay originates from $\mathrm{Re}(\Lambda) = -\infty$ for $\tau \to 0$, and shows the typical nonmonotonic behavior that leads to stability islands for appropriate τ and K. For larger feedback gain K, the eigenvalue originating from the uncontrolled system is no longer separated from those which are generated by the delay as shown in Figs. 3.37c and d. Moreover, one can observe a scaling behavior of the real parts of the eigenvalues for large τ: in Figs. 3.37a–c, there is a single eigenvalue retaining a positive real part, whereas all the other real parts tend to zero for large τ. The insets show that the eigenvalues in fact accumulate along the imaginary axis. This observation will be studied in detail in the following.

The scaling behavior of eigenvalues of general linear delay-differential equations for large delay τ has been analyzed in Ref. [60]. In particular, it turns out that one can distinguish the strongly unstable eigenvalues and a pseudocontinuous spectrum. Note that a pseudocontinuous spectrum was also found in the context of delay-coupled oscillators which exhibit destabilization via a competition-cooperation mechanism [61]. This is similar to the dynamics in an Eckhaus instability. See, for instance, Ref. [62], where the correspondence between Stuart–Landau oscillators with delay and the amplitude equation of the complex Ginzburg–Landau model was demonstrated.

The strongly unstable eigenvalues Λ_s have positive real parts that do not tend to zero with increasing τ, i.e., $\Lambda_s \to$ const. and $\mathrm{Re}(\Lambda_s) \geq \delta$ for some finite quantity $\delta > 0$ as τ tends to infinity.

The pseudocontinuous spectrum has eigenvalues Λ_p with real parts that scale as $1/\tau$, i.e., $\Lambda_p = \frac{1}{\tau}\gamma + i\left(\Omega + \frac{1}{\tau}\varphi\right) + \mathcal{O}\left(\frac{1}{\tau^2}\right)$ with some γ, Ω, and φ. A spectrum with this scaling behavior and positive real part leads to so-called weak instabilities. For more details, see Refs. [60, 63].

In order to obtain the strongly unstable eigenvalues, one can insert $\Lambda_s =$ const. into Eq. 3.6.2 and assume $\tau \to \infty$. Since $\mathrm{Re}(\Lambda_s) > \delta$, the exponential term vanishes and one arrives at the expression for Λ_s

$$\Lambda_s = \lambda - K \pm i\omega \tag{3.6.4}$$

which holds for $\lambda - K > 0$. Thus the following statement holds:

For $K < \lambda$, there exist two eigenvalues of the controlled stationary state, Λ_{s1} and its complex conjugate Λ_{s2}, such that

$$\lim_{\tau\to\infty} \Lambda_{s1} = \lambda - K + i\omega. \tag{3.6.5}$$

The real parts of these eigenvalues are positive and, hence, the stationary state is strongly unstable as depicted in Figs. 3.37a–c.

In order to obtain the asymptotic expression for the remaining pseudocontinuous part of the spectrum, one has to insert the scaling $\Lambda_p = \frac{1}{\tau}\gamma + i\left(\Omega + \frac{1}{\tau}\varphi\right)$ into the characteristic equation (3.6.2). Up to the leading order one obtains the equation

$$i\Omega + K\left(1 - e^{-\gamma}e^{-i\varphi}\right) = \lambda \pm i\omega, \tag{3.6.6}$$

and the additional condition $\Omega = \Omega^{(m)} = 2\pi m/\tau$, $m = \pm 1, \pm 2, \pm 3, \ldots$. Equation 3.6.6 can be solved with respect to $\gamma(\Omega)$

$$\gamma(\Omega) = -\frac{1}{2}\ln\left[\left(1 - \frac{\lambda}{K}\right)^2 + \left(\frac{\Omega \pm \omega}{K}\right)^2\right]. \tag{3.6.7}$$

The fact that $\mathrm{Re}(\Lambda_p) \approx \gamma(\Omega)/\tau$ and $\mathrm{Im}(\Lambda_p) \approx \Omega$ up to the leading order means that the eigenvalues Λ_p accumulate in the complex plane along curves $(\gamma(\Omega), \Omega)$, provided that the real axis is scaled as $\mathrm{Re}(\Lambda)\tau$. The actual positions of the eigenvalues on the curves can be obtained by evaluating Ω at points $\Omega^{(m)} = 2\pi\, m/\tau$. With increasing τ, the eigenvalues cover the curves densely [60]. Hence, a second statement holds:

The fixed point of system (3.6.1a, 3.6.1b) has a set of eigenvalues which behave asymptotically as $\Lambda_p(\Omega^{(k)}) = \frac{1}{\tau}\gamma(\Omega^{(k)}) + i\left(\Omega^{(k)} + \frac{1}{\tau}\varphi(\Omega^{(k)})\right)$ with $\gamma(\Omega)$ given by Eq. 3.6.7. There exists weak instability if the maximum of $\gamma(\Omega)$ is positive, i.e.,

$$\gamma_{\max} = \max_{\Omega} \gamma(\Omega) = -\ln\left|1 - \frac{\lambda}{K}\right| > 0 \tag{3.6.8}$$

which is the case for $K > \lambda/2$.

Figure 3.38 illustrates the spectrum of the fixed point of system (3.6.1a, 3.6.1b) for $\tau = 20$. One can clearly distinguish the two types of eigenvalues. For $K < \lambda/2$ as in Fig. 3.38a, the fixed point has a pair of strongly unstable eigenvalues, whereas the pseudocontinuous spectrum is stable. Note that the red symbols show the spectrum computed numerically from the full eigenvalue equation, whereas the dashed lines are the curves $(\gamma(\Omega), \Omega)$ from the asymptotic approximation where the pseudocontinuous spectrum accumulates for large τ. At $K = \lambda/2$ as depicted in Fig. 3.38b, the pseudocontinuous spectrum touches the imaginary axis resulting in the appearance of a weak instability for $K > \lambda/2$. This leads to the coexistence of strong and weak instabilities for $\lambda/2 < K < \lambda$ as shown in Fig. 3.38c with $K = 0.75$. At $K = \lambda$, the strongly unstable eigenvalues disappear, being absorbed by the pseudocontinuous spectrum, which develops a singularity at this moment as in Fig. 3.38d. Finally, for $K > \lambda$ in Fig. 3.38e, there occurs only a weak instability induced by the pseudocontinuous spectrum.

After inspecting all possibilities given in Fig. 3.38, one can conclude that stabilization by the feedback control scheme (3.6.1a, 3.6.1b) always has an upper limit τ_c such that for $\tau > \tau_c$ it fails. Additionally, note that for $K < \lambda$ and large delay, the stationary state is strongly unstable with the complex conjugate eigenvalues $\Lambda_{1,2} = \lambda - K \pm i\omega$, and for $K > \lambda$ weakly unstable with a large number of unstable eigenvalues given by Eq. 3.6.6, the real parts of which scale as $1/\tau$.

For the fixed point which is close to the Hopf bifurcation, let us assume $K > \lambda$, and, hence it has an unstable pseudocontinuous spectrum, as shown in Fig. 3.38e. As λ stays fixed, with increasing τ the curve of the pseudocontinuous spectrum will be densely filled with the eigenvalues ($\Omega^{(m)} = 2\pi m/\tau$). The only possibility for the fixed point to become stable is to assume that λ is also scaled with increasing τ. Particularly, I will show that in order to achieve control it have to be scaled as $\lambda = \lambda_0 \varepsilon^2$ with fixed λ_0 where small parameter $\varepsilon = 1/\tau$ is introduced for notational convenience.

Figure 3.39 illustrates the part of the curve $\gamma(\Omega)$ which may induce an instability in the system. More precisely, the interval of unstable frequencies is $\Omega_1 < \Omega < \Omega_2$, where Ω_1 and Ω_2 are given by the zeros of $\gamma(\Omega)$:

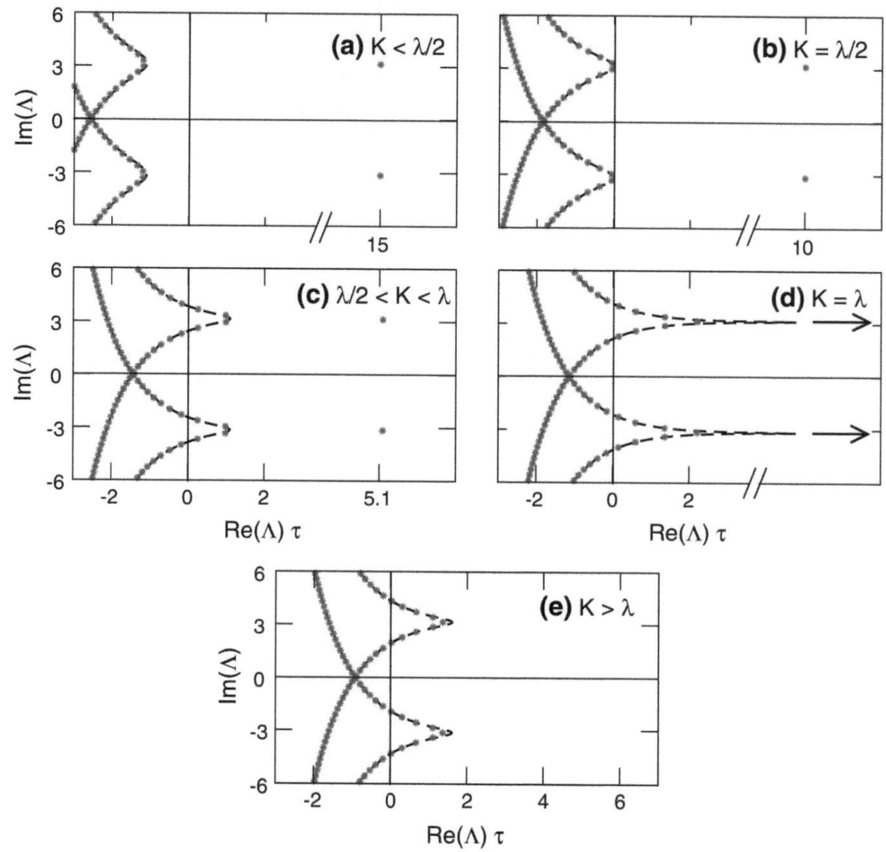

Fig. 3.38 Numerically computed spectrum of eigenvalues for $\tau = 20$ (*red asterisks*). The *dashed lines* depict the asymptotic pseudocontinuous spectrum. Panel **a** shows strong instability for $K = 0.25$ ($K < \lambda/2$). Panel **b** refers to the critical case of $K = 0.5 = \lambda/2$ at which the weak instability occurs in addition to the strong one. **c** displays strong and weak instability for $K = 0.75$ which is in the range of $\lambda/2 < K < \lambda$). In panel **d**, a strong instability disappears via the singularity of the pseudocontinuous spectrum for $K = 1.0 = \lambda$ and panel **e** depict the case of weak instability for $K = 1.25$ ($K > \lambda$). System's parameters as in Fig. 3.37

$$\Omega_{1,2} = \omega \pm K\sqrt{1 - \left(1 - \frac{\lambda}{K}\right)^2}. \tag{3.6.9}$$

For a derivation of this equation see Eq. 3.2.16a. For small λ one can approximate this as

$$\Omega_{1,2} = \omega \pm \sqrt{2\lambda K}. \tag{3.6.10}$$

Thus, the length of the interval $\Delta\Omega$ of unstable frequencies is given by $\Delta\Omega = \Omega_2 - \Omega_1 = 2\sqrt{2\lambda K}$.

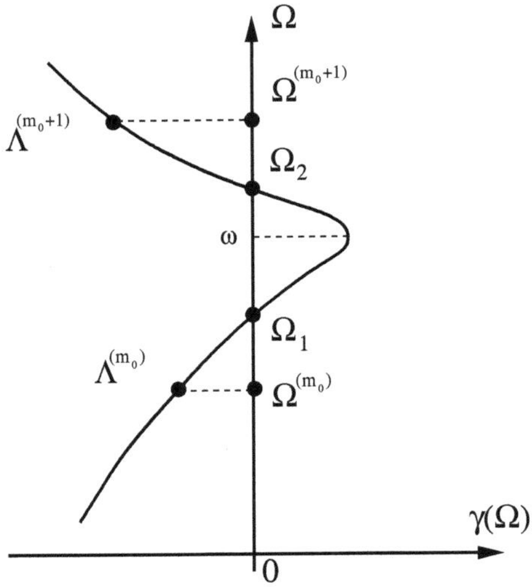

Fig. 3.39 Curve of the pseudocontinuous spectrum. The actual position of the complex eigenvalues $\Lambda = \frac{1}{\tau}\gamma + i[\Omega + \mathcal{O}(\tau^{-1})]$ on the curve corresponds to $\Omega^{(m)} = 2\pi m\varepsilon, m = \pm1, \pm2, \pm3, \ldots, \varepsilon = 1/\tau$

Note that the actual position of the eigenvalues on the curve corresponds to the values of $\Omega^{(m)} = 2\pi m\varepsilon$ with any integer m and $\varepsilon = \tau^{-1}$. It is easy to see that the distance between the frequencies of neighboring eigenvalues $\Omega^{(m+1)} - \Omega^{(m)} = 2\pi\varepsilon$ scales as ε. Therefore, the control can be successful if $\lambda = \lambda_0\varepsilon^2$. In this case the length of the unstable interval is $\Delta\Omega = 2\varepsilon\sqrt{2\lambda_0 K}$ and scales also as ε. The control can be achieved if the length is smaller than the distance between neighboring eigenvalues, i.e., $\Delta\Omega = 2\varepsilon\sqrt{2\lambda_0 K} < 2\pi\varepsilon$, leading to

$$K < \frac{\pi^2}{2\lambda_0}. \tag{3.6.11}$$

Equation 3.6.11 gives a necessary condition for successful control. The fixed point is stable if the imaginary parts of the eigenvalues are outside of the interval $\Omega_1 < \Omega < \Omega_2$. Such a case with $\Omega^{(m_0)} < \Omega_1 < \Omega_2 < \Omega^{(m_0+1)}$ is illustrated in Fig. 3.39, in which the leading eigenvalues $\Lambda^{(m_0)}$ and $\Lambda^{(m_0+1)}$ have negative real parts.

The relative phase of the delay plays an additional important role. Depending on this phase, control occurs periodically with τ. In order to quantify this effect, let us introduce $\omega_\tau = 2\pi/\tau$ to be the frequency associated with the delay. Then the ratio of the internal frequency ω and ω_τ is given by $\omega/\omega_\tau = \gamma_\tau \bmod 1$. Here $0 < \gamma_\tau < 1$ measures the detuning from the resonance between the internal frequency and the delay-induced one. Using this notation and Eq. 3.6.10, one can rewrite

$$\Omega_{1,2} = m_0\omega_\tau + \gamma_\tau\omega_\tau \pm \varepsilon\sqrt{2\lambda_0 K} = \Omega^{(m_0)} + \varepsilon\left(2\pi\gamma_\tau \pm \sqrt{2\lambda_0 K}\right) \tag{3.6.12}$$

with some integer number m_0.

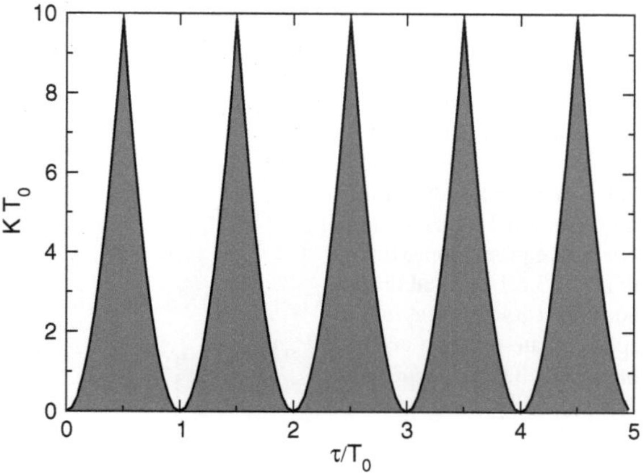

Fig. 3.40 *Shaded region*: domain of control in the (τ, K) plane for the fixed point close to the Hopf bifurcation, given by the asymptotic formula Eq. 3.6.14 for $\lambda = \lambda_0/\tau^2$. *Parameters* $\lambda_0 = 1$, $\omega = \pi (T_0 = 2)$

The necessary and sufficient condition for the stability is $\Omega^{(m_0)} < \Omega_1 < \Omega_2 < \Omega^{(m_0+1)}$ as shown in Fig. 3.39. This leads to

$$\sqrt{2\lambda_0 K} < 2\pi \min\{\gamma_\tau, 1 - \gamma_\tau\} \tag{3.6.13}$$

or equivalently

$$K < \frac{2\pi^2}{\lambda_0} (\min\{\gamma_\tau, 1 - \gamma_\tau\})^2 = \frac{2\pi^2}{\lambda_0} \left(\min\left\{ \left[\frac{\omega\tau}{2\pi}\right]_f, 1 - \left[\frac{\omega\tau}{2\pi}\right]_f \right\} \right)^2, \tag{3.6.14}$$

where $[\omega\tau/(2\pi)]_f$ is the fractional part of $\omega\tau/(2\pi)$. Figure 3.40 displays the domain of control according to this formula for $\lambda = \lambda_0/\tau^2$.

In order to return to unscaled parameters, one has to substitute $\lambda_0 = \lambda/\varepsilon^2 = \lambda\tau^2$. Figure 3.41a shows the obtained domain of control for fixed small $\lambda = 0.01$. The maximum allowed values of K decrease as $1/\tau^2$. More precisely, the maximum feedback gain $K_{\max}$ is given by

$$K_{\max}(\tau) = \frac{\pi^2}{2\lambda\tau^2}. \tag{3.6.15}$$

The application of the asymptotic analysis allows to reveal many essential features and mechanisms of the stabilization control scheme (3.6.1a, 3.6.1b) for large delay τ. On the other hand, the obtained approximations are valid as soon as K is much larger than λ. Figure 3.41 shows a comparison of the boundaries of the stability domain, which are given by the asymptotic methods and exact analytical formulas already derived in Eqs. 3.2.18a, 3.2.18b, 3.2.18c of Sect. 3.2. Very close to the Hopf bifurcation, e.g., $\lambda = 0.01$, the agreement is excellent even at small values of τ as in Fig. 3.41a, while for larger λ shown in Fig. 3.41b the deviations become more visible. In addition, the approximate solution does not give the lower boundary of the control domain for small K which only shows up in Fig. 3.42.

As it is shown in Ref. [13], the pairs of curves derived as Eqs. 3.2.18a and 3.2.18b in Sect. 3.2 form the boundaries of the control domains in the (τ, K)

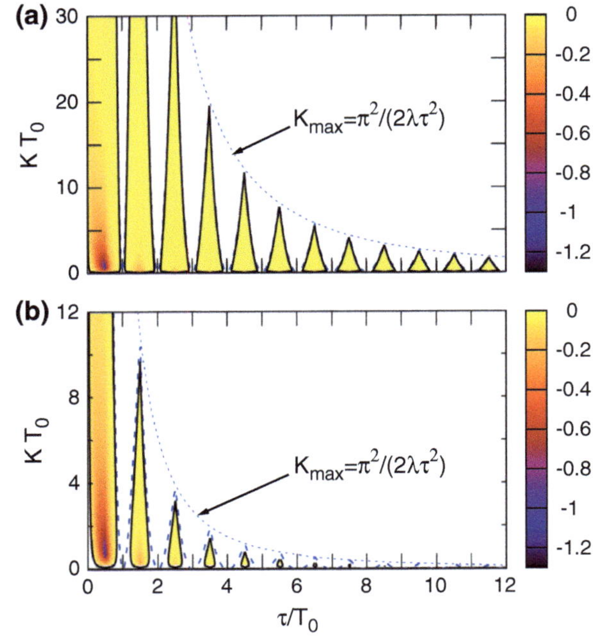

Fig. 3.41 Domain of control in the (τ, K) plane, and largest negative real part of the complex eigenvalues $\Lambda(K, \tau)$ (in *color code*) calculated from the characteristic equation using the Lambert function [Eq. 3.6.3]. *Dashed blue lines* asymptotic approximation Eq. 3.6.14 of stability boundary; *dotted blue line* approximate maxima according to Eq. 3.6.15. *Solid lines* exact stability boundaries, cf. Eqs. 3.2.18a and 3.2.18b. *Parameters* $\lambda = 0.01$, $\omega = \pi (T_0 = 2)$ in panel (**a**) and $\lambda = 0.1$, $\omega = \pi (T_0 = 2)$ in panel (**b**)

parameter plane, as depicted by solid lines in Figs. 3.41 and 3.42. These islands become smaller for increasing n and the corresponding values for K are confined by

$$K_{\min} \leq K \leq K_{\max}(n), \tag{3.6.16}$$

where the maximum feedback gain $K_{\max}(n)$ is given by an intersection point of the two branches $\tau_1(K, n)$ and $\tau_2(K, n)$. These intersection points correspond to double-Hopf points of codimension two. They are given by solutions of the transcendental equation

$$\arccos \frac{\lambda - K}{K} = \frac{(2n+1)\pi}{\omega} \sqrt{(2K - \lambda)\lambda}. \tag{3.6.17}$$

The corresponding values of τ are given by

$$\tau_{\max}(n) = \tau_{\min}(n) = \frac{(2n+1)\pi}{\omega}. \tag{3.6.18}$$

Note that the condition (3.6.17) is satisfied also for $K = K_{\min}$. The stability domain vanishes if $K_{\min}$ and $K_{\max}$ coincide. Forming the derivative of Eq. 3.6.17 with respect to K yields

$$\frac{1}{K} = \frac{(2n+1)\pi}{\omega}. \tag{3.6.19}$$

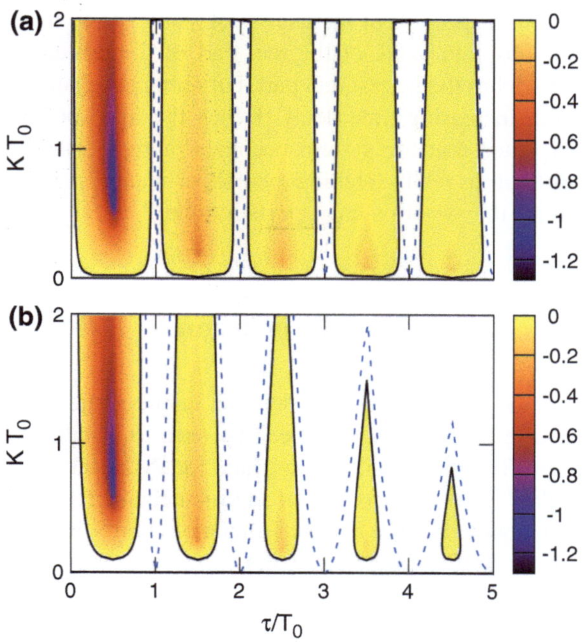

Fig. 3.42 Enlargement of Fig. 3.41: Deviation of the asymptotic results (*dashed*) from the exact stability boundary (*solid*) for small K or large λ. Parameters as in Fig. 3.41

Inserting the value of the minimum feedback gain $K = K_{\min} = \lambda/2$ finally gives the relation

$$\omega = \frac{(2n+1)\pi\lambda}{2}. \tag{3.6.20}$$

If this relation is satisfied, there is a resonant double-Hopf point of codimension three. Since n has to be an integer, this happens only for particular choices of λ and ω. Otherwise, the integer part of the value n obtained from this relation gives the number of nondegenerate stability islands.

Using Eq. 3.6.18, the maximum delay time τ_c which allows for stabilization is obtained as

$$\tau_c = \frac{2}{\lambda}. \tag{3.6.21}$$

Note that this boundary is sharp only if τ is an odd integer multiple of π/ω. For

$$\frac{\omega}{\lambda} < \frac{\pi}{2}, \tag{3.6.22}$$

even the first stability island vanishes and stabilization is not possible. This analytical approach also allows to identify the peaks of the control domains, which occur at $\tau_{\max} = (2n+1)\pi/\omega, n = 0, 1, 2, \ldots$, as double-Hopf bifurcation points. The critical time delay, above which control fails, is given by $\tau_c = 2/\lambda$.

An inspection of the islands of stabilization in Figs. 3.41 and 3.42 reveals that the absolute value of the real part of the critical eigenvalue, i.e. the eigenvalue which has the largest real part, but remains negative within those islands, decreases with increasing time delay. Hence the fixed point becomes less stable, and it is expected that the system becomes more sensitive to noise, and it will be more difficult to realize stabilization experimentally, if the delay time is chosen several times the system's characteristic time T_0.

3.7 Intermediate Conclusion

Let me finish this chapter with an intermediate summary. I have shown that time-delayed feedback control can be used to stabilize unstable steady states of focus type. This has been demonstrated for Pyragas control involving only a single time delay τ as well as for extended time-delayed feedback with integer multiples of τ. I calculated numerically the domain of control in various one- and two-dimensional projections of the space spanned by the control parameters. In addition, I derived analytical results, for instance, on the shape of this domain, the value of the minimum feedback gain, and solutions of the characteristic equation using the multivalued Lambert function.

Furthermore, I have discussed effects of non-zero latency times and nondiagonal phase-dependent coupling which are both relevant in an experimental realization of time-delayed feedback control. I found that a proper adjustment of the time delay is able to compensate for reduced effectiveness of the control method, for instance, due to latency or feedback phase. In addition, I have demonstrated that time-delayed feedback control fails to stabilize saddle points in the realization both in the diagonal and phase-dependent coupling.

Finally, I have elaborated various analytical conditions for successful control of a fixed point of focus type. By asymptotic expansion methods for large delay, and a detailed comparison with exact solutions I have established the parameter ranges for successful control. Thereby I have not only obtained the precise shape of the islands of control in the parameter plane spanned by time delay and feedback gain, but have also gained insight into the mechanism of control by analyzing the eigenvalue spectrum of the fixed point of the delay-differential equation which consists of a pseudocontinuous spectrum and up to two strongly unstable complex eigenvalues. Although our analysis has shown that the standard control scheme generally fails for large delay, I have found that if the uncontrolled system is sufficiently close to its instability threshold, i.e., a Hopf bifurcation, control does work even for relatively large delay times, compared to an intrinsic oscillation period.

After the control of unstable steady states, I will consider time-delayed feedback applied to unstable periodic orbits in the next chapter. Specifically, I will show that a topological limitation of time-delayed feedback control known as odd number limitation theorem does not hold in general. This limitation refers to the case of an unstable periodic orbit with an odd number of real Floquet multipliers larger than unity [44, 45] and states that this class of periodic orbits cannot be stabilized by time-delayed feedback. I will refute this theorem following Refs. [46–48, 64, 65].

References

1. Gills Z, Iwata C, Roy R, Schwartz IB, Triandaf I (1992) Tracking unstable steady states: extending the stability regime of a multimode laser system. Phys Rev Lett 69:3169
2. Bielawski S, Bouazaoui M, Derozier D, Glorieux P (1993) Stabilization and characterization of unstable steady states in a laser. Phys Rev A 47:3276
3. Parmananda P, Rhode MA, Johnson GA, Rollins RW, Dewald HD, Markworth AJ (1994) Stabilization of unstable steady states in an electrochemical system using derivative control. Phys Rev E 49:5007
4. Parmananda P, Madrigal R, Rivera M, Nyikos L, Kiss IZ, Gáspár V (1999) Stabilization of unstable steady states and periodic orbits in an electrochemical system using delayed-feedback control. Phys Rev E 59:5266
5. Ahlborn A, Parlitz U (2004) Stabilizing unstable steady states using multiple delay feedback control. Phys Rev Lett 93:264101
6. Rosenblum MG, Pikovsky AS (2004) Controlling synchronization in an ensemble of globally coupled oscillators. Phys Rev Lett 92:114102
7. Just W, Bernard T, Ostheimer M, Reibold E, Benner H (1997) Mechanism of time-delayed feedback control. Phys Rev Lett 78:203
8. Just W, Reibold E, Benner H, Kacperski K, Fronczak P, Holyst J (1999) Limits of time-delayed feedback control. Phys Lett A 254:158
9. Pyragas K (2002) Analytical properties and optimization of time-delayed feedback control. Phys Rev E 66:26207
10. von Loewenich C, Benner H, Just W (2004) Experimental relevance of global properties of time-delayed feedback control. Phys Rev Lett 93:174101
11. Balanov AG, Janson NB, Schöll E (2005) Delayed feedback control of chaos: bifurcation analysis. Phys Rev E 71:016222
12. Pyragas K (1995) Control of chaos via extended delay feedback. Phys Lett A 206:323
13. Hövel P, Schöll E (2005) Control of unstable steady states by time-delayed feedback methods. Phys Rev E 72:046203
14. Yanchuk S, Wolfrum M, Hövel P, Schöll E (2006) Control of unstable steady states by long delay feedback. Phys Rev E 74:026201
15. Dahms T, Hövel P, Schöll E (2007) Control of unstable steady states by extended time-delayed feedback. Phys Rev E 76:056201
16. Gjurchinovski A, Urumov V (2008) Stabilization of unstable steady states by variable delay feedback control. Europhys Lett 84:40013
17. Strogatz SH (1994) Nonlinear dynamics and chaos. Westview Press, Cambridge, MA
18. Schöll E (2001) Nonlinear spatio-temporal dynamics and chaos in semiconductors. Nonlinear science series, vol 10. Cambridge University Press, Cambridge
19. Schuster HG, Stemmler MB (1997) Control of chaos by oscillating feedback. Phys Rev E 56:6410
20. Wright EM (1949) The linear difference-differential equation with constant coefficients. Proc R Soc Edinb A Math Phys Sci 62:387
21. Wright EM (1955) A non-linear difference-differential equation. J Reine Angew Math 194:66
22. Bellmann R, Cooke KL (1963) Differential-difference equations. Academic Press, New York

23. Hale JK (1971) Functional differential equations. Applied mathematical sciences, vol 3. Springer, New York
24. Asl FM, Ulsoy AG (2003) Analysis of a system of linear delay differential equations. ASME J Dyn Syst Meas Control 125:215
25. Amann A, Schöll E, Just W (2007) Some basic remarks on eigenmode expansions of time-delay dynamics. Physica A 373:191
26. Hövel P (2004) Effects of chaos control and latency in time-delay feedback methods. Master's thesis, Technische Universität Berlin
27. Hale JK (1977) Theory of functional differential equations. Springer, New York
28. Hale JK, Verduyn Lunel SM (1993) Introduction to functional differential equations. Springer, New York
29. Just W, Reibold E, Kacperski K, Fronczak P, Holyst JA, Benner H (2000) Influence of stable Floquet exponents on time-delayed feedback control. Phys Rev E 61:5045
30. Just W (2000) On the eigenvalue spectrum for time-delayed Floquet problems. Physica D 142:153
31. Beck O, Amann A, Schöll E, Socolar JES, Just W (2002) Comparison of time-delayed feedback schemes for spatio-temporal control of chaos in a reaction-diffusion system with global coupling. Phys Rev E 66:016213
32. Just W, Popovich S, Amann A, Baba N, Schöll E (2003) Improvement of time-delayed feedback control by periodic modulation: analytical theory of Floquet mode control scheme. Phys Rev E 67:026222
33. Janson NB, Balanov AG, Schöll E (2004) Delayed feedback as a means of control of noise-induced motion. Phys Rev Lett 93:010601
34. Hinz R (2009) Transient behaviour in systems with time-delayed feedback. Master's thesis, Technische Universität Berlin
35. Socolar JES, Sukow DW, Gauthier DJ (1994) Stabilizing unstable periodic orbits in fast dynamical systems. Phys Rev E 50:3245
36. Baba N (2001) Stabilisierung instabiler räumlicher Muster durch zeitverzögerte Rückkopplung mit räumlichen Filtern. Master's thesis, Technische Universität Berlin
37. Dahms T (2007) Stabilisierung von Fixpunkten durch zeitverzögerte Rückkopplung in Lasern. Master's thesis, Technische Universität Berlin
38. Blakely JN, Illing L, Gauthier DJ (2004) Controling fast chaos in delay dynamical systems. Phys Rev Lett 92:193901
39. Schikora S, Hövel P, Wünsche HJ, Schöll E, Henneberger F (2006) All-optical noninvasive control of unstable steady states in a semiconductor laser. Phys Rev Lett 97:213902
40. Sukow DW, Bleich ME, Gauthier DJ, Socolar JES (1997) Controlling chaos in a fast diode resonator using time-delay autosynchronisation: experimental observations and theoretical analysis. Chaos 7:560
41. Just W, Reckwerth D, Reibold E, Benner H (1999) Influence of control loop latency on time-delayed feedback control. Phys Rev E 59:2826
42. Hövel P, Socolar JES (2003) Stability domains for time-delay feedback control with latency. Phys Rev E 68:036206
43. Wünsche HJ, Schikora S, Henneberger F (2008) Noninvasive control of semiconductor lasers by delayed optical feedback. In: Schöll E, Schuster HG (eds) Handbook of chaos control. Wiley-VCH, Weinheim (second completely revised and enlarged edition)
44. Nakajima H (1997) On analytical properties of delayed feedback control of chaos. Phys Lett A 232:207
45. Nakajima H, Ueda Y (1998) Limitation of generalized delayed feedback control. Physica D 111:143
46. Fiedler B, Flunkert V, Georgi M, Hövel P, Schöll E (2007) Refuting the odd number limitation of time-delayed feedback control. Phys Rev Lett 98:114101
47. Just W, Fiedler B, Flunkert V, Georgi M, Hövel P, Schöll E (2007) Beyond odd number limitation: a bifurcation analysis of time-delayed feedback control. Phys Rev E 76:026210

48. Fiedler B, Yanchuk S, Flunkert V, Hövel P, Wünsche HJ, Schöll E (2008) Delay stabilization of rotating waves near fold bifurcation and application to all-optical control of a semiconductor laser. Phys Rev E 77:066207
49. Dahms T, Hövel P, Schöll E (2008) Stabilizing continuous-wave output in semiconductor lasers by time-delayed feedback. Phys Rev E 78:056213
50. Fischer A, Andersen O, Yousefi M, Stolte S, Lenstra D (2000) Experimental and theoretical study of filtered optical feedback in a semiconductor laser. IEEE J Quantum Electron 36:375
51. Rosenblum MG, Pikovsky AS (2004) Delayed feedback control of collective synchrony: an approach to suppression of pathological brain rhythms. Phys Rev E 70:041904
52. Flunkert V, Schöll E (2007) Suppressing noise-induced intensity pulsations in semiconductor lasers by means of time-delayed feedback. Phys Rev E 76:066202
53. Schikora S, Wünsche HJ, Henneberger F (2008) All-optical noninvasive chaos control of a semiconductor laser. Phys Rev E 78:025202
54. Pyragas K (1992) Continuous control of chaos by self-controlling feedback. Phys Lett A 170:421
55. Yanchuk S (2005) Discretization of frequencies in delay coupled oscillators. Phys Rev E 72:036205
56. Yanchuk S, Wolfrum M (2004) Instabilities of stationary states in lasers with long-delay optical feedback. Rep Weierstraß Inst Appl Anal Stoch 962:1
57. Yanchuk S (2005) Properties of stationary states of delay equations with large delay and applications to laser dynamics. Math Methods Appl Sci 28:363
58. Yanchuk S, Wolfrum M (2005) Synchronous and asynchronous instabilities of two lasers with a long delayed coupling. In: van Campen DH, Lazurko MD, van den Oever WPJM (eds) Proceedings of 5th EUROMECH nonlinear dynamics conference ENOC-2005:Eindhoven. Eindhoven University of Technology, Eindhoven, Netherlands), pp 2069–2073, eNOC Eindhoven (CD ROM). ISBN 90 386 2667 3
59. Lang R, Kobayashi K (1980) External optical feedback effects on semiconductor injection laser properties. IEEE J Quantum Electron 16:347
60. Yanchuk S, Wolfrum M (2005) Instabilities of equilibria of delay-differential equations with large delay. In: In: van Campen DH, Lazurko MD, van den Oever WPJM (eds) Proceedings of 5th EUROMECH Nonlinear Dynamics Conference ENOC-2005:Eindhoven. Eindhoven University of Technology, Eindhoven, Netherlands, pp 1060–1065, eNOC Eindhoven (CD ROM). ISBN 90 386 2667 3
61. Yanchuk S, Wolfrum M (2008) Destabilization patterns in chains of coupled oscillators. Phys Rev E 77:26212
62. Wolfrum M, Yanchuk S (2006) Eckhaus instability in systems with large delay. Phys Rev Lett 96:220201
63. Lepri S, Giacomelli G, Politi A, Arecchi FT (1994) High-dimensional chaos in delayed dynmical-systems. Physica D 70:235
64. Fiedler B, Flunkert V, Georgi M, Hövel P, Schöl E (2008) Beyond the odd number limitation of time-delayed feedback control. In: Schöll E, Schuster HG (eds) Handbook of chaos control. Wiley-VCH, Weinheim, pp 73–84 (second completely revised and enlarged edition)
65. Fiedler B, Flunkert V, Georgi M, Hövel P, Schöll E (2008) Delay stabilization of rotating waves without odd number limitation. In: Schuster HG (ed) Reviews of nonlinear dynamics and complexity, vol 1. Wiley-VCH, Weinheim, pp 53–68

Chapter 4
Refuting the Odd Number Limitation Theorem

2 is the oddest prime.

(unknown humorous mathematician)

In the previous chapter, I have discussed the control of an unstable steady state by time-delayed feedback. This method, however, was originally developed by K. Pyragas to stabilize periodic orbits where the period of the target orbit determined the choice of the time delay [1]. Therefore, this chapter deals with the stabilization of a special class of periodic orbits which have an odd number of real Floquet multipliers greater than unity. It has been contended that periodic orbits with this property cannot be stabilized by time-delayed feedback control of Pyragas type. This severe limitation is known as odd number limitation theorem which was stated by H. Nakajima for both simple and extended time-delayed feedback [2, 3]. In this chapter, I will refute this often invoked theorem by a counterexample which consists of a normal form of a subcritical Hopf bifurcation. Thus, the general limitation for orbits with an odd number of real unstable Floquet multipliers greater than unity does not hold, but stabilization may be possible for suitable choices of the control parameters given by the time delay and a complex feedback gain [4–8].

The structure of this chapter is the following: At first, I will briefly review the odd number limitation theorem by Nakajima. For this, I will sketch the main steps of the proof and present also modifications of time-delayed feedback control that were employed to overcome this theorem. In Sect. 4.2, I will introduce the model equations of the counterexample including Pyragas control. The structure of the domain of control and the mechanism of stabilizing an odd number orbit will be topic of Sect. 4.3. In addition, I will extend the theoretical analysis to systems which are described rotating waves which can occur, for instance, in a fold bifurcation in Sects. 4.4 and 4.5. I will concluded this chapter with an intermediate summary in Sect. 4.6.

P. Hövel, *Control of Complex Nonlinear Systems with Delay*,
Springer Theses, DOI: 10.1007/978-3-642-14110-2_4,

4.1 Review of the Odd Number Limitation Theorem

Before the discussion of the counterexample, I will briefly review the main findings of the odd number limitation theorem and sketch its proof according to Ref. [2] where I will use the notation of the previous chapters. At the end of this section, I will summarize some modifications of the original Pyragas controller which were proposed to overcome this severe limitation.

The systems under investigation is given in general form including time-delayed feedback by the following equation

$$\frac{d}{dt}\mathbf{x}(t) = \mathbf{f}(\mathbf{x}(t)) + \mathbf{K}[\mathbf{x}(t-\tau) - \mathbf{x}(t)], \tag{4.1.1}$$

where $\mathbf{x}(t) \in \mathbb{R}^n$ is a n-dimensional vector and the control parameters are denoted by the time delay τ and feedback gain $\mathbf{K}$. In general, the feedback gain $\mathbf{K} \in \mathbb{R}^{n\times n}$ is a $n \times n$ matrix. For details concerning the coupling of the feedback term see Sect. 2.3. The system without control is assumed to have an unstable periodic solution $\mathbf{x}^*(t)$ with an odd number of real Floquet multipliers greater than unity. The time delay coincides with the period of this target solution $\mathbf{x}^*(t)$ which guarantees the noninvasive property of time-delayed feedback control.

The odd number limitation theorem states:

Theorem 4.1.1 *If the linear variational equation around* $\mathbf{x}^*(t)$ *of the uncontrolled system which is given by Eq.* 4.1.1 *for* $\mathbf{K} = 0$ *has an odd number of real Floquet multipliers greater than unity, the unstable periodic orbit* $\mathbf{x}^*(t)$ *can never be stabilized by time-delayed feedback of the form* (4.1.1) *with any value of the feedback gain* $\mathbf{K}$ [2].

After a linearization of system (4.1.1) around the target orbit $\mathbf{x}^*(t)$, Floquet theory is employed and using the fundamental matrix leads to a characteristic equation. The location of the roots of this equation is discussed with the restriction of an odd number of real Floquet multipliers greater than unity in the absence of a control force. By this assumption, it is concluded that the system (4.1.1)

> has at least one unstable solution, i.e., there does not exist any gain matrix $\mathbf{K}$ with which the unstable periodic orbit $\mathbf{x}^*(t)$ is stabilized [2].

Later on in Ref. [2], Theorem 3 considers the special case of a diagonal feedback gain matrix $\mathbf{K}$.

In the following, I will sketch the main steps of the proof on the basis of Ref. [2], but in the notation of this thesis:

Let Φ be the fundamental matrix of the uncontrolled, linearized system

$$\frac{d}{dt}\delta\mathbf{x}(t) = \mathbf{A}(t)\delta\mathbf{x}(t) \tag{4.1.2}$$

and Ψ the respective fundamental matrix of the controlled, linearized systems

$$\frac{d}{dt}\delta\mathbf{x}(t) = \left[\mathbf{A}(t) + \left(e^{-\Lambda\tau} - 1\right)\mathbf{K}\right]\delta\mathbf{x}(t), \tag{4.1.3}$$

where $\delta\mathbf{x}(t)$ denotes the deviation from the periodic orbit $\mathbf{x}^*(t)$ with period τ, Λ is the Floquet exponent, and $\mathbf{A}(t)$ describes the Jacobian matrix of the uncontrolled system. With these definitions, the following equations for Φ and Ψ hold

$$\frac{d}{dt}\Phi(t) = \mathbf{A}(t)\Phi(t) \tag{4.1.4a}$$

$$\frac{d}{dt}\Psi(t) = \left[\mathbf{A}(t) + \left(e^{-\Lambda\tau} - 1\right)\mathbf{K}\right]\Psi(t) \tag{4.1.4b}$$

with the normalizations $\Phi(0) = \mathbf{Id}$ and $\Psi(0) = \mathbf{Id}$, i.e., Φ and Ψ are equal to the $n \times n$ identity matrix for $t = 0$. Furthermore, Φ and Ψ are related by the following equations

$$\Psi(t) = \Phi(t)\left\{\Phi(0) + \int_0^t \Phi(-s)\left[\left(e^{-\Lambda\tau} - 1\right)\mathbf{K}\right]\Psi(s)ds\right\}, \tag{4.1.5}$$

which can be seen by inserting this expression into Eq. 4.1.4b. Starting at the left-hand side of Eq. 4.1.4b yields substituting Eq. 4.1.4a for the time derivative of Φ

$$\begin{aligned}\frac{d}{dt}\Psi(t) = \left(\frac{d}{dt}\Phi(t)\right)&\left\{\Phi(0) + \int_0^t \Phi(-s)\left[\left(e^{-\Lambda\tau} - 1\right)\mathbf{K}\right]\Psi(s)ds\right\} \\ &+ \Phi(t)\left\{\Phi(-t)\left[\left(e^{-\Lambda t} - 1\right)\mathbf{K}\right]\Psi(t)\right\}\end{aligned} \tag{4.1.6a}$$

$$\begin{aligned}&= \mathbf{A}(t)\Phi(t)\left\{\Phi(0) + \int_0^t \Phi(-s)\left[\left(e^{-\Lambda\tau} - 1\right)\mathbf{K}\right]\Psi(s)ds\right\} \\ &\quad + \left[\left(e^{-\Lambda\tau} - 1\right)\mathbf{K}\right]\Psi(t)\end{aligned} \tag{4.1.6b}$$

$$= \left[\mathbf{A}(t) + \left(e^{-\Lambda\tau} - 1\right)\mathbf{K}\right]\Psi(t). \tag{4.1.6c}$$

For $t = \tau$, Eq. 4.1.5 becomes

$$\Psi(\tau) = \Phi(\tau)\left\{\Phi(0) + \int_0^\tau \Phi(-s)\left[\left(e^{-\Lambda\tau} - 1\right)\mathbf{K}\right]\Psi(s)ds\right\}. \tag{4.1.7}$$

The Floquet exponent Λ is a root of the characteristic equation

$$F(\mu) = \det(\mu\mathbf{Id} - \Psi(\tau)) = 0 \tag{4.1.8a}$$

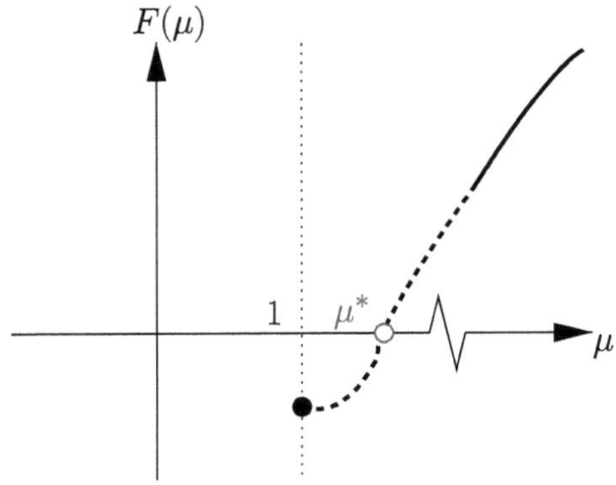

Fig. 4.1 Schematic diagram of the characteristic function $F(\mu)$. The *black (filled) dot and solid curve* are given by Eqs. 4.1.10 and 4.1.11, respectively. The *red (open) dot* marks a root $\mu^*$of $F(\mu)$

$$F(\mu) = \prod_{i=1}^{n} (\mu - \mu_i) \tag{4.1.8b}$$

which is written in terms of the Floquet multipliers $\mu_i = \exp(\Lambda_i \tau)$ with $i = 1, 2, \ldots, n$. $F(\mu)$ is called the characteristic function. The discussion of Eq. 11, especially the structure of $F(\mu)$, is key point of the proof.

Inserting $\mu = 1$ leads to the following expression using Eq. 4.1.7

$$F(1) = \det(\mathbf{Id} - \Psi(\tau)) \tag{4.1.9a}$$

$$= \det\left(\mathbf{Id} - \Phi(\tau)\left\{\Phi(0) + \left(\mu^{-1} - 1\right)\Big|_{\mu=1} \int_0^{\tau} \Phi(-s)\mathbf{K}\Psi(s)ds\right\}\right) \tag{4.1.9b}$$

$$= \det(\mathbf{Id} - \Phi(\tau)) \tag{4.1.9c}$$

which recovers the characteristic equations of the uncontrolled, linearized system given by Eq. 4.1.2. The initial assumption that the uncontrolled system has an odd number of real Floquet multipliers yields

$$F(\mu)\big|_{\mu=1} < 0. \tag{4.1.10}$$

Since the Jacobian matrix in the controlled case of Eq. 4.1.3 is bounded, the integral in the characteristic function $F(\mu)$ inherits this boundedness. See Eqs. 4.1.8a and 4.1.9 and one obtains for large μ

$$\lim_{\mu \to \infty} F(\mu) = \infty. \tag{4.1.11}$$

The results concerning the structure of the characteristic function $F(\mu)$ until this point are summarized in Fig. 4.1 as black (filled) dot and solid curve.

Since $F(\mu)$ is continuous there exists a Floquet multiplier μ^* with $F(\mu^*) = 0$ which is real and greater than unity. This multiplier is shown as red (open) dot in Fig. 4.1 and corresponds to a real positive Floquet exponent. Thus, Nakajima concludes that the system given by Eq. 4.1.8a remains unstable. A similar proof was given for the control by extended time-delayed feedback [9] in Ref. [3].

The error of the proof is the inequality (4.1.10). In systems which have, for instance, a rotational symmetry, there exists a Floquet multiplier at $\mu = 1$ corresponding to a Floquet exponent $\Lambda = 0$. This multipliers arise from the fact that solutions of the system's equation are invariant with respect to a phase shift of the orbit. This will demonstrated in the next section. See, for instance, Eqs. 4.2.6a and 4.2.6b. In fact, the proof fails for any autonomous system.

The mode corresponding to the Floquet multiplier at $\mu = 1$ is sometimes called Goldstone mode. If the time delay is equal to the period of a periodic orbit in autonomous systems, this mode persists also by construction of the feedback term in the case when a Pyragas control force is applied. Thus, one can have a root of the characteristic equation pinned at $\mu = 1$ and hence, the inequality (4.1.10) and its consequences do not hold anymore. Note that a similar proof as for the Pyragas case is given in Ref. [3] for the extended version of time-delayed feedback. The proof remains valid for orbits without a Floquet multiplier at $\mu = 1$, i.e., non-autonomous systems.

The paper [2] which states the odd number limitation theorem had a large impact on the nonlinear dynamics community. It was cited more than 130 times according the *web of science* (http://www.apps.isiknowledge.com) until March 2009. There is a number of subsequent papers which introduce modifications of the original Pyragas control scheme in order to stabilize odd number orbits [10–14]. One of these works is done by K. Pyragas himself. In Ref. [10], he introduced the concept of an unstable controller. The crucial point of this extension is to add one unstable dimension to the system under control such that the odd number limitation theorem cannot be applied [11]. This is done by an addition differential equation yielding the following control system

$$\frac{d}{dt}\mathbf{x}(t) = \mathbf{f}(\mathbf{x}(t)) - K\mathbf{F}(t), \tag{4.1.12}$$

where the time-delayed control force $\mathbf{F}(t)$ is given by

$$\mathbf{F}(t) = \tilde{\mathbf{F}}(t) + w(t) \tag{4.1.13a}$$

$$\frac{d}{dt}w(t) = aw(t) + b\tilde{\mathbf{F}}(t) \tag{4.1.13b}$$

with positive constant $a > 0$, negative constant $b < 0$, and the usual Pyragas feedback force $\tilde{\mathbf{F}}(t) = \mathbf{g}(\mathbf{x}(t)) - \mathbf{g}(\mathbf{x}(t-\tau))$. The function $\mathbf{g}(\mathbf{x}(t))$ determines which components of the system enter the control force. For details see Chap. 2. Since the constant a is positive, the variable $w(t)$ adds one unstable dimension to the system.

The discussion of the success of this method can be found in Refs. [1, 15] where it is applied to the Lorenz system and an electrochemical oscillator. The mechanism is the following: As the feedback gain K increases, the Floquet exponent corresponding to the additional equation (4.1.13b) collides with the positive exponent of Eq. 4.1.12. They form a complex conjugate pair which crosses the imaginary axis and enters the half-plane of negative real part. Thus, the troublesome mode of the odd number orbit is stabilized. Note that this extension has been realized in an electronic experiment by an additional unstable control loop [14].

Schuster and Stemmler suggested a periodic modulation of the feedback gain K as an extension of the original Pyragas controller to overcome the odd number limitation [16]. Then, the feedback gain changes periodically with period 2τ between zero and a constant value as follows

$$K = 0 \quad \text{for} \quad n\tau \leq t < (n+1)\tau \tag{4.1.14a}$$

and

$$K \neq 0 \quad \text{for} \quad (n+1)\tau \leq t < (n+2)\tau. \tag{4.1.14b}$$

This method was successfully applied to the Kuramoto–Sivashinksy equation [17] which describes irregular flow of a liquid film down a vertical plane and exhibits spatio-temporal chaos. Starting from initial conditions, the homogeneous state was stabilized via this oscillating feedback.

Furthermore, in dynamical systems with symmetry, half-period feedback control can be used to stabilize hyperbolic periodic orbits [18]. The controlled system becomes

$$\frac{d}{dt}\mathbf{x}(t) = \mathbf{f}(\mathbf{x}(t)) - \mathbf{h}[K\{\mathbf{g}(\mathbf{x}(t)) + \mathbf{g}(\mathbf{x}(t - T/2))\}], \tag{4.1.15}$$

where T denotes the period of the target orbit. The success of this control method was demonstrated for the Duffing oscillator and the Lorenz system [19] in Ref. [18]. In the first case, the non-autonomous dynamical systems including half-period feedback is given by

$$\frac{d}{dt}x(t) = y(t) - K[x(t - T/2) + x(t)] \tag{4.1.16a}$$

$$\frac{d}{dt}y(t) = -\delta y(t) - \alpha x(t) - x(t)^3 + B\cos(t) - K[y(t - T/2) + y(t)] \tag{4.1.16b}$$

with constants δ, α, and B [18]. The uncontrolled system shows the following symmetry: If $(x(t), y(t))$ is a solution, then $(-x(t - T/2), -y(t - T/2))$ is also a solution of the uncontrolled equation where the period T is $T = 2\pi$ for the present scaling of time. If the half-period feedback is applied to the Lorenz system, the dynamical equation become

$$\frac{d}{dt}x(t) = -\sigma[x(t) - y(t)] - K[x(t - T/2) + x(t)] \tag{4.1.17a}$$

$$\frac{d}{dt}y(t) = rx(t) - y(t) - x(t)z(t) - K[y(t - T/2) + y(t)] \tag{4.1.17b}$$

$$\frac{d}{dt}z(t) = x(t)y(t) - bz(t) - K[z(t) - z(t - T/2)] \tag{4.1.17c}$$

which shows chaotic dynamics in the uncontrolled case for $\sigma = 10$, $r = 28$, and $b = 8/3$. Note that the uncontrolled Lorenz system has the symmetry that a solution $(x(t), y(t), z(t))$ leads to another solution $(-x(t), -y(t), z(t))$. Thus, the condition $(x(t), y(t), z(t)) = (-x(t - T/2), -y(t - T/2), z(t - T/2))$ yields a T-periodic solution. In both case, a periodic solution with period T was stabilized using half-period feedback [18]. Due to the special symmetry of the models, the control remains noninvasive, i.e., the feedback vanishes if the orbit is stabilized.

Other suggestions to overcome the topological limitation involve adaptive control involving low-pass filtering [12]. In Ref. [13], a two-step algorithm was proposed to enhance the performance of the controller. This algorithm was then realized in an experiment with an electronic circuit. Recently, a similar scheme to overcome the odd-number limitation in the case of steady states was proposed and successfully applied to the Lorenz system [20]. Stabilization was achieved by variable time-delayed feedback control, where the time delay τ became time-dependent. This time dependence was implemented by a sawtooth function $\tau(t) = \tau_0 + \varepsilon[2(at \bmod 1) - 1]$, a random distribution $\tau(t) = \tau_0 + \varepsilon\xi(t)$ where $\xi(t)$ denotes a random numbers in the interval $[-1, 1]$, and a periodic modulation $\tau(t) = \tau_0 + \varepsilon \sin(at)$ [20].

After this summary of the odd number limitation theorem, I will introduce the equations of the counterexample in the next section.

4.2 Model Equations of the Counterexample

In this section, I will introduce the equations of the counterexample to refute the odd number limitation theorem [2, 3] following Refs. [4, 5, 6]. This counterexample consists of the normal form of a subcritical Hopf bifurcation subject to time-delayed feedback control. At first, I will investigate the system's dynamics and then, I will discuss the strategy to stabilize the subcritical orbits.

The system's equation are given in complex notation by the following equation

$$\frac{dz(t)}{dt} = \left[\lambda + i + (1 + i\gamma)|z(t)|^2\right]z(t) + b[z(t - \tau) - z(t)] \tag{4.2.1}$$

with complex variable $z(t)$ and real parameters λ and γ. The control parameters are denoted by the time delay τ and the complex feedback gain $b = b_0 e^{i\beta} = \mathrm{Re}(b) + i\mathrm{Im}(b)$. In the absence of a control force, Eq. 4.2.1 becomes in amplitude and phase notation, i.e., $z(t) = r(t)e^{i\theta(t)}$,

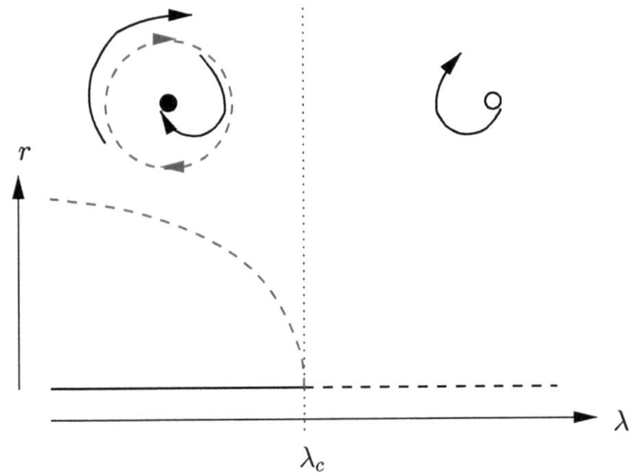

Fig. 4.2 *Lower part*: Schematic bifurcation diagram of the subcritical Hopf bifurcation, i.e., dependence of the amplitude r on the bifurcation parameter λ. At the critical value λ_c a bifurcation occurs. The *upper part* depicts schematically the periodic orbit and steady states of the system. *Solid lines and filled dots* denote stability, *dashed lines and empty dots* show instability. The *thick arrows* are example trajectories

$$\frac{dr(t)}{dt} = \left[\lambda + r(t)^2\right] r(t) \tag{4.2.2a}$$

$$\frac{d\theta(t)}{dt} = 1 + \gamma r(t)^2. \tag{4.2.2b}$$

Figure 4.2 shows a schematic bifurcation diagram of the subcritical Hopf bifurcation in the absence of time-delayed feedback as given by Eqs. 4.2.2a and 4.2.2b. The upper part of the figure depicts schematically the periodic orbit and steady states of the system. Solid lines and filled dots denote stability, dashed lines and empty dots indicate instability. The thick arrows show example trajectories. The lower part displays the characteristic square root dependence of the amplitude r on the bifurcation parameter λ. This dependence can be derived from Eq. 4.2.2a whose time-independent solutions are

$$r(t) \equiv 0 \quad \text{and} \quad r(t) \equiv \sqrt{-\lambda}. \tag{4.2.3}$$

The first solution corresponds to the fixed point at the origin. The latter solution refers to the periodic orbit shown as red dashed curve in Fig. 4.2. In the present case, the critical value of the bifurcation parameter is $\lambda_c = 0$.

The period of the subcritical orbit can be derived from the equation of the phase (4.2.2b). Inserting $r = \sqrt{-\lambda}$ yields for $\theta(t)$

$$\frac{d\theta(t)}{dt} = 1 - \gamma\lambda \tag{4.2.4a}$$

$$\Leftrightarrow \quad \theta(t) = (1 - \gamma\lambda)t + \theta_0 \tag{4.2.4b}$$

with some initial phase θ_0 which is set to zero for convenience. Thus, the period T of one oscillation is given by

$$T = \frac{2\pi}{1 - \gamma\lambda}. \tag{4.2.5}$$

Note that this subcritical periodic orbit belongs to the class of odd number orbits and thus, should not be feasible for stabilization by time-delayed feedback of Pyragas type. The Floquet multiplier can be derived by linearization of the uncontrolled Eqs. 4.2.2a and 4.2.2b around the orbit $r_0 = \sqrt{-\lambda}$ with a corresponding frequency $\omega = 1 - \gamma\lambda$ which will be elaborated in the following. Consider small deviations $\delta r(t)$ and $\delta\theta(t)$ from the periodic orbit, i.e., $r(t) = r_0 + \delta r(t)$ and $\theta(t) = \omega\, t + \delta\theta(t)$. Linearizing Eqs. 4.2.2a and 4.2.2b yields

$$\begin{pmatrix} \frac{d\delta r(t)}{dt} \\ \frac{d\delta\theta(t)}{dt} \end{pmatrix} = \begin{pmatrix} \lambda - 3r^2 & 0 \\ 2\gamma r & 0 \end{pmatrix}\Bigg|_{r=r_0} \begin{pmatrix} \delta r(t) \\ \delta\theta(t) \end{pmatrix} \tag{4.2.6a}$$

$$= \begin{pmatrix} -2\lambda & 0 \\ 2\gamma\sqrt{-\lambda} & 0 \end{pmatrix} \begin{pmatrix} \delta r(t) \\ \delta\theta(t) \end{pmatrix}. \tag{4.2.6b}$$

The Floquet exponents are given by the roots of the characteristic equation

$$0 = \det\begin{pmatrix} -2\lambda - \Lambda & 0 \\ 2\gamma\sqrt{-\lambda} & -\Lambda \end{pmatrix} \tag{4.2.7a}$$

$$0 = \Lambda(2\lambda + \Lambda). \tag{4.2.7b}$$

Thus, one Floquet exponent is unity, i.e., $\Lambda = 0$, which reflects the rotational symmetry of the rotating wave solution. The other Floquet exponent is given as $\Lambda = -2\lambda$. For $\lambda < 0$ in the range, where the subcritical orbit exists, the leads to a positive, real Floquet exponent or equivalently to a real Floquet multiplier $\mu = \exp(\Lambda\tau)$ greater than unity. Therefore, the subcritical periodic orbit belongs to the class of odd number orbits.

The period is the subcritical orbit given by Eq. 4.2.2a and 4.2.5 determines the value of the time delay τ in Eq. 4.2.1 which is chosen as an integer multiple of T, i.e., $\tau = nT$ with $n = 1, 2, \ldots$. Therefore, Eq. 4.2.5 fixes curves in the (λ, τ) plane which correspond to the period of the target orbit. I will call these curves Pyragas curves τ_P

$$\tau_P(\lambda) = \frac{2\pi n}{1 - \gamma\lambda}, \quad n = 1, 2, \ldots \tag{4.2.8}$$

There is another set of curves which will become important in the following. These Hopf curves indicate the change of stability of the fixed point located at the origin. They are characterized by purely imaginary solutions of the characteristic equation which results from the linearization of Eq. 4.2.1 at the steady state $z(t) = z^* = 0$. This characteristic equation is given by

$$\eta = \lambda + i + b(e^{-\eta\tau} - 1), \tag{4.2.9}$$

where η denotes the eigenvalue of the fixed point. Note that this equation can be solved using the multi-valued Lambert function W which is the inverse function of $f(z) = ze^z$ for complex z. Thus, the solution can be written as

$$\eta\tau = W\left(b\tau e^{-(\lambda+i)\tau - b\tau}\right) + (\lambda + i)\tau - b\tau. \tag{4.2.10}$$

For details on the Lambert function see Sect. 3.2, e.g., Eq. 3.2.8. In order to derive a dependence of the time delay τ on the system's parameter λ as for the Pyragas curves (4.2.8), it is helpful to consider real and imaginary party of Eq. 4.2.9 separately at the point where the stability of the fixed point changes and the eigenvalue becomes purely imaginary, i.e., $\eta = i\omega$. Then, the characteristic equation reads

$$0 = \lambda + b_0[\cos(\beta - \omega\tau) - \cos\beta] \tag{4.2.11a}$$

$$\omega - 1 = b_0[\sin(\beta - \omega\tau) - \sin\beta], \tag{4.2.11b}$$

where b_0 and β denote the amplitude and phase of the complex feedback gain b. The Hopf curves are calculated as follows. Equation 4.2.11a leads to an expression for the quantity $\omega\tau$

$$\omega\tau = \pm \arccos\left(\frac{b_0 \cos\beta - \lambda}{b_0}\right) + \beta + 2\pi n \tag{4.2.12}$$

with an integer number n that indicates the multi-valued character of the arccos function. As a next step, one can derive a formula for the imaginary part ω of the eigenvalue η. This is done by isolating the trigonometric functions in Eqs. 4.2.11a and 4.2.11b and exploiting the identity $\cos^2(\cdot) + \sin^2(\cdot) = 1$

$$\left(\frac{b_0 \cos\beta - \lambda}{b_0}\right)^2 = \cos(\beta - \omega\tau)^2 \tag{4.2.13a}$$

$$\left(\frac{\omega - 1 + b_0 \sin\beta}{b_0}\right)^2 = \sin(\beta - \omega\tau)^2 \tag{4.2.13b}$$

which yields the following polynomial equation in terms of ω

$$b_0^2 = (b_0 \cos\beta - \lambda)^2 + (\omega - 1 + b_0 \sin\beta)^2 \tag{4.2.14a}$$

$$\begin{aligned} \Leftrightarrow \quad 0 &= \omega^2 + 2\omega(b_0 \sin\beta - 1) \\ &\quad - 2b_0\lambda\cos\beta + \lambda^2 - 2b_0 \sin\beta + 1. \end{aligned} \tag{4.2.14b}$$

The solution of this equation are given by

$$\begin{aligned} \omega &= -(b_0 \sin\beta - 1) \pm \sqrt{(b_0 \sin\beta - 1)^2 + 2b_0\lambda\sin\beta - \lambda^2 + 2b_0 \sin\beta - 1} \\ &= -(b_0 \sin\beta - 1) \pm \sqrt{\lambda(2b_0\cos\beta - \lambda) + b_0^2 \sin^2\beta}. \end{aligned} \tag{4.2.15}$$

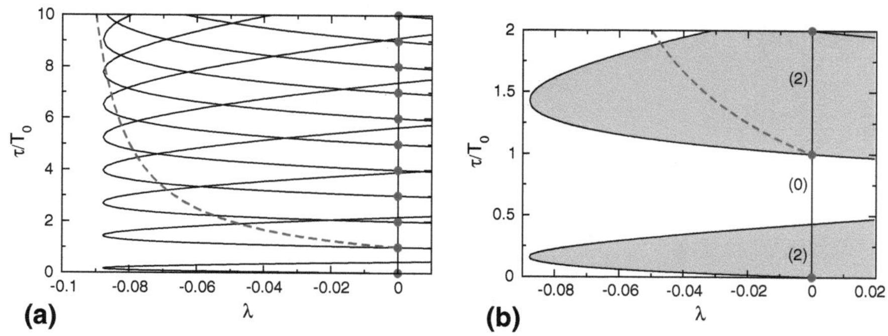

Fig. 4.3 Pyragas (*red dashed*) and Hopf (*black solid*) curves in the (λ, τ) plane according to Eqs. 4.2.8 and 4.2.16: Panel (**a**) shows the first 10 Hopf bifurcation curves. Panel (**b**) is an enlargement in the range of small time delays τ. The *yellow* (*shaded*) *area* marks the domain where the steady state at the origin is unstable, and the numbers in parentheses indicate the dimension of the unstable manifold at $z = 0$. The time delay τ is given in units of the intrinsic timescale T_0 of the trivial fixed point, i.e., $T_0 = 2\pi$. The red dots mark special values of the time delay on the λ axis. Parameters: $\gamma = -10$, $b_0 = 0.3$ and $\beta = \pi/4$

Inserting this value of ω into Eq. 4.2.12 leads to an explicit expression for the Hopf curves $\tau_H(\lambda)$

$$\tau_H(\lambda) = \frac{\pm \arccos\left(\frac{b_0 \cos\beta - \lambda}{b_0}\right) + \beta + 2\pi n}{1 - b_0 \sin\beta \mp \sqrt{\lambda(2b_0\cos\beta - \lambda) + b_0^2 \sin^2\beta}}. \tag{4.2.16}$$

Note that $\tau_H(\lambda)$ is not defined in case of $\beta = 0$ and $\lambda < 0$. Thus complex b is a necessary condition for the existence of the Hopf curve in the subcritical regime $\lambda < 0$. Figure 4.3 displays the family of Hopf and Pyragas curves in the (λ, τ) plane according to Eqs. 4.2.16 and 4.2.8, respectively. Panel (a) shows the first 10 Hopf bifurcation curves, i.e., $n = 1, 2, \ldots, 10$ in Eq. 4.2.16, and panel (b) is an enlargement in the range of small time delays τ. The yellow shading marks the domain where the steady state at the origin is unstable, and the numbers in parentheses indicate the dimension of the unstable manifold of $z = 0$. Note that the time delay τ is given in units of the intrinsic timescale T_0 of the trivial fixed point, i.e., $T_0 = 2\pi$. This reflects the imaginary part of the eigenvalue in the absence of control as can be derived from the characteristic equation (4.2.9) for $b = 0$.

By construction of time-delayed feedback control, the time delay τ is equal to the period of the target orbit on the Pyragas curve because the control force vanishes for this choice of τ if the orbit is stabilized. Therefore, the strategy to control the subcritical orbit is clear. At least for small values of $|\lambda|$, one can stabilize the orbit of the corresponding point if the Pyragas curve lies inside the area where the fixed point is unstable. Then, the orbits inherits the stability of the steady state which in turn becomes unstable. Note that an unstable steady state at $z = 0$ is not a sufficient condition for stabilization of the subcritical orbit since

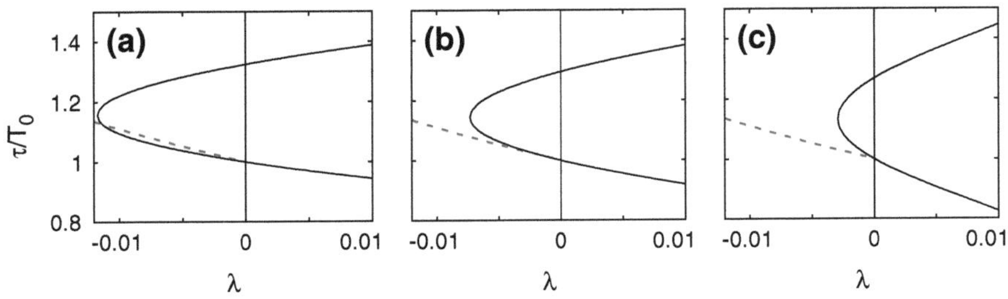

Fig. 4.4 Region of unstable steady state in the (λ, τ) plane in dependence on the amplitude b_0 of the feedback gain where the time delay is given in units of the intrinsic timescale $T_0 = 2\pi$. Panels (**a**), (**b**), and (**c**) correspond to $b_0 = 0.04$, 0.025, and 0.01, respectively. *Solid and dashed curves* refer to the Hopf and Pyragas curves according to Eqs. 4.2.16 and 4.2.8, respectively. Other parameters as in Fig. 4.3

other, e.g., global, bifurcations must be considered as well. These bifurcations will be discussed later in Sect. 4.3.

Figure 4.4 shows how the regime of unstable fixed point at $z = 0$ changes if the feedback amplitude b_0 varies. Panels (a), (b), and (c) correspond to a value of $b_0 = 0.04$, 0.025, and 0.01, respectively. As in Fig. 4.3, the dashed and solid curves correspond to the Pyragas and Hopf curves given by Eqs. 4.2.8 and 4.2.16, respectively. While panel (a) is similar to Fig. 4.3b because the Pyragas curves lies partly inside the unstable domain of the trivial fixed point, panel (b) shows the threshold case of a tangential Pyragas curve on this domain. For smaller feedback amplitudes as displayed in panel (c), no stabilization is expected because the Pyragas and Hopf curves do not cross.

Next, I will analyze the conditions under which stabilization of the subcritical periodic orbit is possible. From Figs. 4.3b and 4.4 it is evident that the Pyragas curve must lie inside the yellow region, i.e., the Pyragas and Hopf curves emanating from the point $(\lambda, \tau) = (0, 2\pi)$ must locally satisfy the inequality $\tau_H(\lambda) < \tau_P(\lambda)$ for $\lambda < 0$. More generally, let me investigate the eigenvalue crossings of the Hopf eigenvalues $\eta = i\omega$ along the τ-axis of Fig. 4.3. In particular I will derive conditions for the unstable dimensions of the trivial steady state near the Hopf bifurcation point $\lambda = 0$ in our model equation (4.2.1). On the τ-axis $(\lambda = 0)$, the characteristic equation (4.2.9) for $\eta = i\omega$ is reduced to

$$\eta = i + b(e^{-\eta\tau} - 1) \tag{4.2.17a}$$

$$\Leftrightarrow \quad i\omega = i + b_0(e^{i(\beta-\omega\tau)} - e^{\beta}). \tag{4.2.17b}$$

This leads to two series of Hopf points given by

$$0 \leq \tau_n^A = 2\pi n \tag{4.2.18a}$$

$$0 < \tau_n^B = \frac{2\beta + 2\pi n}{1 - 2b_0 \sin\beta} \tag{4.2.18b}$$

with $n = 0, 1, 2, \ldots$. The corresponding Hopf frequencies are $\omega^A = 1$ and $\omega^B = 1 - 2b_0 \sin\beta$, respectively.

A derivation of the expressions for τ_n^A and τ_n^B is presented in the following. Note that series A consists of all Pyragas points, since

$$\tau_n^A = nT = \left.\frac{2\pi n}{1-\gamma\lambda}\right|_{\lambda=0} = \frac{2\pi n}{\omega^A}. \tag{4.2.19}$$

In the series B the integers n have to be chosen such that the delay $\tau_n^B \geq 0$. The case $b_0 \sin\beta = 1/2$, only, corresponds to $\omega^B = 0$ and does not occur for finite delays τ. The series is given by Eq. 4.2.16 for vanishing λ

$$\tau_H(\lambda)|_{\lambda=0} = \left.\frac{\arccos\left(\frac{b_0\cos\beta-\lambda}{b_0}\right)+\beta+2\pi n}{1-b_0\sin\beta-\sqrt{\lambda(2b_0\cos\beta-\lambda)+b_0^2\sin^2\beta}}\right|_{\lambda=0} \tag{4.2.20a}$$

$$\tau_H(\lambda)|_{\lambda=0} = \frac{2\beta+2\pi n}{1-2b_0\sin\beta} \tag{4.2.20b}$$

which gives the equation for τ_n^B of Eq. 4.2.18b.

Next, I will evaluate the crossing directions of the critical Hopf eigenvalues along the positive τ-axis and for both series in Figs. 4.3 and 4.4. The crossing direction is given by sign(Re($\partial\eta/\partial\tau$)). Thus, a loss of stability corresponds to negative derivative of the eigenvalue with respect to τ in Figs. 4.3 and 4.4 and a recovery of stability to a positive value of this quantity. For purely, imaginary eigenvalues $\eta = i\omega$ as given at the threshold of stability, the characteristic equation (4.2.17a) can be rewritten as

$$i(\omega-1) = b_0\left(e^{i(\beta-\omega\tau)} - e^{\beta}\right) \tag{4.2.21a}$$

$$= b_0\{\cos(\beta-\omega\tau) - \cos\beta + i[\sin(\beta-\omega\tau) - \sin\beta]\}. \tag{4.2.21b}$$

Implicit differentiation of Eq. 4.2.21b with respect to τ at $\eta = i\omega$ implies

$$\operatorname{sign}\left[\operatorname{Re}\left(\frac{\partial\eta}{\partial\tau}\right)\right] = -\operatorname{sign}(\omega)\operatorname{sign}(\sin(\omega\tau-\beta)). \tag{4.2.22}$$

One is interested specifically in the Pyragas–Hopf points of series A as marked by red dots in Fig. 4.3 which satisfy $\tau = \tau_n^A = 2\pi n$ and $\omega = \omega^A = 1$. Inserting these values into Eq. 4.2.22 yields

$$\operatorname{sign}\left[\operatorname{Re}\left(\frac{\partial\eta}{\partial\tau}\right)\right] = -\operatorname{sign}(\omega^A)\operatorname{sign}(\sin(\omega^A\tau_n^A-\beta)) \tag{4.2.23a}$$

$$= -\operatorname{sign}(\sin(2\pi n-\beta)) \tag{4.2.23b}$$

$$= \operatorname{sign}(\beta). \tag{4.2.23c}$$

Indeed $\mathrm{sign}(\mathrm{Re}(\partial\eta/\partial\tau)) = \mathrm{sign}(\sin\beta) > 0$ holds, under the assumption $0 < \beta < \pi$, i.e., for positive imaginary part of the feedback gain $\mathrm{Im}(b) > 0$. This condition alone, however, is not sufficient to guarantee stability of the steady state for $\tau < 2\pi n$. One also has to consider the crossing direction $\mathrm{sign}(\mathrm{Re}(\partial\eta/\partial\tau))$ along series B because the fixed point at the origin can gain unstable dimensions at the Pyragas–Hopf points of series B. Inserting the expressions for ω^B and τ_n^B as given by Eq. 4.2.18b

$$\omega^B = 1 - 2b_0 \sin\beta \tag{4.2.24a}$$

$$\Rightarrow \tau_n^B = \frac{2\beta + 2\pi n}{\omega^B} = \frac{2\beta + 2\pi n}{1 - 2b_0 \sin\beta} \tag{4.2.24b}$$

into Eq. 4.2.22 leads to

$$\begin{aligned}\mathrm{sign}\left[\mathrm{Re}\left(\frac{\partial\eta}{\partial\tau}\right)\right] &= -\mathrm{sign}(\omega^B)\mathrm{sign}\Big(\sin(\omega^B\tau_n^B - \beta)\Big)\\ &= -\mathrm{sign}(1 - 2b_0\sin\beta)\mathrm{sign}\left[\sin\left(\omega^B\frac{2\beta + 2\pi n}{\omega^B} - \beta\right)\right]\end{aligned} \tag{4.2.25a}$$

$$= -\mathrm{sign}(1 - 2b_0\sin\beta)\mathrm{sign}(\sin(-\beta)) \tag{4.2.25b}$$

$$= \mathrm{sign}[(2b_0\sin\beta - 1)\sin\beta]. \tag{4.2.25c}$$

To compensate for the destabilization of $z = 0$ upon each crossing of any point $\tau_n^A = 2\pi n$, one must require stabilization corresponding to the condition

$$\mathrm{sign}\left[\mathrm{Re}\left(\frac{\partial\eta}{\partial\tau}\right)\right] < 0 \tag{4.2.26}$$

at each point τ_n^B of series B. This requires with the constraint $0 < \beta < \pi$

$$0 < \beta < \arcsin\left(\frac{1}{2b_0}\right) \quad \text{or} \quad \pi - \arcsin\left(\frac{1}{2b_0}\right) < \beta < \pi \tag{4.2.27}$$

as can be derived from Eq. 4.2.25c. The distance between two successive points τ_n^B and τ_{n+1}^B is $2\pi/\omega^B > 2\pi$. Therefore, there is at most one τ_n^B between any two successive Hopf points of series A. Stabilization requires exactly one such τ_n^B, specifically: $\tau_{k\text{-}1}^A < \tau_{k\text{-}1}^B < \tau_k^A$ for all $k = 1, 2, \ldots, n$.

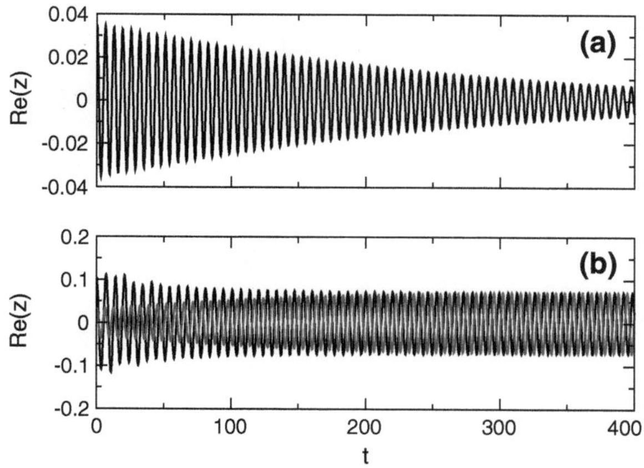

Fig. 4.5 Time series of system (4.2.1): Panel (**a**) shows the time evolution of the real part of z in the absence of control. Panel (**b**) depicts the time series with control for two different initial conditions. The control parameters are chosen as $b_0 = 0.3$ and $\beta = \pi/4$. System's parameters: $\lambda = -0.005$ and $\gamma = -10$

At the end of this section, Fig. 4.5 displays the time series of the real part Re(z) according to the counterexample given by Eq. 4.2.1. While panel (a) shows the time evolution in the absence of a control force, panel (b) depicts the case of two different initial conditions with control. The control parameters are chosen as $b_0 = 0.3$ and $\beta = \pi/4$. One can see that the system approaches the fixed point at the origin if no control force is applied. With time-delayed feedback, however, the trivial fixed point is unstable and the system evolves to the periodic orbit at $\sqrt{-\lambda}$.

After the discussion of the general strategy to stabilize the periodic odd-number orbit, I will investigate the domain of control in the next section. There, I will also explain the mechanism which causes the stabilization.

4.3 Domains of Control

In the previous section, I have demonstrated that the odd number limitation theorem does not hold in general. By example, I have shown that time-delayed feedback can eventually be used to stabilize periodic orbits which have an odd number of real Floquet multipliers greater than unity. I will investigate in this section the underlying mechanism and calculate the domain of control in dependence on the complex feedback gain.

The stability of the subcritical odd number orbit and the trivial steady state is determined by the Floquet exponents Λ and the eigenvalues η, respectively. The latter can be calculated by solving the characteristic equation (4.2.9). In analogy, the Floquet exponents Λ are solutions of the characteristic equation of the periodic orbit on the Pyragas curve since the time delay is fixed at $\tau = \tau_P = 2\pi/(1-\gamma\lambda)$. I will derive this characteristic equation in the following.

The Floquet exponents Λ of the Pyragas orbit can be calculated explicitly by rewriting Eq. 4.2.1 in polar coordinates $z(t) = r(t)e^{i\theta(t)}$

$$\frac{dr(t)}{dt} = [\lambda + r^2(t)]r(t) \\ + b_0[\cos(\beta + \theta(t-\tau) - \theta(t))r(t-\tau) - r(t)\cos\beta] \tag{4.3.1a}$$

$$\frac{d\theta(t)}{dt} = 1 + \gamma r^2(t) \\ + b_0\left[\sin(\beta + \theta(t-\tau) - \theta(t))\frac{r(t-\tau)}{r(t)} - \sin\beta\right] \tag{4.3.1b}$$

and linearizing around the periodic orbit according to $r(t) = r_0 + \delta r(t)$ and $\theta(t) = \omega\, t + \delta\theta(t)$, with $r_0 = \sqrt{-\lambda}$ and $\omega = 1 - \gamma\lambda$ as given by Eqs. 4.2.3 and 4.2.4b. Note that this solution $z(t) = r_0\exp(i\omega t)$ has the form of a rotating wave which will be discussed in more detail in Sect. 4.4. Inserting the expressions for $r(t)$ and $\theta(t)$ yields for the deviations $\delta r(t)$ and $\delta\theta(t)$

$$\begin{pmatrix} \frac{d\delta r(t)}{dt} \\ \frac{d\delta\theta(t)}{dt} \end{pmatrix} = \begin{bmatrix} -2\lambda - b_0\cos\beta & b_0 r_0 \sin\beta \\ 2\gamma r_0 - b_0 \sin\beta r_0^{-1} & -b_0\cos\beta \end{bmatrix} \begin{pmatrix} \delta r(t) \\ \delta\theta(t) \end{pmatrix} \\ + \begin{bmatrix} b_0\cos\beta & -b_0 r_0 \sin\beta \\ b_0 \sin\beta r_0^{-1} & b_0\cos\beta \end{bmatrix} \begin{pmatrix} \delta r(t-\tau) \\ \delta\theta(t-\tau) \end{pmatrix}. \tag{4.3.2}$$

With the ansatz

$$\begin{pmatrix} \delta r(t) \\ \delta\theta(t) \end{pmatrix} = u\exp(\Lambda t), \tag{4.3.3}$$

where u is a two-dimensional vector, one obtains the autonomous linear equation

$$\begin{bmatrix} -2\lambda + b_0\cos\beta(e^{-\Lambda\tau} - 1) - \Lambda & -b_0 r_0 \sin\beta(e^{-\Lambda\tau} - 1) \\ 2\gamma r_0 + b_0 r_0^{-1}\sin\beta(e^{-\Lambda\tau} - 1) & b_0\cos\beta(e^{-\Lambda\tau} - 1) - \Lambda \end{bmatrix} u = 0. \tag{4.3.4}$$

The condition of vanishing determinant then gives the transcendental characteristic equation

$$0 = \left[-2\lambda + b_0\cos\beta(e^{-\Lambda\tau} - 1) - \Lambda\right]\left[b_0\cos\beta(e^{-\Lambda\tau} - 1) - \Lambda\right] \\ + b_0 r_0 \sin\beta(e^{-\Lambda\tau} - 1)\left[2\gamma r_0 + b_0 r_0^{-1}\sin\beta(e^{-\Lambda\tau} - 1)\right] \tag{4.3.5a}$$

$$= \Lambda^2 - 2\left[{r_0}^2 + b_0\cos\beta\left(e^{-\Lambda\tau} - 1\right)\right]\Lambda \\ + 2{r_0}^2(\cos\beta + \gamma\sin\beta)b_0\left(e^{-\Lambda\tau} - 1\right) + b_0^2\left(e^{-\Lambda\tau} - 1\right)^2 \tag{4.3.5b}$$

for the Floquet exponents Λ which can be solved numerically. Note that the parameter λ is substituted by the amplitude r_0 via $r_0 = \sqrt{-\lambda}$.

This is shown in Fig. 4.6 which depicts solutions Λ of the characteristic equation (4.3.5a) of the periodic solution on the Pyragas curve. Panel (a) (top) shows the dependence of the real part of the critical Floquet exponent Λ on the

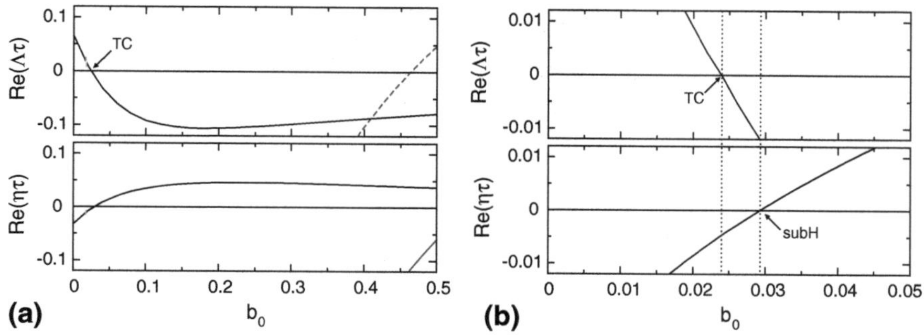

Fig. 4.6 Real part of Floquet exponents Λ of the periodic orbit and real part of the eigenvalue η of the steady state in dependence on the feedback amplitude b_0. The time delay is fixed at $\tau = 2\pi/(1 - \gamma\,\lambda)$. Panel (**b**) is an enlargement for small b_0. The abbreviations TC and subH refer to a transcritical and subcritical Hopf bifurcation, respectively. Parameters: $\lambda = -0.005$, $\gamma = -10$, and $\beta = \pi/4$

amplitude of the feedback gain b_0. The largest real part is positive for $b_0 = 0$. Thus the periodic orbit is unstable. The abbreviations TC and subH refer to a transcritical and a subcritical Hopf bifurcation, respectively, and will be discussed later. As the amplitude of the feedback gain increases, the largest real part of the exponent becomes smaller and eventually changes sign. Hence the periodic orbit is stabilized. Note that an infinite number of Floquet exponents are created by the control scheme; their real parts tend to negative infinity in the limit of vanishing b_0, and some of them may cross over to positive real parts for larger b_0 as indicated by the blue curve, terminating the stability of the periodic orbit. The largest real part of the eigenvalue η of the fixed point at the origin shows the opposite behavior. It starts with a negative value because the steady state is stable in the absence of control. As the feedback amplitude b_0 increases, if becomes positive which corresponds to a loss of stability.

Panel (b) is an enlargement in the range of small b_0. The vertical dotted lines indicate values of the feedback amplitude for which the periodic orbit and the steady state change their stability. Note that these changes of stability occur at different values of the feedback amplitude. At first, the periodic orbit becomes stable. Then, the trivial steady state loses its stability. This distinct behavior will become important for the understanding of the underlying mechanism which will be analyzed in the following.

Figure 4.7 depicts the dependence of the real part of the Floquet exponents on the feedback amplitude b_0. Panels (a) and (b) correspond to a value of $\lambda = -0.064$ and -0.068, respectively. The time delay τ is chosen on the Pyragas curve according to Eq. 4.2.8. The black curve originates at the Floquet exponent of the uncontrolled case. The red dashed curves show the branches which perform an exchange of stability with the system's branch. Additionally, four other modes are depicted as blue (dotted) curves. The two choices of system parameter λ show the cases where the periodic orbit is just stabilized and where the stabilization fails.

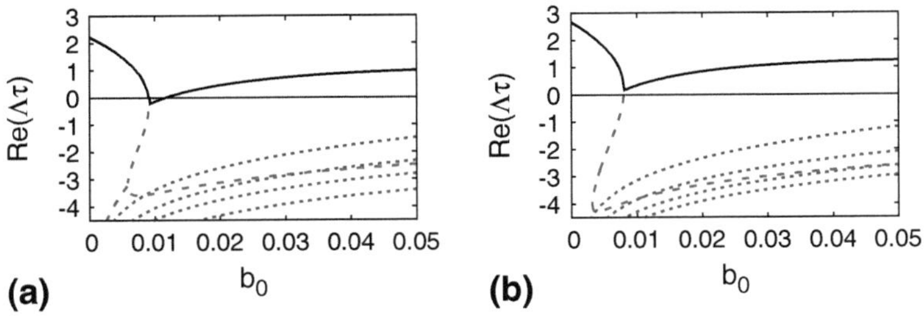

Fig. 4.7 Floquet spectrum, i.e., real part of the exponents, in dependence on the control amplitude b_0 and two different values of $\lambda = -0.064$ and -0.068 in panels (**a**) and (**b**), respectively. The time delay is chosen on the Pyragas curve, i.e., $\tau = 2\pi/(1 - \gamma\lambda)$. Other parameters as in Fig. 4.6

This is indicated by the largest Floquet exponent which becomes slightly negative in the first case and remains positive in the latter case.

In order to understand the mechanism of stabilization of the subcritical odd-number orbit, it is helpful to calculate all periodic orbits of the system in the presence of time-delayed feedback as given by Eqs. 4.2.1 or in polar coordinates (4.3.1a, 4.3.1b). Due to the feedback term, there exist next to the subcritical Hopf orbit of Eqs. 4.2.3 and 4.2.5 some delay-induced orbits with periods different from the subcritical orbit. Clearly, these orbits are influenced by the control force as well because their periods are different from the time delay. One can calculate these orbits which still have the form of rotating waves by the ansatz $z(t) = r_0 e^{i\omega t}$ with constant radius r_0 and frequency ω. A linearization around these orbits follows the derivation of the characteristic equation of the odd-number orbit given in Eq. 4.3.2. To calculate possible orbits of the delayed system, however, one needs to take the nonlinear terms into account. Inserting the ansatz into Eqs. 4.3.1a and 4.3.1b yields

$$0 = [\lambda + r_0^2]r_0 + b_0 r_0[\cos(\beta - \omega\tau) - \cos\beta] \tag{4.3.6a}$$

$$\omega = 1 + \gamma r_0^2 + b_0[\sin(\beta - \omega\tau) - \sin\beta]. \tag{4.3.6b}$$

Eliminating r_0^2 in Eq. 4.3.6a and inserting the resulting expression into the phase equation (4.3.6b) leads to a transcendental equation for the phase ω which can be written as

$$r_0^2 = -\lambda - b_0[\cos(\beta - \omega\tau) - \cos\beta] \tag{4.3.7a}$$

$$\Rightarrow \omega = 1 - \gamma\lambda + b_0[-\gamma\cos(\beta - \omega\tau) + \gamma\cos\beta + \sin(\beta - \omega\tau) - \sin\beta] \tag{4.3.7b}$$

$$\Rightarrow \omega = 1 - \gamma\lambda + 2b_0\left[-\gamma\sin\left(\frac{\omega\tau}{2}\right)\sin\left(\beta - \frac{\omega\tau}{2}\right) + \cos\left(\frac{\omega\tau}{2}\right)\sin\left(\beta - \frac{\omega\tau}{2}\right)\right]. \tag{4.3.7c}$$

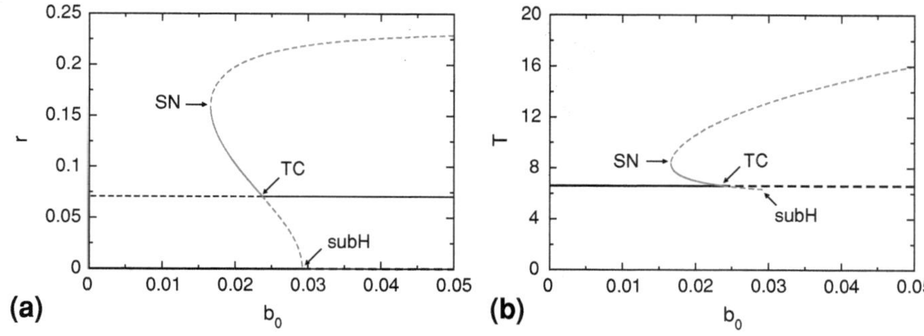

Fig. 4.8 Periods and radii of orbits in dependence on the feedback amplitude b_0 in panels (**a**) and (**b**), respectively. The *black curves* refer to the subcritical periodic orbit and trivial fixed point. The *green curves* correspond to delay-induced orbits. The *solid curves* denotes stability and the *dashed curve* indicates instability. The abbreviations TC, subH, and SN refer to a transcritical, subcritical Hopf, and saddle-node bifurcations, respectively. Other parameters as in Fig. 4.6

To obtain all possible periodic orbits of the system with time-delayed feedback control, one has to solve Eq. 4.3.7b numerically. Then, Eq. 4.3.7a gives the corresponding radii of the periodic orbit. For some frequencies ω obtained from Eq. 4.3.7b the right-hand side of Eq. 4.3.7a may become negative. These spurious solutions can be omitted.

Figure 4.8 shows which states are present in the delayed system (4.2.1) as the feedback amplitude b_0 varies. The time delay τ is chosen on the Pyragas curve, i.e., $\tau = \tau_P = 2\pi/(1 - \gamma\lambda)$, and the feedback phase is fixed at $\beta = \pi/4$. The black curves correspond to the steady states and rotating waves already present in the uncontrolled system. The red curves refer to delay-induced orbits. Panel (a) depicts the corresponding radii as well as the trivial steady state at $r = 0$ and panel (b) shows the period T of the periodic orbits. Solid and dashed curves denote stability and instability, respectively.

For small feedback amplitudes b_0, there exists only the subcritical orbit which is still unstable and the stable trivial steady state. As b_0 increases, a saddle-node bifurcation occurs indicated by SN and a pair of an unstable and a stable periodic orbit is born. Note that the periods of these delay-induced orbits are different from the target orbit. For further increase of b_0, the stable orbit approaches the subcritical orbit, eventually collides with it, and they exchange their stability in a transcritical bifurcation labeled by TC. If the feedback amplitude increases even more, the now unstable orbit meets the fixed point at the origin in a subcritical Hopf bifurcation indicated by subH and ceases to exist. Above this subcritical Hopf bifurcation, the fixed point is unstable. The result is a stabilized subcritical odd number orbit and a destabilized trivial steady state.

The stabilization of the odd number orbit is also displayed in Fig. 4.9 in terms of the Floquet multipliers $\mu = \exp(\Lambda\tau)$. This figure shows the behavior of the Floquet multipliers μ in the complex plane with the increasing amplitude of the feedback gain b_0 in the directions of propagation marked by arrows. Note that

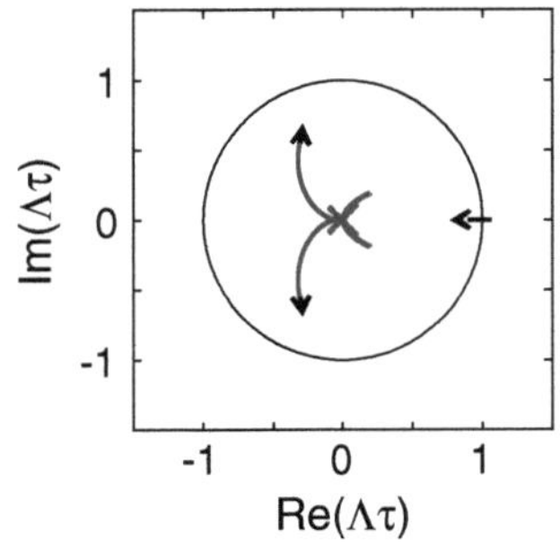

Fig. 4.9 Floquet multipliers $\mu = \exp(\Lambda\tau)$ for increasing feedback amplitude $b_0 \in [0, 0.3]$ in the complex plane. The *arrows* indicate the propagation direction. Other parameters as in Fig. 4.6

there is an isolated real multiplier crossing the unit circle at $\mu = 1$, in contrast to the result stated in [2].

The mechanism of stabilization of the Pyragas orbit by a transcritical bifurcation relies upon the possible existence of such delay-induced periodic orbits with $T \neq \tau$, which was overlooked, e.g., in Ref. [2]. Technically, the proof of the odd number limitation theorem in Ref. [2] fails because the trivial Floquet multiplier $\mu = 1$ corresponding to the Goldstone mode of the system was neglected there; $F(1)$ in Eq. 14 in Ref. [2] is thus zero and not less than zero, as assumed. See also the detailed discussion in Sect. 4.1. At the transcritical bifurcation TC, where a second Floquet multiplier crosses the unit circle, this results in a Floquet multiplier $\mu = 1$ of algebraic multiplicity two.

At the end of the Sect. 4.2, I derived conditions for stabilization in terms of an odering of the Pyragas–Hopf points. These conditions are satisfied if, and only if, the feedback phase β obey the following inequality

$$0 < \beta < \beta_n^*, \tag{4.3.8}$$

where $0 < \beta_n^* < \pi$ is the unique solution of the transcendental equation

$$\frac{1}{\pi}\beta_n^* + 2nb_0 \sin\beta_n^* = 1. \tag{4.3.9}$$

This holds because the condition $\tau_{n\text{-}1}^A < \tau_{n\text{-}1}^B < \tau_n^A$ first fails when $\tau_{n\text{-}1}^B = \tau_n^A$. This condition can be used to derive Eq. 4.3.9. Evaluating the expressions for τ_n^A and τ_n^B as given by Eqs. 4.2.18a and 4.2.18b the condition $\tau_{n\text{-}1}^B = \tau_n^A$ becomes

$$\frac{2\beta^* + 2\pi n - 2\pi}{1 - 2b_0 \sin\beta^*} = 2\pi n \tag{4.3.10a}$$

$$\Leftrightarrow \quad 2\beta + 2\pi n - 2\pi = (1 - 2b_0 \sin\beta^*)2\pi n \tag{4.3.10b}$$

$$\Leftrightarrow \quad \frac{\beta_n^*}{\pi} + 2nb_0 \sin\beta_n^* = 1. \tag{4.3.10c}$$

Equation (4.3.8) represents a necessary but not sufficient condition that the Pyragas choice $\tau_P = nT$ for the delay time will stabilize the periodic orbit.

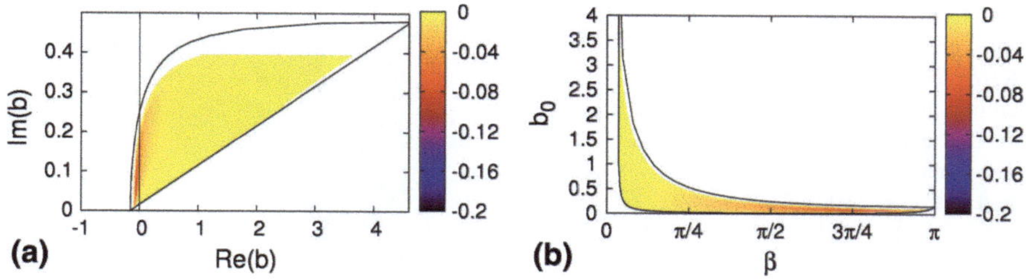

Fig. 4.10 Domain of control in the (Re(b), Im(b)) and (β, b_0) planes in panels (**a**) and (**b**), respectively. The *solid lines* depict the reference boundary of the domain of control for the limit $\lambda \to 0$ according to Eqs. 4.3.13 and 4.3.14. The *color code* shows the largest real part of the Floquet exponent Λ where only negative values are displayed. Other parameters as in Fig. 4.6

To evaluate the second condition, $\tau_H < \tau_P$ near $(\lambda, \tau) = (0, 2\pi)$, an expansion of the exponential in the characteristic equation (4.2.9) for $\omega\tau \approx 2\pi n$ is useful to obtain the approximate Hopf curve for small $|\lambda|$:

$$\tau_H(\lambda) \approx 2\pi n - \frac{1}{\mathrm{Im}(b)}[2\pi n \mathrm{Re}(b) + 1)]\lambda. \tag{4.3.11}$$

Recalling Eq. 4.2.8, the Pyragas stabilization condition $\tau_H(\lambda) < \tau_P(\lambda)$ is therefore satisfied for small values of $|\lambda|$ with $\lambda < 0$ if, and only if,

$$\frac{1}{\mathrm{Im}(b)}\left(\mathrm{Re}(b) + \frac{1}{2\pi n}\right) < -\gamma. \tag{4.3.12}$$

Equation 4.3.12 defines a domain in the plane of the complex feedback gain $b = \mathrm{Re}(b) + i\mathrm{Im}(b) = b_0 e^{i\beta}$ bounded from below for $\gamma < 0 < \mathrm{Im}(b)$ by the straight line

$$\mathrm{Im}(b) = \frac{1}{-\gamma}\left(\mathrm{Re}(b) + \frac{1}{2\pi n}\right). \tag{4.3.13}$$

Equation 4.3.9 represents a curve $b_0(\beta)$, i.e.,

$$b_0(\beta) = \frac{1}{2n \sin\beta}\left(1 - \frac{\beta}{\pi}\right) \tag{4.3.14}$$

which forms the upper boundary of a domain given by the inequality (4.3.8). Thus, Eqs. 4.3.13 and 4.3.14 describe the boundaries of the domain of control in the complex plane of the feedback gain b in the limit of small λ.

Figure 4.10 depicts the domain of control in dependence on the complex feedback gain b. Panels (a) and (b) display this domain in the (Re(b), Im(b)) and (b_0, β) plane, respectively. The solid lines depict as a reference the boundary of the domain of control in the limit of vanishing λ, i.e., $\lambda \to 0$, according to Eqs. 4.3.13 and 4.3.14. The color code shows the largest real part of the Floquet exponent Λ of

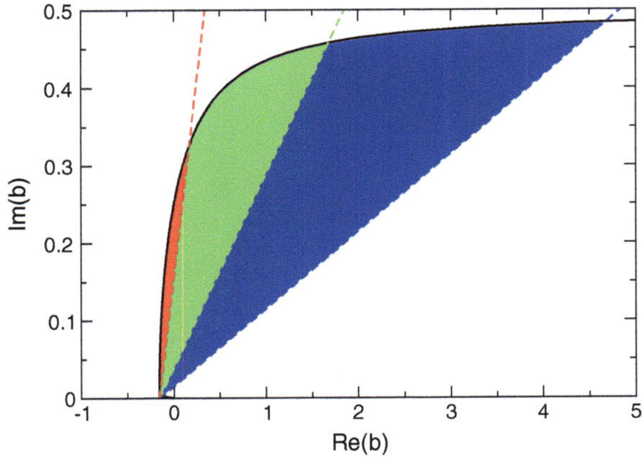

Fig. 4.11 Asymptotic control domain in the (Re(b), Im(b)) plane for different values of γ. The *shaded areas* above the *dashed blue, green*, and *red curves* correspond to $\gamma = -10, -4$, and -1, respectively. The *black curve* refers to the upper boundary of the domain according to Eq. 4.3.14 and the lower γ dependent boundary are given by Eq. 4.3.13. Other parameters as in Fig. 4.6

the subcritical periodic orbit. Only negative values are displayed. Outside the color shaded areas the subcritical orbit is not stabilized.

The effects of the system parameter γ can be seen in Fig. 4.11 which depicts the domain of control in the asymptotic case of vanishing λ according to according to Eqs. 4.3.13 and 4.3.14. The shaded areas above the dashed blue, green, and red curves correspond to $\gamma = -10, -4$, and -1, respectively. The domains are bounded from above by the black curve which is given by Eq. 4.3.14. Note that the domains become smaller as γ approaches zero.

Figure 4.12 displays an enlargement of the domain of control in the (Re(b), Im(b)) plane in the range of small feedback amplitude. As in Fig. 4.10, the solid lines depict as a reference the boundary of the domain of control in the limit of vanishing λ according to Eqs. 4.3.13 and 4.3.14. The red dots refer to three values of the feedback amplitude $b_0 = 0.04, 0.25$, and 0.01 with constant feedback phase $\beta = \pi/4$ and correspond to the Pyragas and Hopf curves of Fig. 4.4a, b and c, respectively. Note that the red dot at point (b) in Fig. 4.12 lies on the boundary

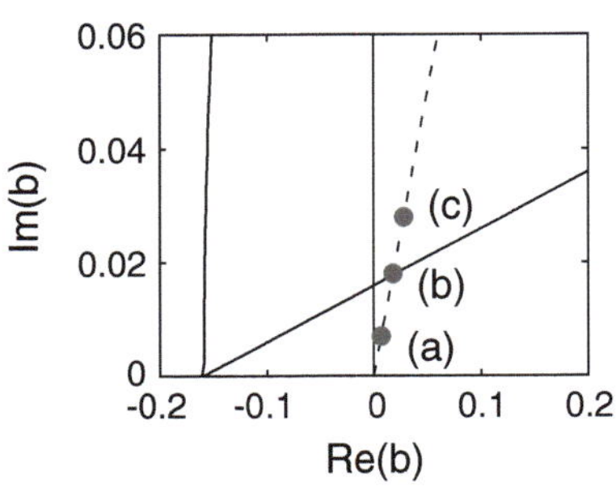

Fig. 4.12 Enlargement of the domain of control in the (Re(b), Im(b)) plane with three values of the feedback $b = 0.04\exp(i\pi/4)$, $0.025\exp(i\pi/4)$, and $0.01\exp(i\pi/4)$ which corresponds to panels (**a**), (**b**), and (**c**) of Fig. 4.4, respectively. The *solid lines* depict as a reference the boundary of the domain of control in the limit of vanishing λ, i.e., $\lambda \to 0$, according to Eqs. 4.3.13 and 4.3.14. Other parameters as in Fig. 4.6

of the domain of control and thus, corresponds to the threshold case of Fig. 4.4b where to Pyragas curve becomes tangent to the Hopf curve.

Note that for real feedback gain b, i.e., $\beta = 0$, no stabilization occurs at all. Therefore, stabilization fails if the feedback term is coupled via a multiple of the identity matrix. The value of the feedback gain at the threshold of control as approximately given in point (b) of Fig. 4.12 can also be derived analytically. For fixed feedback phase β, it is determined by the intersection of the lower bound of the domain of control given by Eq. 4.3.13 and the linear curve

$$\mathrm{Im}(b) = \mathrm{Re}(b)\tan\beta \tag{4.3.15}$$

which parameterizes the dashed line in Fig. 4.12. This leads to the following condition for the critical feedback amplitude b_{crit}

$$\mathrm{Re}(b_{\mathrm{crit}})\tan\beta = \frac{1}{-\gamma}\left(\mathrm{Re}(b_{\mathrm{crit}}) + \frac{1}{2\pi n}\right) \tag{4.3.16a}$$

$$\Leftrightarrow \quad \mathrm{Re}(b_{\mathrm{crit}}) = -\frac{1}{2\pi n(\gamma\tan\beta + 1)}. \tag{4.3.16b}$$

Equivalently, the corresponding feedback amplitude $b_{0,\mathrm{crit}}$ is given by

$$b_{0,\mathrm{crit}} = \sqrt{\mathrm{Re}(b_{\mathrm{crit}})^2 + \mathrm{Im}(b_{\mathrm{crit}})^2} \tag{4.3.17a}$$

$$= \sqrt{\tan^2\beta + 1}\,|\mathrm{Re}(b_{\mathrm{crit}})| \tag{4.3.17b}$$

$$= \frac{\sqrt{\tan^2\beta + 1}}{|2\pi n(\gamma\tan\beta + 1)|}. \tag{4.3.17c}$$

This results for a feedback phase of $\beta = \pi/4$ as in Fig. 4.12 in a value of $b_0 \approx 0.025$ as chosen for point (b).

If the parameter λ becomes more negative, i.e., for increasing $|\lambda|$, the domain of control shrinks which is shown in Fig. 4.13. This figure depicts the domain of control for values of $\lambda = -0.005$, -0.01, and -0.015. The black lines indicate the analytical boundary of the domain as reference case with the limit $\lambda \to 0$. As in Fig. 4.10, the color code indicates the value of the largest real part of the Floquet exponents Λ. Note that the deviation from the linear approximation given in Eq. 4.3.11 becomes larger as $|\lambda|$ increases, i.e., further away from the bifurcation point.

At the end of this section, I will discuss the dynamics of the subcritical Hopf bifurcation system (4.2.1) for large feedback amplitudes b_0. It can be seen already in Fig. 4.6a that delay-induced mode can alter the stability of the system. See, for instance, the additional branch displayed as blue dashed curve in Fig. 4.6a for large b_0. I will show in the following that the stabilized Pyragas orbit becomes unstable in a torus bifurcation which is also known as Neimark–Sacker bifurcation.

The dashed blue curve in Fig. 4.6a indicates how the real part $\mathrm{Re}(\Lambda)$ of a delay induced Floquet exponent of the Pyragas orbit $z(t)$ crosses zero when the control

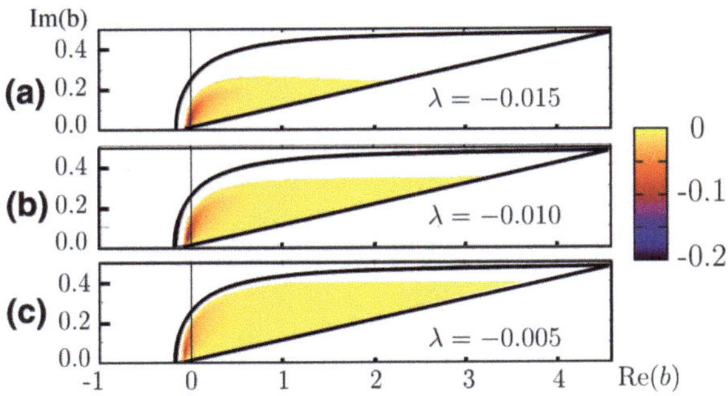

Fig. 4.13 Domain of control in the (Re(b), Im(b)) plane for different values of λ. Panels (**a**), (**b**), and (**c**) correspond to $\lambda = -0.005$, -0.01, and 0.015, respectively. The *solid lines* depict the reference boundary of the domain of control for the limit $\lambda \to 0$ according to Eqs. 4.3.13 and 4.3.14. The *color code* shows the largest real part of the Floquet exponent Λ where only negative values are displayed. Other parameters as in Fig. 4.6

Fig. 4.14 Minimum and maximum radius as *solid black curve* obtained from simulations and radius of the Pyragas orbit $r = \sqrt{-\lambda}$ as *dashed red curve* in dependence on the control amplitude b_0. Parameters as in Fig. 4.6

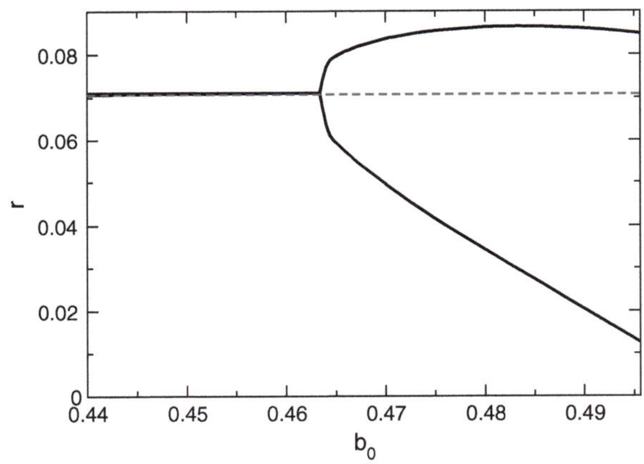

amplitude b_0 is increased. Figures 4.14 and 4.15 show the resulting dynamic consequences for a particular choice of parameters.

Both figures depict the minimum and maximum radius calculated from simulations. The reference radius of the subcritical orbit is given in red color. Note that Fig. 4.14 shows the radius in dependence on the feedback amplitude b_0 while Fig. 4.15 displays the dependence on the system parameter λ for fixed $b_0 = 0.3$. The inset in Fig. 4.15 shows an enlargement at the bifurcation point. As the feedback amplitude increases in Fig. 4.14 or as the parameter λ decreases, the radius does not remain a constant value at $r = \sqrt{\lambda}$, but oscillations occur.

These observed amplitude oscillations can be explained by a supercritical Neimark–Sacker torus bifurcation scenario. Indeed the nontrivial purely imaginary Floquet exponent Λ causes the associated nonreal Floquet multiplier $\mu = e^{\Lambda T}$ to cross the complex unit circle and destabilize the periodic Pyragas orbit in spite of an increasing control amplitude as shown in Fig. 4.14. The resulting bifurcation is

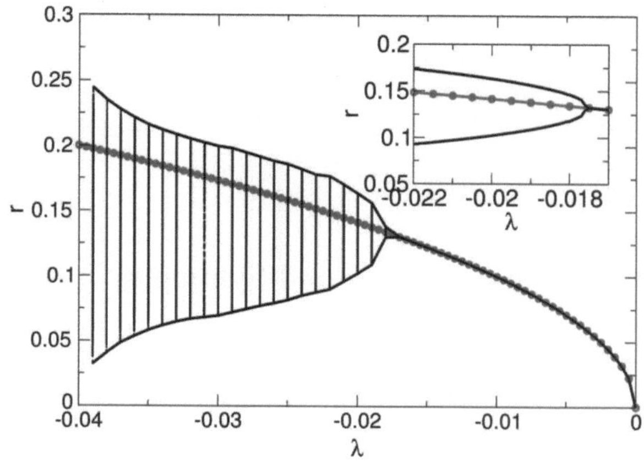

Fig. 4.15 Minimum and maximum radius as *solid black curve* obtained from simulations and radius of the Pyragas orbit $r = \sqrt{-\lambda}$ as *dotted red curve* in dependence on the bifurcation parameter λ. The inset shows an enlargement at the bifurcation point. The feedback amplitude is fixed at $b_0 = 0.3$. Other parameters as in Fig. 4.6

supercritical in our example as indicated in Fig. 4.15. The bifurcating 2-torus inherits stability, taking it away from the Pyragas orbit by exchange of stability.

Equivariance under rotations causes a slight improvement over standard torus bifurcation. The rotation number along the bifurcating torus branch, which would behave in a devil's staircase manner in the general case, in fact becomes a smooth function of the amplitude of the resulting oscillations of $r(t) = |z(t)|$.

In the case of Pyragas control one can conclude that stabilization and destabilization of Pyragas orbits occurs, either by transcritical bifurcations of non-Pyragas periodic orbits, or else by Neimark–Sacker torus bifurcation with nonresonant, smooth dependence of the rotation number.

To conclude this section on the stabilization of the periodic orbit generated by a subcritical Hopf bifurcation, Fig. 4.16 depicts the phase portrait of the variable z in the complex plane for large feedback amplitudes. The feedback phase is fixed at $\beta = \pi/4$. Panels (a) and (b) corresponds to values of $b_0 = 0.46$ and 0.47, respectively. The figure shows variable z for $t \in [4500, 5000]$ such that the transient is omitted. Note that Fig. 4.16a shows the case where the system is still below the torus bifurcation and thus, the periodic orbit at $|z| = \sqrt{-\lambda}$ is stabilized.

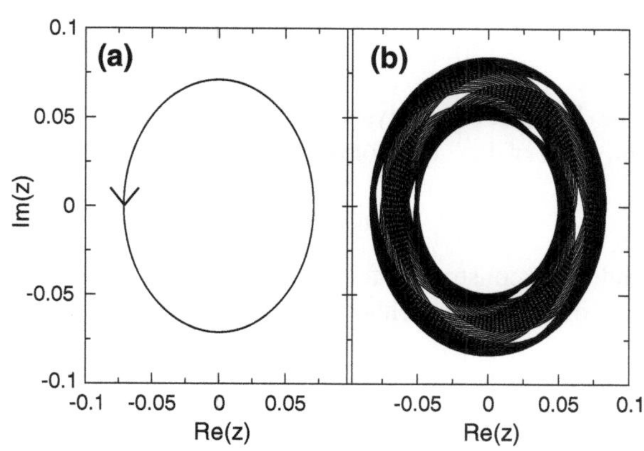

Fig. 4.16 Phase portrait of system (4.2.1) in the complex plane for large feedback amplitudes: Panel (**a**) shows the trajectory for $b_0 = 0.46$. Panel (**b**) depicts the time series for $b_0 = 0.47$. The *arrow* in panel (**a**) indicates the direction of rotation. The time series $z(t)$ is shown for a time interval $t \in [4500, 5000]$. Other parameters as in Fig. 4.14

In contrast, the torus can be seen in Fig. 4.16b. Compare also to Fig. 4.14 which depicts the amplitude $|z|$ in dependence on b_0.

After the discussion of a counterexample to refute the odd number limitation theorem, I will investigate in the next section some aspects of rotating wave solutions and symmetry arguments.

4.4 Rotating Waves and Symmetry

In this section, I will conceptually summarize—from a mathematical perspective—the above calculations on stabilization of the subcritical Hopf model given in Eq. 4.2.1 by Pyragas control following Refs. [6, 7]. Specifically, I will address S^1-equivariance and co-rotating coordinates $z(t) = e^{i\omega t}\zeta(t)$, rotating waves and their bifurcations, the role of center manifolds, and bifurcation to non-rotating waves.

For any fixed real ω, the transformation $z(t) = e^{i\omega t}\zeta(t)$ to co-rotating complex coordinates $\zeta(t)$ transforms Eq. 4.2.1 into the equivalent delay equation

$$\begin{aligned}\frac{d\zeta(t)}{dt} &= [\lambda + (1-\omega)i]\zeta(t) + (1+i\gamma)|\zeta(t)|^2\zeta(t) \\ &\quad + b_0 e^{i\beta}\left[e^{-i\omega\tau}\zeta(t-\tau) - \zeta(t)\right].\end{aligned} \tag{4.4.1}$$

The co-rotating equation (4.4.1) is autonomous since $e^{i\vartheta}z(t)$ solves Eq. 4.2.1 for any fixed real ϑ, whenever $z(t)$ does. Steady states $d\zeta(t)/dt = 0$ of Eq. 4.4.1 are precisely the rotating waves of Eq. 4.2.1, i.e., solutions of the form $z(t) = e^{i\omega t}\zeta_0$ with nonzero rotation frequency ω and nonzero $\zeta_0 \in \mathbb{C}$. The minimal period of such solutions $z(t)$ is of course given by $T = 2\pi/|\omega|$. Note that the Pyragas curves are determined by $\tau = mT = 2\pi m/|\omega|$ for positive integers $m = 1, 2, 3, \ldots$. The derivation of a characteristic equation is given in Sect. 4.3. The roots of this equation are the Floquet exponents of the rotating wave solution. See, for instance, Eq. 4.3.5a and Fig. 4.6 for details.

Rotating waves are nontrivial steady states $\zeta(t) = \zeta_0$ of Eq. 4.4.1, i.e., solutions of

$$0 = \lambda + (1-\omega)i + (1+i\gamma)|\zeta_0|^2 + b_0 e^{i\beta}\left(e^{-i\omega\tau} - 1\right). \tag{4.4.2}$$

The bifurcation analysis of Sects. 4.2 and 4.3 was concerned with solutions $(\omega, |\zeta_0|^2)$ of Eq. 4.4.2 only. One can solve this equation for real parts

$$|\zeta_0|^2 = -\lambda - b_0[\cos(\beta - \omega\tau) - \cos(\beta)] \tag{4.4.3}$$

under the constraint of a positive right hand side. Substituting into the imaginary part of Eq. 4.4.2 yields the real equation

$$1 - \omega + b_0[\sin(\beta - \omega\tau) - \sin(\beta)] - \gamma\{\lambda + b_0[\cos(\beta - \omega\tau) - \cos(\beta)]\} = 0. \tag{4.4.4}$$

One can seek solutions ω, depending on the five real parameters λ, γ, b_0, β, τ. This is elaborated in Eqs. 4.3.7a, 4.3.7b and 4.3.7c. Inserting the solution for ω into the equation of the real part (4.4.3) yields possible rotating wave solutions of the system (4.2.1). See also Fig. 4.8 of Sect. 4.3. The degenerate case $\omega = 0$ which corresponds to a circle of equilibria, alias a frozen wave of vanishing angular velocity ω, arises only for $\gamma\lambda = 1$ [21].

Note how the fold description, in two parameters (λ, τ), of the transcritical bifurcation along the Pyragas curve $\tau = 2\pi/|\omega|$ follows from Eq. 4.4.4 considering ω (or $|\zeta_0|^2$) as a function of (λ, τ), for fixed suitable γ, b_0, β. Indeed, all this follows if one explicitly solves Eq. 4.4.4 for λ, as a function of γ, b_0, β, τ, ω, and then project the resulting graph onto whatever parameter plane seems desirable.

At the Hopf bifurcation there exists a simple pair of purely imaginary eigenvalues, and no other imaginary eigenvalues. Therefore, the center manifold is two dimensional at the Hopf bifurcation [22]. Dimension two also holds at the transcritical bifurcation of rotating waves. Moreover, the center manifold can be chosen to be invariant with respect to the S^1-action $z \mapsto e^{i\vartheta}z$ [23]. In polar coordinates the dynamics in any two-dimensional center manifold is therefore given by a system of the general form

$$\frac{dr(t)}{dt} = g(r(t)^2, \underline{\mu})r(t) \tag{4.4.5a}$$

$$\frac{d\theta(t)}{dt} = h(r(t)^2, \underline{\mu}). \tag{4.4.5b}$$

In the counterexample of the previous section, the functions g and h are given in the absence of time-delayed feedback control by the real and imaginary parts of the dynamic function

$$f(z(t)) = \left[\lambda + i + (1 + i\gamma)|z(t)|^2\right]z(t) \tag{4.4.6}$$

and the system's parameters are $\underline{\mu} = (\lambda, \gamma)$. Here, at the Hopf bifurcation, one seeks a system of ordinary differential equations of the form (4.4.5a, 4.4.5b) whose dynamics match with the delayed system. Thus, in the case of the counterexample in the previous sections, one has to extend the parameter set as $\underline{\mu} = (\lambda, \gamma, b_0, \beta, \tau)$. Note that θ does not enter the equation for $d\,r(t)/dt$ or $d\theta(t)/dt$. Indeed, $(r, \theta + \varphi)$ must be a solution for any fixed φ, by S^1-equivariance, whenever (r, θ) is. Also note that Eqs. 4.4.5a and 4.4.5b is a system of differential equations which does not involve time delayed arguments. Rather, the original time delay τ enters as one parameter among others.

To determine g one can observe that $g(r(t)^2, \underline{\mu}) = 0$ defines rotating (or frozen) waves with $|\zeta_0| = r$, and thus must be equivalent to Eq. 4.4.2 with $h(r^2, \underline{\mu}) = \omega$. The solution set $(r^2, \omega, \underline{\mu})$ is therefore given by Eqs. 4.4.3 and 4.4.4, and defines the zero set of g. Again, $g(r^2, \underline{\mu}) = 0$ if, and only if, $(\omega, \underline{\mu})$ solve Eq. 4.4.4 and r^2 is given by Eqs. 4.4.5a and 4.4.5b.

To determine the stability of the rotating waves within the center manifold it remains to determine the sign of g outside the zero set. That sign is known at the trivial equilibrium $r = 0$, by standard exchange of stability at nondegenerate Hopf bifurcations. Normally hyperbolic rotating waves correspond to simple zeros of g in the r-direction, i.e., $\partial f/\partial r \neq 0$. This allows to determine the sign of g in all bifurcation diagrams. The (in-)stability properties of all rotating waves within the two-dimensional center manifold follows. By spectral analysis at the Hopf bifurcation or, equivalently, at the transcritical bifurcation of rotating waves, stability or instability in the full delay system (4.2.1) follows from the center manifold analysis without ever computing the manifold itself.

Rotating waves $z(t) = e^{i\omega t}\zeta_0$ are periodic solutions of Eq. 4.2.1. But not all periodic solutions need to be rotating waves. Any bifurcation from rotating waves, however, must be visible in co-rotating coordinates $\zeta(t) = e^{-i\omega t}z(t)$ as well. Since rotating waves of Eq. 4.2.1 are equilibria of Eq. 4.4.1 any such bifurcation must be accompanied by purely imaginary eigenvalues Λ of the characteristic equation associated to ζ_0.

It may be useful to insert $\Lambda = i\omega$ here and check for the resulting curve for b_0 and β. Such a curve would discover torus bifurcations from rotating waves or, equivalently, Hopf bifurcations from (circles of) nontrivial equilibria ζ_0 of Eq. 4.4.1. Due to S^1-equivariance such tori would exhibit neither phase-locking nor devil's staircases, but smooth dependence of rotation numbers on parameters.

At the rotating wave equilibrium $r_0 = \zeta_0 > 0$ of the co-rotating system Eq. 4.4.1, the characteristic equation for Floquet multipliers $\mu = e^{\Lambda\tau}$ becomes

$$\begin{aligned} 0 &= \Lambda^2 - 2\left[{r_0}^2 + b_0\cos\beta\left(e^{-\Lambda\tau} - 1\right)\right]\Lambda \\ &\quad + 2{r_0}^2(\cos\beta + \gamma\sin\beta)b_0\left(e^{-\Lambda\tau} - 1\right) + b_0^2\left(e^{-\Lambda\tau} - 1\right)^2. \end{aligned} \tag{4.4.7}$$

See Eq. 4.3.5a and its derivation for details.

In particular, the usual phase locking or resonance phenomena at rational rotation numbers on invariant tori do not occur. This can be seen, for example, by an analysis in suitable Palais coordinates along the relative equilibrium to the S^1-action which is given by the destabilized Pyragas orbit. Palais introduced helpful coordinates near group orbits of group actions. Examples of the construction of these coordinates can be found in Refs. [24, 25]. Eliminating the S^1-action, the Neimark–Sacker bifurcation then becomes a nondegenerate relative Hopf bifurcation from the relative equilibrium $\zeta_0 = r_0$. The Hopf eigenvalues are provided by the purely imaginary Floquet exponent Λ. See Ref. [26] and the survey [27] for details of such a setting. A similar phenomenon occurs, albeit in a reaction-diffusion setting, when rigidly rotating spiral waves in excitable media destabilize to meandering motions [27, 28]. For the likewise related destabilization of wavy vortices to modulated wavy vortices in Taylor–Couette fluid flow between rotating concentric cylinders see Ref. [29] and the references there. Intensity pulsations in laser systems provide yet another experimental source of the same phenomenon of smooth parameter dependence of rotation numbers [30, 31]. Unlike the present case, though, these phenomena do not involve delayed feedback control.

4.5 Fold Bifurcation

In the previous sections, I demonstrated along the lines of an example consisting of a subcritical Hopf bifurcation that the odd number limitation theorem does not hold in general. In fact, the subcritical periodic orbit can be stabilized by a delay-induced saddle-node bifurcation in combination with a transcritical bifurcation yielding an exchange of stability. The subcritical Hopf bifurcation, however, is not the only type of systems with an odd number of real Floquet multipliers greater that unity that can be stabilized by time-delayed feedback. In this section, I will stabilization of periodic orbits which occur due to a fold bifurcation following Ref. [8].

This bifurcation is sometimes called saddle-node bifurcation of limit cycles because the periodic orbit generated at the bifurcation point is repulsive for points outside the limit cycle and attracting for points from the inside. Therefore, it is a saddle-node bifurcation of fixed points in the Poincaré section. Above the bifurcation point, it splits into an unstable and a stable orbit. This can be seen in Fig. 4.17 which show a schematic bifurcation diagram of the fold bifurcation. Dashed lines and empty dots correspond to instability and solid lines indicate stability. The bifurcation is set to $\lambda = 0$.

Let me stress that the periodic orbits generated in a fold bifurcation are particularly important for applications to optical systems. One such system, a three-section semiconductor laser, was discussed recently in Ref. [8]. In this work, numerical bifurcation analysis confirmed that an all-optical delayed feedback control can successfully stabilize rotating waves close to a fold bifurcation in this system. All-optical control exploits the advantage of delayed feedback control, as well as simplicity and inherent high-speed operation. Note that all-optical control of unstable steady states close to a supercritical Hopf bifurcation of the same system has been reported in Refs. [32, 33, 34]. In optical systems, time-delayed feedback can also be exploited to suppress relaxation oscillations excited by noise due to spontaneous emission.

At first, I will introduce the model equation and investigate the dynamics in the absence of a control force. In the next step, I will discuss a delay-induced bifurcation

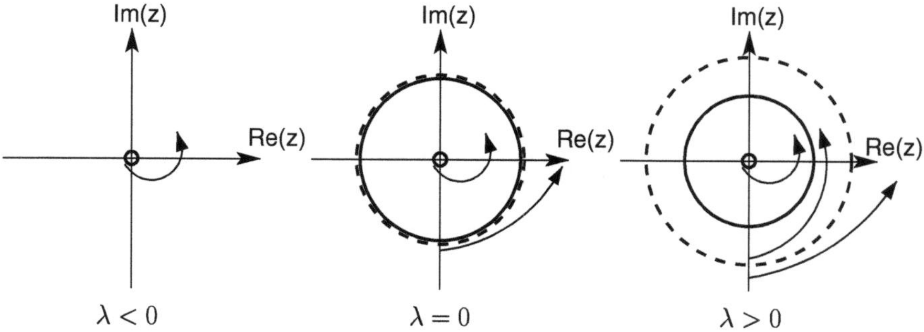

Fig. 4.17 Schematic bifurcation diagram of the fold bifurcation. *Solid lines* indicate stability. *Dashed lines* and *empty dots* refer to instability. The *arrows* show example trajectories

of the trivial fixed point and finally show how the periodic orbit becomes stable. Throughout this section, I will consider only periodic orbits which have the special form of a rotating wave. This is particularly important for applications to optical systems and, in addition, allows detailed analytical treatment.

A fold bifurcation of rotating waves can be described by the following complex equation

$$\frac{dz(t)}{dt} = f(|z(t)|^2)z(t) \tag{4.5.1a}$$

$$= \left[g(|z(t)|^2) + ih(|z(t)|^2)\right]z(t), \tag{4.5.1b}$$

where $z(t)$ denotes a complex variable and g and h are the real and imaginary parts of the function f, respectively. Since f depends on the absolute value of z, no phase-dependence is present such that the system exhibits S^1-equivariance. In fact, $e^{i\vartheta}z(t)$ is a solution whenever $z(t)$ is for any fixed $e^{i\vartheta}$ in the unit circle. Equations 4.5.1a and 4.5.1b can be rewritten in polar coordinates using $z(t) = r(t)e^{i\theta(t)}$ which leads to

$$\frac{dr(t)}{dt} = g(r(t)^2)r(t) \tag{4.5.2a}$$

$$\frac{d\theta(t)}{dt} = h(r(t)^2). \tag{4.5.2b}$$

The S^1-invariance becomes again clear from the last equations because the phase does not enter the right-hand side. Thus, a phase shift $\theta \mapsto \theta + \vartheta$ leaves the equation unchanged. Solutions of Eqs. 4.5.2a and 4.5.2b are rotating waves which can be written as

$$z(t) = r_0 e^{i\omega t} \tag{4.5.3}$$

with nonzero real constants r_0 and ω. Inserting this solution into Eqs. 4.5.2a and 4.5.2b leads to the following requirements

$$0 = g\left(r_0^2\right) \quad \text{and} \quad \omega = h\left(r_0^2\right). \tag{4.5.4}$$

Throughout this section, I will consider the following nonlinear functions

$$g(r(t)^2) = \left[r(t)^2 - 1\right]^2 - \lambda \tag{4.5.5a}$$

$$h(r(t)^2) = \gamma\left[r(t)^2 - 1\right] + \omega_0 \tag{4.5.5b}$$

with positive constants γ and ω_0. The parameter λ will serve as a bifurcation parameter. The form of g and h is chosen such that a nondegenerate fold bifurcation occurs at $r(t)^2 \equiv 1$ for vanishing λ.

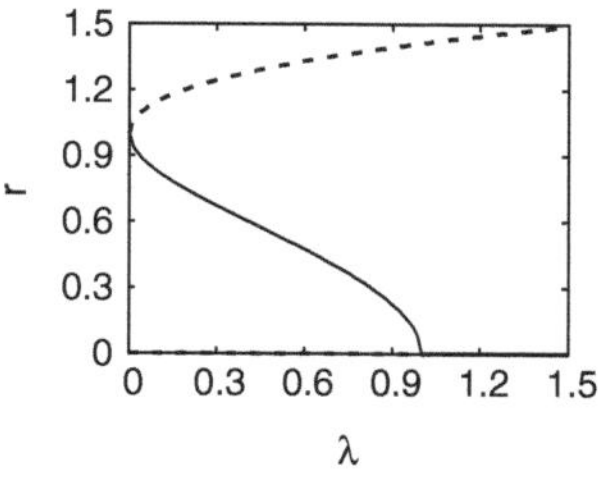

Fig. 4.18 Bifurcation diagram of rotating waves: The *solid* and *dashed curves* correspond to stability and instability, respectively, of the orbits and fixed points of Eqs. 4.5.2a, 4.5.2b, 4.5.5a and 4.5.5b

Figure 4.18 depicts a bifurcation diagram of Eqs. 4.5.2a, 4.5.2b, 4.5.5a and 4.5.5b depending on the system's parameter λ. At $\Lambda = 0$, a pair of an unstable and stable periodic orbit is generated. The radius of the stable orbit becomes smaller for increasing λ and eventually collides with the fixed point at the origin in a reverse supercritical Hopf bifurcation. The analytical formula for the radius $r(t)$ can be inferred from Eqs. 4.5.4 and 4.5.5a

$$0 = g\left(r_0^2\right) \tag{4.5.6a}$$

$$\Leftrightarrow \quad 0 = \left[r(t)^2 - 1\right]^2 - \lambda \tag{4.5.6b}$$

$$\Leftrightarrow \quad r(t)^2 = 1 \pm \sqrt{\lambda}. \tag{4.5.6c}$$

Similarly, the corresponding frequency is given by Eqs. 4.5.4 and 4.5.5b

$$\omega = h\left(r_0^2\right) \tag{4.5.7a}$$

$$\Leftrightarrow \quad \omega = \gamma\left[r(t)^2 - 1\right] + \omega_0 \tag{4.5.7b}$$

$$\Leftrightarrow \quad \omega = \omega_0 \pm \gamma\sqrt{\lambda}. \tag{4.5.7c}$$

Note that the plus and minus signs belong to the unstable and stable orbit in Fig. 4.18, respectively. The goal of control is to stabilize the unstable orbit.

Applying time-delayed feedback to system (4.5.1a, 4.5.1b) leads to

$$\frac{dz(t)}{dt} = f\left(|z(t)|^2\right)z + b_0 e^{i\beta}[z(t-\tau) - z(t)], \tag{4.5.8}$$

where the complex feedback gain is already written as amplitude b_0 and phase β. In order to satisfy the noninvasive property of time-delayed feedback, the time delay τ is chosen as an integer multiple of the target orbit's period T. This minimum period is given by $T = 2\pi/\omega$ for the rotating wave $z(t) = re^{i\omega t}$ which leads to the following formula for the time delay τ using Eq. 4.5.7c

$$\tau = \tau(\lambda) = \frac{2\pi n}{\omega_0 + \gamma\sqrt{\lambda}} \tag{4.5.9}$$

with $n = 1, 2, \ldots$ where only the unstable orbit, i.e., plus sign in Eq. 4.5.6c, is taken into account. As in the Sect. 4.2, this leads to a set of Pyragas curves in the

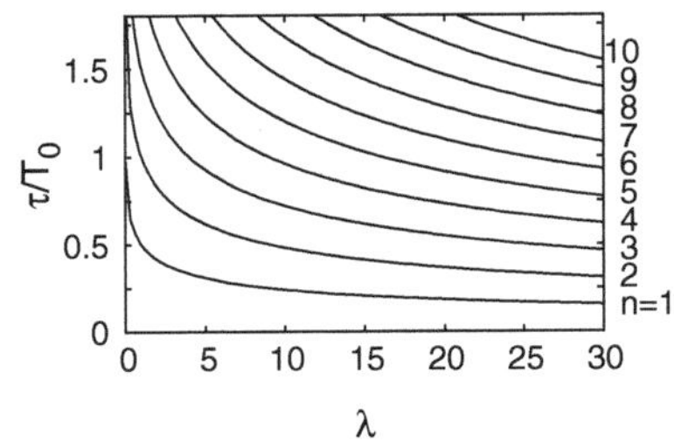

Fig. 4.19 Pyragas curves $\tau(\lambda)$, corresponding to the unstable branch in Fig. 4.18, in the parameter plane (λ, τ) according to Eq. 4.5.10. The time delay is given in units of timescale T_0 of the fixed point at the origin, i.e., $T_0 = 2\pi$. Parameters: $\gamma = \omega_0 = 1$

(λ, τ) plane. The dependence of λ on the time delay τ can be extracted from Eq. 4.5.9 which yields

$$\lambda = \lambda(\tau) = \left(\frac{2\pi n - \omega_0 \tau}{\gamma\tau}\right)^2 \tag{4.5.10}$$

with $n = 1, 2, \ldots$.

Figure 4.19 displays the first 10 Pyragas curves in the (λ, τ) plane for which the control scheme is noninvasive. Note that the time delay is given in units of timescale T_0 of the fixed point at the origin, i.e., $T_0 = 2\pi$. Vanishing parameter λ corresponds to a time delay $\tau = 2\pi n/\omega_0$ which recovers for $n = 1$ the period of the uncontrolled orbits. Due to the relation between λ and τ by Eqs. 4.5.9 and 4.5.10, I will consider τ as bifurcation parameter in the following.

Before exploring the domain of control for stabilization of the periodic orbit, let me discuss the case of b_0 first. Inserting the formula $\lambda(\tau)$ of Eq. 4.5.10 yields for the radius and frequency of the rotating wave solution given by Eqs. 4.5.6c and 4.5.7c

$$r^2 = 1 \pm \frac{2\pi n - \omega_0 \tau}{\gamma\tau} \tag{4.5.11a}$$

$$\omega = \omega_0 \pm \left(\frac{2\pi n - \omega_0 \tau}{\tau}\right). \tag{4.5.11b}$$

The two resulting branches form a transcritical bifurcation at $\tau = 2\pi n/\omega_0$ or equivalently $\lambda = 0$ as discussed in Fig. 4.18. This transcriticality looks like an artefact caused by the change of bifurcation parameter from λ to τ on the Pyragas curve. Note, however, that the two orbits in Eqs. 4.5.11a and 4.5.11b have different periods away from the transcritical point. In fact, only one of the two crossing branches features minimum period T such that the Pyragas condition $\tau = nT$ holds. This happens along the branch

$$r^2 = 1 + \frac{2\pi n - \omega_0 \tau}{\gamma\tau} \tag{4.5.12}$$

with the frequency $\omega = 2\pi n/\tau$. I will call this branch which corresponds to the plus sign in Eqs. 4.5.11a and 4.5.11b the Pyragas branch. The other branch has minimum period T with

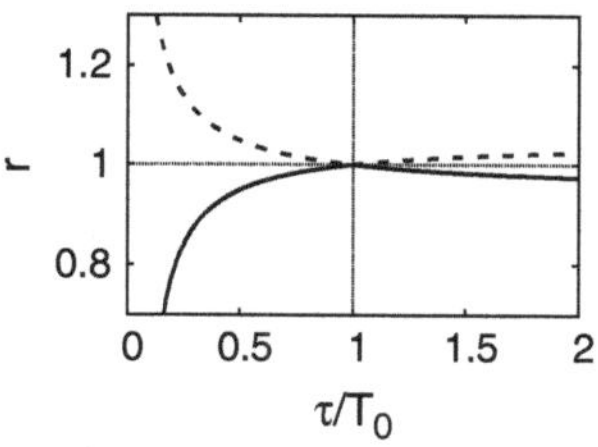

Fig. 4.20 Bifurcation diagram of rotating waves of Eq. 4.5.8 at vanishing control amplitude $b_0 = 0$. Parameters: $T_0 = 2\pi$, $\omega_0 = 1$, $\gamma = 10$

$$nT = \frac{\pi n}{\omega_0 \tau - \pi n}\tau \neq \tau, \tag{4.5.13}$$

except at the crossing point $\omega_0 \tau = 2\pi n$. The minus branch therefore violates the Pyragas condition for noninvasive control, even though it has admittedly been generated from the same fold bifurcation.

Figure 4.20 shows a bifurcation diagram of the rotating wave solutions of Eq. 4.5.8 at vanishing control amplitude b_0 in dependence on the time delay τ which is given in units of the timescale of the trivial fixed point $T_0 = 2\pi$. The solid and dashed curves indicate stability and instability, respectively.

The strategy for Pyragas control of the unstable part of the Pyragas branch is now simple. For a nonzero control amplitude b_0, the Pyragas branch persists without change, due to the noninvasive property $\tau = nT$ along the Pyragas curve $\lambda = \lambda(\tau)$ given by Eq. 4.5.10. The minus branch in Eqs. 4.5.11a and 4.5.11b, however, will be perturbed slightly for small $b_0 \neq 0$. If the resulting perturbed transcritical bifurcation $\tau = \tau_c$ moves to the left, i.e., below $2\pi n/\omega_0$, then the stability region of the Pyragas branch has invaded the unstable region of the fold bifurcation. This again refutes the odd number limitation theorem.

Let $\tau = \tau_c$ denote the transcritical bifurcation point on the Pyragas curve $\lambda = \lambda(\tau)$ according to Eq. 4.5.10. Let $z(t) = r_c e^{i\omega_c t}$ denote the corresponding rotating wave. I will derive in the following conditions for the transcritical bifurcation in Eq. 4.5.8.

In order to derive these conditions, one could proceed by brute force. This means a linearization of the control system (4.5.8) along the Pyragas branch, in polar coordinates, a derivation of the characteristic equation in a co-rotating coordinate frame, and an elimination of the trivial zero characteristic root to determine critical values $\tau = \tau_c$, $r = r_c$, and $b_0 = b_c$ such that a nontrivial zero characteristic root remains. Instead, I will proceed locally in a two-dimensional center manifold of the fold, following the arguments in Just et al. [7]. Any periodic solution in the center manifold of Eq. 4.5.8 is a rotating wave $z(t) = re^{i\omega t}$. Hence, one can compute the rotating waves of the system with time-delayed feedback control (4.5.8), globally. Substituting $z(t) = re^{i\omega t}$ into Eq. 4.5.8 and decomposing into real and imaginary parts yields similar to Eq. 4.5.4

$$0 = g(r^2) + b_0 r[\cos(\beta - \omega\tau) - \cos\beta] \tag{4.5.14a}$$

$$= \left(r^2 - 1\right)^2 r + 2b_0 r \sin\frac{\omega\tau}{2} \sin\left(\beta - \frac{\omega\tau}{2}\right) \tag{4.5.14b}$$

$$\omega = h\left(r^2\right) + b_0\left[\sin(\beta - \omega\tau) - \sin\beta\right] \tag{4.5.14c}$$

$$= \left(r^2 - 1\right) - 2b_0 \sin\frac{\omega\tau}{2} \cos\left(\beta - \frac{\omega\tau}{2}\right). \tag{4.5.14d}$$

Note that the equations above can also be derived by transformation of the system with time-delayed feedback as given in Eq. 4.5.8 to co-rotating coordinates as shown in the following. In polar coordinates Eq. 4.5.8 becomes

$$\frac{dr(t)}{dt} = g(r^2(t))r(t) + b_0[\cos(\beta + \theta(t - \tau) - \theta(t))r(t - \tau) - r(t)\cos\beta] \tag{4.5.15a}$$

$$\frac{d\theta(t)}{dt} = h(r^2(t)) + b_0\left[\sin(\beta + \theta(t - \tau) - \theta(t))\frac{r(t - \tau)}{r(t)} - \sin\beta\right], \tag{4.5.15b}$$

where r and θ denote the amplitude and phase z, respectively. Equations 4.5.14a and 4.5.14c are then recovered by passing to co-rotating coordinates $\psi(t) = \theta(t) - \omega t$ and searching for equilibria of the resulting equations

$$\frac{dr(t)}{dt} = g(r^2(t))r(t) + b_0[\cos(\beta + \psi(t - \tau) - \psi(t) - \omega\tau)r(t - \tau) - r(t)\cos\beta] \tag{4.5.16a}$$

$$\frac{d\theta(t)}{dt} = h(r^2(t)) - \omega + b_0\left[\sin(\beta + \psi(t - \tau) - \psi(t) - \omega\tau)\frac{r(t - \tau)}{r(t)} - \sin\beta\right]. \tag{4.5.16b}$$

The equilibria are given for $r = const.$ and $\psi = 0$, i.e., $\theta(t) = \omega t$ and have the form of the rotating waves $z(t) = re^{i\omega t}$ considered above.

With the abbreviation $\varepsilon = r^2 - 1$ which is introduced for notational convenience and the special choices for the functions g and h given in Eqs. 4.5.5a and 4.5.5b, the Eqs. 4.5.14a and 4.5.14c become

$$0 = \varepsilon^2 - \lambda(\tau) + 2b_0 \sin\frac{\omega\tau}{2} \sin\left(\beta - \frac{\omega\tau}{2}\right) \tag{4.5.17a}$$

$$\omega = \gamma\varepsilon + \omega_0 - 2b_0 \sin\frac{\omega\tau}{2} \cos\left(\beta - \frac{\omega\tau}{2}\right). \tag{4.5.17b}$$

For small enough b_0, one can solve Eq. 4.5.17b for $\omega = \omega(\varepsilon)$ and insert into Eq. 4.5.17a which leads to

$$0 = G(\tau, \varepsilon). \tag{4.5.18}$$

Here $G(\tau, \varepsilon)$ abbreviates the right hand side of Eq. 4.5.17a with $\omega = \omega(\varepsilon)$ substituted for ω. The condition for a transcritical bifurcation in the system with control then reads

$$0 = \frac{\partial}{\partial \varepsilon} G(\tau, \varepsilon) \tag{4.5.19}$$

in addition to Eq. 4.5.18. It simplifies matters significantly that this calculation has to be performed along the Pyragas branch only, where $\omega\tau = 2\pi\tau/T = 2\pi n$. Therefore, Eq. 4.5.19 becomes

$$0 = \frac{\partial}{\partial \varepsilon} G(\tau, \varepsilon) \tag{4.5.20a}$$

$$= \frac{\partial}{\partial \varepsilon}\left[\varepsilon^2 - \lambda(\tau) + 2b_0 \sin\frac{\omega(\varepsilon)\tau}{2} \sin\left(\beta - \frac{\omega(\varepsilon)\tau}{2}\right)\right] \tag{4.5.20b}$$

$$\begin{aligned} &= 2\varepsilon + 2b_0\left[\frac{\tau}{2}\frac{\partial\omega(\varepsilon)}{\partial\varepsilon}\cos(n\pi)\sin(\beta - n\pi)\right. \\ &\quad \left. - \sin(n\pi)\frac{\tau}{2}\frac{\partial\omega(\varepsilon)}{\partial\varepsilon}\cos(\beta - n\pi)\right] \end{aligned} \tag{4.5.20c}$$

$$= 2\varepsilon + b_0\tau\frac{\partial\omega(\varepsilon)}{\partial\varepsilon}\cos(n\pi)\sin(\beta - n\pi) \tag{4.5.20d}$$

$$= 2\varepsilon + b_0\tau\frac{\partial\omega(\varepsilon)}{\partial\varepsilon}\sin\beta. \tag{4.5.20e}$$

To obtain the derivative of ω with respect to ε, one has to differentiate Eq. 4.5.17b implicitly, at $\omega\tau = 2n\pi$

$$\frac{\partial\omega(\varepsilon)}{\partial\varepsilon} = \gamma - b_0\tau\frac{\partial\omega(\varepsilon)}{\partial\varepsilon}\cos\beta. \tag{4.5.21}$$

Solving for values of this derivative, for small b_0, yields

$$\frac{\partial\omega(\varepsilon)}{\partial\varepsilon} = \frac{\gamma}{1 + b_0\tau\cos\beta} = \frac{\gamma}{1 + b_0\frac{2n\pi}{\omega_0 + \gamma\varepsilon}\cos\beta}, \tag{4.5.22}$$

where $\omega\tau = 2k\pi$ and $\omega = \omega_0 + \gamma\varepsilon$ is employed. Plugging Eq. 4.5.22 into Eq. 4.5.20e, the control amplitude b_0 enters linearly which leads to

$$0 = \varepsilon(\omega_0 + \gamma\varepsilon)\left(1 + b_0\frac{2n\pi}{\omega_0 + \gamma\varepsilon}\cos\beta\right) + b_0 n\pi\gamma\sin\beta \tag{4.5.23a}$$

$$= \varepsilon(\omega_0 + \gamma\varepsilon + b_0 2n\pi\cos\beta) + b_0 n\pi\gamma\sin\beta. \tag{4.5.23b}$$

Solving for b_0, one finally arrives at a relation between the control amplitude b_c at which the transcritical bifurcation occurs and ε given by

$$b_c(\varepsilon) = -\varepsilon \frac{\omega_0 + \gamma\varepsilon}{n\pi(\gamma \sin\beta + 2\varepsilon \cos\beta)}. \tag{4.5.24}$$

Equivalently, one can derive a dependence of b_c on the time delay τ by substitution of $\varepsilon = -\sqrt{\lambda} = r^2 - 1$ and the function $\lambda(\tau)$ according to Eqs. 4.5.11a and 4.5.11b. This yields

$$b_c(\tau) = -\frac{2\pi n - \omega_0\tau_c}{\tau_c\left[\frac{1}{2}\gamma^2\tau_c \sin\beta + (2\pi n - \omega_0\tau_c)\cos\beta\right]}. \tag{4.5.25}$$

As follows from Eqs. 4.5.24 and 4.5.25, for small ε, alias for τ_c near $2n\pi/\omega_0$, the optimal control angle is $\beta = -\pi/2$ in the limit $\varepsilon \to 0$, i.e., $r \to 1$, and for fixed $n, \omega_0, \gamma, \varepsilon$ this control phase β allows for stabilization with the smallest amplitude $|b_c|$. For $\beta = -\pi/2$ the relations Eqs. 4.5.24 and 4.5.25 simplify to

$$b_c(\varepsilon) = \frac{\varepsilon}{n\pi}\left(\frac{\omega_0}{\gamma} + \varepsilon\right) \tag{4.5.26a}$$

$$b_c(\tau) = \frac{2}{(\gamma\tau_c)^2}(2n\pi - \omega_0\tau_c), \tag{4.5.26b}$$

respectively. For small $b_0 > 0$ there is also the expansions

$$\varepsilon = -\left(n\pi\frac{\gamma}{\omega_0}\sin\beta\right)b_0 + \cdots \tag{4.5.27a}$$

$$\tau_c = \frac{2\pi n}{\omega_0} + \left[\frac{1}{2\omega_0}\left(\frac{2n\pi\gamma}{\omega_0}\right)^2 \sin\beta\right]b_0 + \cdots \tag{4.5.27b}$$

for the location of the transcritical bifurcation. In particular, one can see that odd number delay stabilization can be achieved by arbitrary small control amplitudes b_0 near the fold, for $\gamma > 0$ and $\sin\beta < 0$. Note that the stability region of the Pyragas curve increases if $\varepsilon = r_c^2 - 1 > 0$. For vanishing phase of the control, $\beta = 0$, in contrast, delay stabilization cannot be achieved by arbitrarily small control amplitudes b_0, near the fold in our system (4.5.8).

Even far from the fold at $\lambda = 0$, $\tau = 2n\pi/\omega_0$ the above formulas (4.5.24) and (4.5.26b) hold and indicate a transcritical bifurcation from the (global) Pyragas branch of rotating waves of Eq. 4.5.8, along the Pyragas curve $\lambda = \lambda(\tau)$. This follows by analytic continuation. Delay stabilization, however, may fail long before $\tau = \tau_c$ is reached. In fact, nonzero purely imaginary Floquet exponents may arise, which destabilize the Pyragas branch long before $\tau = \tau_c$ is reached.

A more global picture of the orbits involved in the transcritical bifurcation may be obtained by numerical analysis. Rewriting the system's equation (4.5.8) in polar coordinates $z = re^{i\theta}$ yields

$$\frac{dr(t)}{dt} = \left\{\left[r(t)^2 - 1\right]^2 - \lambda\right\} r(t) \\ + b_0[\cos(\beta + \theta(t-\tau) - \theta(t))r(t-\tau) - r(t)\cos\beta] \tag{4.5.28a}$$

$$\frac{d\theta(t)}{dt} = \gamma\left[r(t)^2 - 1\right] + \omega_0 \\ + b_0\left[\sin(\beta + \theta(t-\tau) - \theta(t))\frac{r(t-\tau)}{r(t)} - \sin\beta\right]. \tag{4.5.28b}$$

To find all rotating wave solutions one can follow the strategy of Eq. 4.3.6a of Sect. 4.3 and use the ansatz $r(t) \equiv r_0 = \text{const}$ and $d\theta/dt = \omega = \text{const}$. This leads to

$$0 = \left(r_0^2 - 1\right)^2 r_0 - \lambda + b_0 r_0[\cos(\beta - \omega\tau) - \cos\beta] \tag{4.5.29a}$$

$$\omega = \gamma\left(r_0^2 - 1\right) + \omega_0 + b_0[\sin(\beta - \omega\tau) - \sin\beta]. \tag{4.5.29b}$$

Eliminating r and inserting this expression into Eq. 4.5.29b yields a transcendental equation for ω which reads

$$0 = -\gamma^2\lambda + \gamma^2 b_0[\cos(\beta - \omega\tau) - \cos\beta] \\ + (\omega - \omega_0 - b_0[\sin(\beta - \omega\tau) - \sin\beta])^2. \tag{4.5.30}$$

One can now solve this equation numerically for ω and insert the result into Eq. 4.5.29b, i.e.,

$$r_0 = \sqrt{\frac{\omega - \omega_0}{\gamma} - \frac{b_0}{\gamma}[\sin(\beta - \omega\tau) - \sin\beta] + 1} \tag{4.5.31}$$

to obtain the allowed radii where imaginary radii are discarded.

The orbit which stabilizes the Pyragas branch in the transcritical bifurcation may be the minus-branch of Eqs. 4.5.11a and 4.5.11b or another delay-induced orbit which is born in a fold bifurcation, depending on the parameters. Figure 4.21 displays the different scenarios and the crossover in dependence on the control amplitude b_0 where the dashed and solid lines indicate unstable and stable branches, respectively. The value of γ is chosen as $\gamma = 9$, 10.5, 10.6, and 13 in panels (a), (b), (c), and (d), respectively. The red curves refer to delay-induced solutions. It can be seen that the Pyragas orbit is stabilized by a transcritical bifurcation T_1. As the value of γ increases, a pair of a stable and an unstable orbit generated by a fold bifurcation F_1 approaches the minus-branch as displayed in Fig. 4.21a. On this branch, fold bifurcations labeled by F_2 and F_3 occur as shown in Fig. 4.21b. At $\gamma = 10.6$, the fold points of F_1 and F_2 touch in a transcritical bifurcation T_2 and annihilate. See Fig. 4.21c and d. Thus, for further increase of γ, one is left with the stable minus-branch and the unstable orbit, which was generated at the fold bifurcation F_3. In all panels the radius of the Pyragas orbit is not changed by the

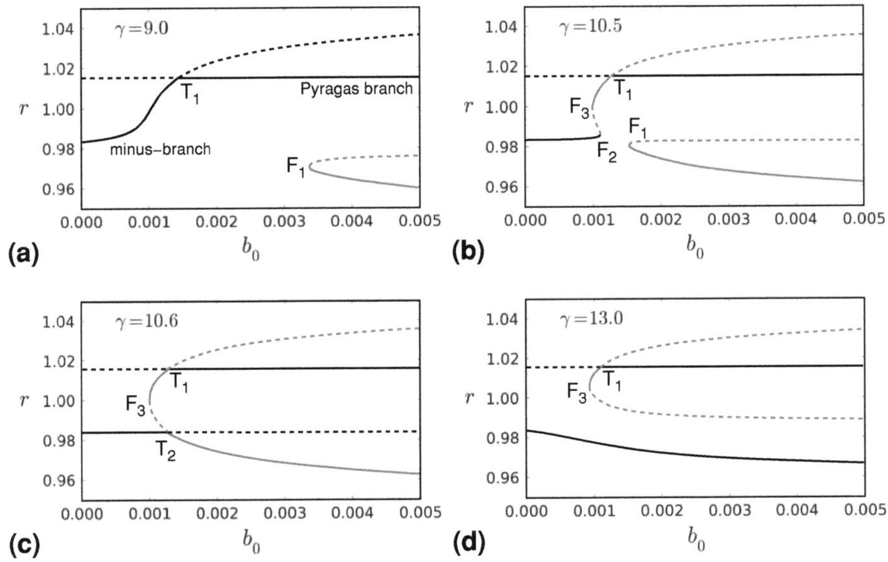

Fig. 4.21 Radii of stable and unstable rotating wave solutions in dependence on the feedback amplitude b_0 for different γ as *solid and dashed curves*, respectively. The *green curves* refer to delay-induced solutions. Panels (**a**), (**b**), (**c**), and (**d**) correspond to a value of $\gamma = 9.0, 10.5, 10.6$, and 13.0, respectively. Other parameters: $\omega_0 = 1$, $\lambda = 0.001$, $\beta = -\pi/2$, and τ according to Eq. 4.5.9

control. The radius of the minus-branch, however, is altered because the delay time does not match orbit period as discussed in Eq. 4.5.13.

One can see from Fig. 4.21 that the mechanism for stabilization can be twofold. In panels (a) and (b) the stable orbit of the saddle-node bifurcation at $\lambda = 0$, i.e., minus-branch, exchanges its stability with the odd-number orbit. Alternatively, a delay-induced saddle-node bifurcation can occur between the minus- and plus-branches as in panels (c) and (d) of Fig. 4.21. In this case, the stable orbit born in this bifurcation becomes larger and eventually exchanges its stability with the odd-number orbit, i.e., the plus-branch, in a transcritical bifurcation.

These two scenarios are schematically depicted in Fig. 4.22 in the complex plane of the variable (z). Panels (a) and (b) correspond to Fig. 4.21a,b and c,d, respectively. Note that solid and dashed curves refer to stability and instability, respectively, and that black color corresponds to orbits and fixed points already present in the uncontrolled system while red color shows delay-induced orbit.

Figure 4.23 depicts the region in the (β, b_0) plane where the Pyragas orbit is stable, for a set of parameters. The color code shows only negative values of the largest real part of the Floquet exponents. One can see that the orbit is most stable for feedback phases $\beta \approx -\pi/2$ which agrees with the previous analytical results for small λ. Figure 4.23 is obtained by linear stability analysis of Eqs. 4.5.28a and 4.5.28b and numerical solution of the transcendental eigenvalue problem for the Floquet exponents. The derivation of the characteristic equation will be elaborated in the following to conclude this section.

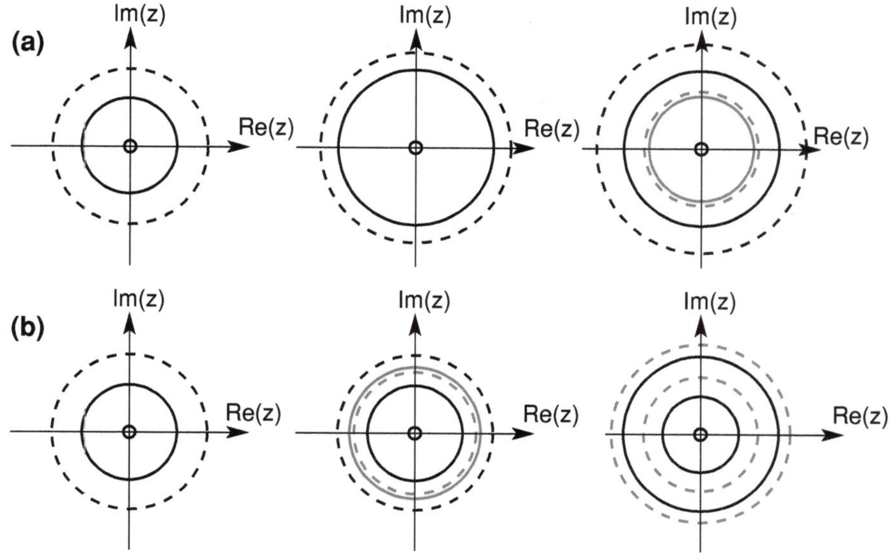

Fig. 4.22 Schematic diagram of the stabilization mechanism in the complex plane. Panels (**a**) and (**b**) correspond to Fig. 4.21a,b and c,d, respectively. The *dashed curves* indicate instability. The *solid curves* refer to stability. *Black color* corresponds to orbits and fixed points already present in the uncontrolled system and *green color* refers to a delay-induced orbit

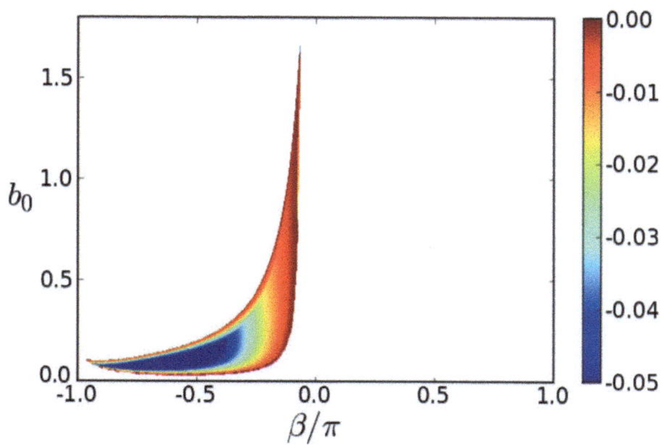

Fig. 4.23 Domain of stability of the Pyragas orbit in the (β, b_0) plane. The *color code* shows only negative values of the largest real part of the Floquet exponents according to the characteristic equation (4.5.33). Parameters: $\omega_0 = 1$, $\lambda = 0.0001$, $\gamma = 0.1$

Linearizing Eqs. 4.5.28a and 4.5.28b around the Pyragas orbit according to $z(t) = (r + \delta r)\exp(i\omega\, t + i\delta\theta)$ yields

$$\frac{d}{dt}\begin{pmatrix}\delta r(t)\\ \delta\theta(t)\end{pmatrix} = \begin{pmatrix}\partial_r g r + g - b_0\cos\beta & r b_0 \sin(\beta - \omega\tau)\\ \partial_r h - b_0\sin(\beta - \omega\tau) r^{-1} & -b_0\cos(\beta - \omega\tau)\end{pmatrix}\begin{pmatrix}\delta r(t)\\ \delta\theta(t)\end{pmatrix} + \begin{pmatrix}b_0\cos(\beta - \omega\tau) & -r b_0\sin(\beta - \omega\tau)\\ b_0\sin(\beta - \omega\tau) r^{-1} & b_0\cos(\beta - \omega\tau)\end{pmatrix}\begin{pmatrix}\delta r(t-\tau)\\ \delta\theta(t-\tau)\end{pmatrix}. \tag{4.5.32}$$

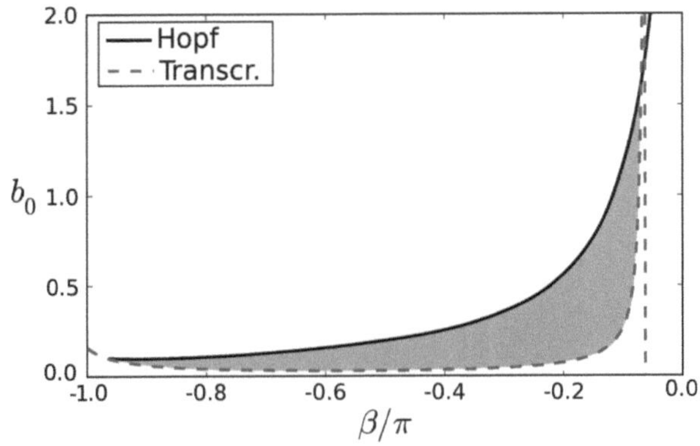

Fig. 4.24 Transcritical and Hopf bifurcation curves as *dashed and solid curves* and the domain of stability as shaded region in the (β, b_0) plane. Parameters as in Fig. 4.23

The delay time τ matches the period of the Pyragas orbit which yields $\omega\tau = 2\pi n$. Using the exponential ansatz $\delta r(t) \sim \exp(\Lambda\, t)$ and $\delta\theta(t) \sim \exp(\Lambda\, t)$ gives a transcendental equation for the Floquet exponents Λ

$$\det\begin{bmatrix} 4(r^2-1)r^2+(r^2-1)^2-\lambda-\Lambda-b_0\cos\beta\left(1-e^{-\Lambda\tau}\right) & rb_0\sin\beta\left(1-e^{-\Lambda\tau}\right) \\ 2\gamma r-(b_0/r)\sin\beta\left(1-e^{-\Lambda\tau}\right) & -\Lambda-b_0\cos\beta\left(1-e^{-\Lambda\tau}\right) \end{bmatrix} = 0. \tag{4.5.33}$$

This equation was numerically solved to obtain Fig. 4.23.

Note that one can find the Hopf bifurcation of the Pyragas orbit in a semi-analytical way by inserting $\Lambda = i\Omega$ into Eq. 4.5.33 and separating the equation into real and imaginary parts

$$\begin{aligned} 0 = &-\Omega^2 - 2\Omega b_0\cos\beta\sin(\Omega\tau) - b_0(cr\sin\beta + a\cos\beta) \\ &[1-\cos(\Omega\tau)] - b_0^2 2[1-\cos(\Omega\tau)]\cos(\Omega\tau) \end{aligned} \tag{4.5.34a}$$

$$\begin{aligned} 0 = &-a\Omega + 2\Omega b_0\cos\beta[1-\cos(\Omega\tau)] - b_0(cr\sin\beta + a\cos\beta)\sin(\Omega\tau) \\ &+ b_0^2 2[1-\cos(\Omega\tau)]\sin(\Omega\tau). \end{aligned} \tag{4.5.34b}$$

One can now use Ω as a parameter and solve the two equations for β and b_0 at each Ω. The resulting Hopf curve and the transcritical bifurcation curve (4.5.24) then form the boundary of the control domain which is shown as Fig. 4.24.

4.6 Intermediate Conclusion

It was long contended that a certain class of periodic orbits, namely those with an odd number of real Floquet multipliers greater than unity cannot be stabilized by time-delayed feedback control. In this chapter, I have shown that this odd number limitation theorem [2, 3] does not hold in general. Along the lines of the normal form of a subcritical Hopf bifurcation which is paradigmatic for a variety of dynamic systems, I have demonstrated that stabilization is possible close to the bifurcation point. For this autonomous system, I have presented analytical conditions for successful control in terms of the complex feedback gain or, equivalently, the feedback amplitude and phase. Non-vanishing feedback phase played a crucial role for this finding.

By solving the characteristic equation, I have calculated the domain of control and discussed the underlying mechanism by analytical and numerical means considering a series of Hopf and Pyragas curves. The former characterize bifurcations of the steady state at the origin. The latter preserved the odd number orbit due to the noninvasiveness of the control force. By exchange of stability with a delay-induced periodic orbit generated in a saddle-node bifurcation, the subcritical orbit transferred its instability to the trivial fixed point via transcritical and a reverse supercritical Hopf bifurcation. The stability of the subcritical odd number orbit was determined from its eigenvalue spectra.

In addition, I have elaborated that also periodic orbits already present in the uncontrolled system can be employed for this exchange of stability. This was realized for a generic model of a fold bifurcation. Similar to the case of a subcritical Hopf bifurcation, I have investigated the domains of control as well as periods and radii of rotating wave solutions in the systems with time-delayed feedback.

From the perspective of applications, these findings are encouraging to realize, e.g., in optical systems. In these systems, the feedback phase is readily accessible and can be adjusted, for instance, in laser systems, where subcritical Hopf bifurcation scenarios are abundant and Pyragas control can be realized via coupling to an external Fabry-Perot resonator in order to stabilize a continuous-wave output [32] or periodic oscillations of the intensity in the GHz frequency range [34]. The feedback phase becomes also important for the stabilization of steady states in the context of neural systems [35], for stabilization of periodic orbits by a time-delayed feedback control scheme using spatio-temporal filtering [36], and for tracking of unstable orbits and bifurcation analysis using time-delayed feedback control [37]. These examples open up fundamental questions as well as a wide range of applications.

References

1. Pyragas K (1992) Continuous control of chaos by self-controlling feedback. Phys Lett A 170:421
2. Nakajima H (1997) On analytical properties of delayed feedback control of chaos. Phys Lett A 232:207
3. Nakajima H, Ueda Y (1998) Limitation of generalized delayed feedback control. Physica D 111:143

4. Fiedler B, Flunkert V, Georgi M, Hövel P, Schöll E (2007) Refuting the odd number limitation of time-delayed feedback control. Phys Rev Lett 98:114101
5. Fiedler B, Flunkert V, Georgi M, Hövel P, Schöll E (2008) Beyond the odd number limitation of time-delayed feedback control. In: Schöll E, Schuster HG (eds) Handbook of chaos control. Wiley-VCH, Weinheim, pp 73–84 (second completely revised and enlarged edition)
6. Fiedler B, Flunkert V, Georgi M, Hövel P, Schöll E (2008) Delay stabilization of rotating waves without odd number limitation. In: Schuster HG (ed) Reviews of nonlinear dynamics and complexity, vol 1. Wiley-VCH, Weinheim, pp 53–68
7. Just W, Fiedler B, Flunkert V, Georgi M, Hövel P, Schöll E (2007) Beyond odd number limitation: a bifurcation analysis of time-delayed feedback control. Phys Rev E 76:026210
8. Fiedler B, Yanchuk S, Flunkert V, Hövel P, Wünsche HJ, Schöll E (2008) Delay stabilization of rotating waves near fold bifurcation and application to all-optical control of a semiconductor laser. Phys Rev E 77:066207
9. Socolar JES, Sukow DW, Gauthier DJ (1994) Stabilizing unstable periodic orbits in fast dynamical systems. Phys Rev E 50:3245
10. Pyragas K (2001) Control of chaos via an unstable delayed feedback controller. Phys Rev Lett 86:2265
11. Pyragas K (2003) Time-delayed feedback control method and unstable controllers. In: Physics and control, 2003. Proceedings 2003 international conference, vol 2, p 456
12. Pyragas K, Pyragas V, Kiss IZ, Hudson JL (2004) Adaptive control of unknown unstable steady states of dynamical systems. Phys Rev E 70:026215
13. Tamaševičius A, Mykolaitis G, Pyragas V, Pyragas K (2007) Delayed feedback control of periodic orbits without torsion in nonautonomous chaotic systems: theory and experiment. Phys Rev E 76:26203
14. Höhne K, Shirahama H, Choe C-U, Benner H, Pyragas K, Just W (2007) Global properties in an experimental realization of time-delayed feedback control with an unstable control loop. Phys Rev Lett 98:214102
15. Pyragas K, Tamaševičius A (1993) Experimental control of chaos by delayed self-controlling feedback. Phys Lett A 180:99
16. Schuster HG, Stemmler MB (1997) Control of chaos by oscillating feedback. Phys Rev E 56:6410
17. Sivashinsky GI, Michelson DM (1980) On irregular wavy flow of a liquid film down a vertical plane. Prog Theor Phys 63:2112
18. Nakajima H, Ueda Y (1998) Half-period delayed feedback control for dynamical systems with symmetries. Phys Rev E 58:1757
19. Lorenz EN (1963) Deterministic nonperiodic flow. J Atmos Sci 20:130
20. Gjurchinovski A, Urumov V (2008) Stabilization of unstable steady states by variable delay feedback control. Europhys Lett 84:40013
21. Fiedler B (1988) Global bifurcation of periodic solutions with symmetry. Springer, Heidelberg
22. Diekmann O, van Gils SA, Verduyn Lunel SM, Walther HO (1995) Delay equations. Springer, New York
23. Vanderbauwhede A (1989) Centre manifolds, normal forms and elementary bifurcations. Dyn Rep 2:89
24. Fiedler B, Turaev D (1998) Normal forms, resonances, and meandering tip motions near relative equilibria of Euclidean group actions. Arch Ration Mech Anal 145:129
25. Fiedler B, Georgi M, Jangle N (2007) Spiral wave dynamics: reacion and diffusion versus kinematics. In: Schimansky-Geier L, Fiedler B, Kurths J, Schöll E (eds) Analysis and control of complex nonlinear processes in physics, chemistry and biology. World Scientific, Singapore, pp 69–114
26. Fiedler B, Sandstede B, Scheel A, Wulf C (1996) Bifurcation from relative equilibria of noncompact group actions: skew products, meanders, and drifts. Documenta Math 1:479

27. Fiedler B, Scheel A (2002) Dynamics of reaction–diffusion patterns. In: Kirkilionis M, Rannacher R, Tomi F (eds) Trends in nonlinear analysis, Festschrift dedicated to Willi Jäger for his 60th birthday. Springer, Heidelberg
28. Fiedler B, Georgi M, Jangle N (2007) Spiral wave dynamics: reaction and diffusion versus kinematics. In: Analysis and control of complex nonlinear processes in physics, chemistry and biology. World Scientific, Singapore
29. Golubitsky M, Stewart I (1988) Singularities and groups in bifurcation theory, vol 2. Applied mathematical sciences, vol 69. Springer, New York
30. Bauer S, Brox O, Kreissl J, Sartorius B, Radziunas M, Sieber J, Wünsche HJ, Henneberger F (2004) Nonlinear dynamics of semiconductor lasers with active optical feedback. Phys Rev E 69:016206
31. Wieczorek S, Krauskopf B, Lenstra D (1999) Unifying view of bifurcations in a semiconductor laser subject to optical injection. Opt Commun 172:279
32. Schikora S, Hövel P, Wünsche HJ, Schöll E, Henneberger F (2006) All-optical noninvasive control of unstable steady states in a semiconductor laser. Phys Rev Lett 97:213902
33. Wünsche HJ, Schikora S, Henneberger F (2008) Noninvasive control of semiconductor lasers by delayed optical feedback. In: Schöll E, Schuster HG (eds) Handbook of chaos control. Wiley-VCH, Weinheim (second completely revised and enlarged edition)
34. Schikora S, Wünsche HJ, Henneberger F (2008) All-optical noninvasive chaos control of a semiconductor laser. Phys Rev E 78:025202
35. Rosenblum MG, Pikovsky AS (2004) Delayed feedback control of collective synchrony: an approach to suppression of pathological brain rhythms. Phys Rev E 70:041904
36. Baba N, Amann A, Schöll E, Just W (2002) Giant improvement of time-delayed feedback control by spatio-temporal filtering. Phys Rev Lett 89:074101
37. Sieber J (2006) Dynamics of delayed relay systems. Nonlinearity 19:2489

Chapter 5
Control of Neutral Delay-Differential Equations

Qu'est-ce que le passé, sinon du présent qui est en retard.[1]

(Pierre Cas)

The topic of this chapter is the control of steady states which arise in neutral systems by means of time-delayed feedback. The crucial difference from systems of ordinary differential equations as considered in the previous chapters is that the uncontrolled system itself contains already delayed variables. Thus, the phase space is infinite dimensional even in the absence of time-delayed feedback and the initial conditions need to be specified over a time interval $[t_0 - \tau, t_0]$, where τ denotes the time delay. To complicate things even further, neutral systems contain the delayed variable in the highest derivative. The numerical treatment and bifurcation analysis of such equations are much more involved than those of regular delay differential equations. For instance, the existing packages for bifurcation analysis of delay differential equations, such as DDE-BIFTOOL [1] and PDDE-CONT [2] currently are unable to perform computations for neutral systems.

This type of systems arise in a variety of physical contexts. Balanov et al. [3], for instance, derived a neutral delay differential equation as a model for torsional waves on a driven drill-string. Another example studied by Blakely and Corron [4] is a model of a chaotic transmission line oscillator, in which a neutral delay differential equations was used to correctly reproduce experimental observations of fast chaotic dynamics. They are also considered in the area of population dynamics [5–7].

Neutral delay-differential equations arise naturally in the context of substructuring and hybrid testing, when part of the dynamics of an experiment is replaced by numerical model [8, 9]. These simulations calculate supplemental forces which

[1] What is past, if it is not the present which is late.

P. Hövel, *Control of Complex Nonlinear Systems with Delay*, Springer Theses, DOI: 10.1007/978-3-642-14110-2_5, are applied to the experimental. In this context, time delays enter the dynamic equations due to latency effects in the interaction between the experimental and numerical part of the system. A mechanical example will be presented in Sect. 5.1.

This chapter is based on Refs. [10, 11] and is organized as follows: In the next Sect. 5.1, I will give an example of substructuring, where neutral delay-differential equations can be found. Section 5.2 is devoted to the introduction of the model which will be used throughout this chapter. I will discuss the domain of stability of the steady state depending on the system's parameters which include already a time delay. Section 5.3 deals with asymptotic properties in the case of large time delays. Time-delayed feedback control will be applied to the system in Sect. 5.4. The center of investigations is the interplay of the different time delays, i.e., the system's delay and the control delay. The chapter concludes with an intermediate summary.

5.1 Substructuring or Hybrid Testing

Neutral delay differential equations are also important in the field of real-time dynamic substructuring or hybrid testing [8, 9]. See, for instance, the example based on Ref. [12] presented in Eqs. 5.1.1a and b and schematically depicted in Fig. 5.1. In this technique, part of a complex structure is transferred to numerical computations whose results are applied back to the remaining complex structure, e.g., by a number of attenuators. This is a helpful procedure if the structure under investigation is large as often considered in civil and aerospace engineering.

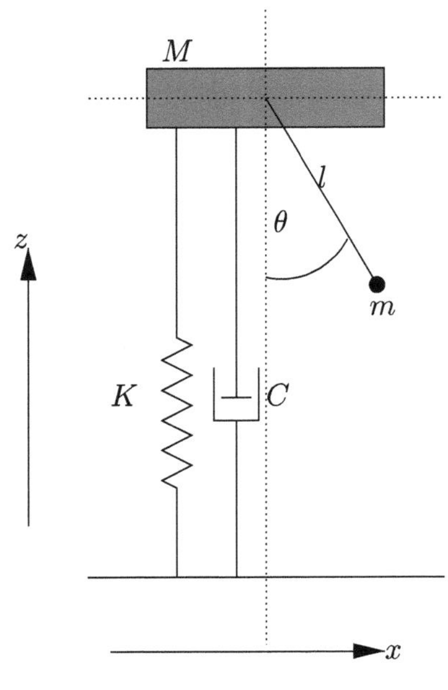

Fig. 5.1 Schematic diagram of the mass-spring-damper-pendulum system described by Eqs. 5.1.1a and b

Delay effects need to be taken into account due to the non-instantaneous transfer system. The delays can be related to a variety of sources such as measurement of the remaining complex structure, transmission to the computational unit, numerical calculations, or application of the numerical results back to the structure. All these sources lead to a total delay [8, 9, 12].

For an example consider the system depicted in Fig. 5.1, which is described by the following equations [12]

$$\begin{aligned} F_{\text{ext}} = {} & M\frac{d^2 z_M(t)}{dt^2} + C\frac{dz_M(t)}{dt} + Kz_M(t) + m\frac{d^2 z_M(t)}{dt^2} \\ & + ml\left[\frac{d^2\theta(t)}{dt^2}\sin\theta(t) - \left(\frac{d\theta(t)}{dt}\right)^2 \cos\theta(t)\right] \end{aligned} \quad (5.1.1a)$$

$$0 = ml^2\frac{d^2\theta(t)}{dt^2} + ml\frac{dz(t)}{dt}\frac{d\theta(t)}{dt}\cos\theta(t) + mgl\sin\theta(t) + ml\frac{d^2 z_M(t)}{dt^2}\sin\theta(t). \quad (5.1.1b)$$

This system consists of a pendulum with mass m and length l which is attached to second mass M. The coordinate $z_M(t)$ denotes the vertical position of the mass M and the angle θ is the angular deflection of the pendulum from the downward position. The mass M is in turn mounted on a spring with a stiffness coefficient K and can only move up and down. In addition, the mass M is subject to viscous damping with damping coefficient C. F_{ext} denotes an external excitation force which could be, for instance, given by a periodic function.

The system's equations 5.1.1a and 5.1.1b are not obvious from the experimental setup. Therefore, I will present a detailed derivation in the following.

The Lagrange function $\mathcal{L}$ of the system is given by

$$\mathcal{L} = \frac{M}{2}\dot{z}_M^2 + \frac{m}{2}(\dot{z}_m^2 + \dot{x}_m^2) - \frac{K}{2}z_M^2 - mgz_m, \quad (5.1.2)$$

where z_M and z_m, x_m denote the positions of the masses M and m, respectively. For notational convenience, the time derivatives are written as, e.g., $\dot{z}$. Using generalized coordinates

$$z_m = z_M - l\cos\theta \quad \text{and} \quad x_m = l\sin\theta \quad (5.1.3)$$

yields

$$\mathcal{L} = \frac{M}{2}\dot{z}_M^2 + \frac{m}{2}(\dot{z}_m^2 + \dot{x}_m^2) - \frac{K}{2}z_M^2 - mgz_m \quad (5.1.4a)$$

$$= \frac{M}{2}\dot{z}_M^2 + \frac{m}{2}\left(\dot{z}_M^2 + l^2\dot{\theta}^2 + 2l\dot{z}_M\dot{\theta}\sin\theta\right) - \frac{K}{2}z_M^2 + mgl\cos\theta. \quad (5.1.4b)$$

The equations of motion (5.1.1a and 5.1.1b) for z_M and θ are given by the respective Lagrange's equation

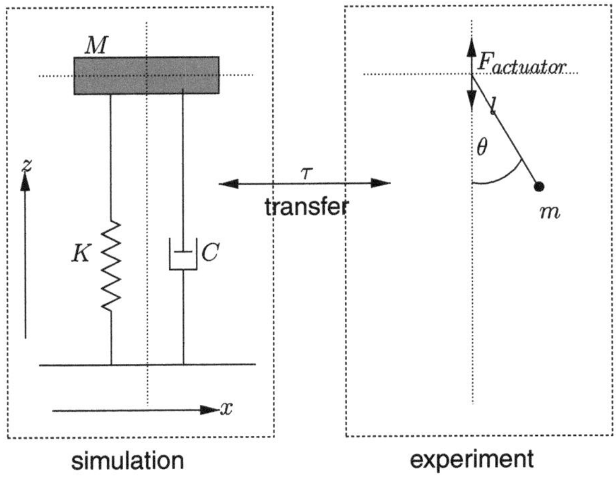

Fig. 5.2 Schematic diagram of the hybrid system described by Eqs. 5.1.8a and b

$$0 = \frac{d}{dt}\left(\frac{\partial \mathcal{L}}{\partial \dot{q}_i}\right) - \frac{\partial \mathcal{L}}{\partial q_i} - \mathcal{R}_i \tag{5.1.5}$$

with $q_i \in \{z_M, \theta\}$ and additional friction terms $\mathcal{R}_i$, which acts only on the mass M as $\mathcal{R}_M = -C \cot z_M$. This leads to the following equation of motion for the mass M

$$0 = \frac{d}{dt}\left(\frac{\partial \mathcal{L}}{\partial \dot{z}_M}\right) - \frac{\partial \mathcal{L}}{\partial z_M} - \mathcal{R}_M \tag{5.1.6a}$$

$$= M\ddot{z}_M + m\ddot{z}_M + ml\ddot{\theta}\sin\theta + ml\dot{\theta}^2\cos\theta + Kz_M + C\dot{z}_M. \tag{5.1.6b}$$

Equivalently, the equations for the angle $\theta(t)$ are given by

$$0 = \frac{d}{dt}\left(\frac{\partial \mathcal{L}}{\partial \dot{\theta}}\right) - \frac{\partial \mathcal{L}}{\partial \theta} \tag{5.1.7a}$$

$$= ml^2\ddot{\theta} + ml\ddot{z}_M\sin\theta + ml\dot{z}_M\dot{\theta}\cos\theta + mgl\sin\theta. \tag{5.1.7b}$$

An additional external force F_{ext} acting on the mass M in the z-direction yields Eqs. 5.1.1a and b.

As a next step, part of the system depicted is replaced by a numerical model as depicted in Fig. 5.2. For instance, the motion of the mass M can be transferred to computation while the pendulum remains in a physical experiment. For this substructuring, suspension point of the pendulum is time-dependent and its position shifted vertically by an actuator by a force F_{actuator} according to the result of the numerics calculating the motion of M. In this numerical simulation, information of the position of the pendulum is measured and transferred to the computer which results in an unavoidable delay in the respective coordinates. Therefore, Eqs. 5.1.1a and b need to be adjusted such that the delay is taken into account:

$$\begin{aligned} F_{\text{ext}} &= M\frac{d^2 z_M(t)}{dt^2} + C\frac{dz_M(t)}{dt} + Kz_M(t) + m\frac{d^2 z_M(t-\tau)}{dt^2} \\ &\quad + ml\left[\frac{d^2\theta(t-\tau)}{dt^2}\sin\theta(t-\tau) - \left(\frac{d\theta(t-\tau)}{dt}\right)^2\cos\theta(t-\tau)\right] \end{aligned} \tag{5.1.8a}$$

$$\begin{aligned} 0 &= ml^2\frac{d^2\theta(t-\tau)}{dt^2} + ml\frac{dz(t-\tau)}{dt}\frac{d\theta(t-\tau)}{dt}\cos\theta(t-\tau) \\ &\quad + mgl\sin\theta(t-\tau) + ml\frac{d^2 z_M(t-\tau)}{dt^2}\sin\theta(t-\tau). \end{aligned} \tag{5.1.8b}$$

Finally, the goal of hybrid testing is to reproduce by the combination of numerical and reduced experimental system the dynamics of the complete experiment as closely as possible. A key question is, for instance, the stability of the model which will be addressed in this chapter.

For simplification, I will consider in the following the case of small angles θ. Then, the Eqs. 5.1.8a, b decouple and Eq. 5.1.8a becomes in the absence of external forcing [12]

$$0 = M\frac{d^2 z_M(t)}{dt^2} + C\frac{dz_M(t)}{dt} + Kz_M(t) + m\frac{d^2 z_M(t-\tau)}{dt^2}. \tag{5.1.9}$$

The investigation of this equations will be the subject of the subsequent Sections of this chapter.

5.2 Model Equations

As a paradigm for the class of neutral delay differential equations, I consider the following model [10]

$$\frac{d^2 z(t)}{dt^2} + 2\zeta\frac{dz(t)}{dt} + z(t) + p\frac{d^2 z(t-\tau)}{dt^2} = 0. \tag{5.2.1}$$

This model was derived in Sect. 5.1 and results from Eq. 5.1.9 by rescaling of parameters which yields non-dimensional variables according to

$$\hat{t} = \omega t, \hat{\tau} = \omega\tau, \omega = \sqrt{\frac{K}{M}}, p = \frac{m}{M}, \zeta = \frac{C}{2\sqrt{MK}}. \tag{5.2.2}$$

This can be seen by inserting into Eq. 5.1.9

$$0 = M\frac{d^2 z_M(t)}{dt^2} + C\frac{dz_M(t)}{dt} + Kz_M(t) + m\frac{d^2 z_M(t-\tau)}{dt^2} \tag{5.2.3a}$$

$$\Leftrightarrow \quad 0 = M\omega^2\frac{d^2 z_M(\hat{t})}{d\hat{t}^2} + C\omega\frac{dz_M(\hat{t})}{d\hat{t}} + Kz_M(\hat{t}) + m\omega^2\frac{d^2 z_M(\hat{t}-\hat{\tau})}{dt^2} \tag{5.2.3b}$$

$$\Leftrightarrow \quad 0 = \frac{M\omega^2}{K}\frac{d^2 z_M(\hat{t})}{d\hat{t}^2} + \frac{C\omega}{K}\frac{dz_M(\hat{t})}{d\hat{t}} + z_M(\hat{t}) + \frac{m\omega^2}{K}\frac{d^2 z_M(\hat{t}-\hat{\tau})}{d\hat{t}^2} \tag{5.2.3c}$$

$$\Leftrightarrow \quad 0 = \frac{d^2 z(\hat{t})}{d\hat{t}^2} + 2\zeta\frac{dz(\hat{t})}{d\hat{t}} + z(\hat{t}) + p\frac{d^2 z(\hat{t}-\hat{\tau})}{d\hat{t}^2} = 0 \tag{5.2.3d}$$

which is identical to Eq. 5.2.1. The parameter ζ denotes a rescaled damping rate. The delayed acceleration term $pd^2z(t - \tau)/dt^2$ can be interpreted as a delayed force acting on the system. The model above was first introduced by Kyrychko et al. [12] in the context of hybrid testing. There it proved to be a good physical model for mass-spring-damper-systems taking into account actuator delays. In this context the parameter p denotes the mass ration between the pendulum and the spring. See also Sect. 5.1 for a detailed analysis.

This nonlinear system consists of a pendulum connected to a mechanical oscillator. When the hybrid testing method is applied, the oscillator dynamics is realized by numerical calculations and information of the pendulums position is transferred by an actuator.

One could also imagine to include a time delay in the first derivative. This would lead to delayed damping or a delayed velocity term which will be considered later in this chapter.

Note that there exists an equivalent form of Eq. 5.2.1 which is derived in the following. Introducing $v(t) = \dot{z}(t)$ and $u(t) = v(t) + pv(t - \tau)$, this equation can be rewritten as a system of differential equations with a shift:

$$\frac{dz(t)}{dt} = u(t) - pv(t-\tau) \tag{5.2.4a}$$

$$\frac{du(t)}{dt} = \frac{d^2z(t)}{dt^2} + p\frac{d^2z(t-\tau)}{dt^2} \tag{5.2.4b}$$

$$= -2\zeta\frac{dz(t)}{dt} - z(t) - p\frac{d^2z(t-\tau)}{dt^2} + p\frac{d^2z(t-\tau)}{dt^2} \tag{5.2.4c}$$

$$= -2\zeta[u(t) - pv(t-\tau)] - z(t) \tag{5.2.4d}$$

$$v(t) = u(t) - pv(t-\tau). \tag{5.2.4e}$$

With initial data $(z(0), u(0)) = (z_0, u_0) \in \mathbb{R} \times \mathbb{R}$ and $v(s) = \phi(s) \in C[-\tau, 0]$, this system can be first solved on $0 \leq t \leq \tau$ interval, then on $\tau \leq t \leq 2\tau$ and so on, provided the following sewing condition is satisfied: $\phi(0) = u_0 - p\phi(-\tau)$. This condition ensures that there are no discontinuities in the solutions at $t = k\tau, k \in \mathbb{Z}$. For arbitrary initial conditions the sewing condition does not hold, and leads to jumps in the derivative of the solution [13].

At this point, I will investigate the stability of the steady state $z^* = 0$ of Eq. 5.2.1. As mentioned in the previous chapters, the stability of a fixed point is determined by the largest real part of the eigenvalues $\lambda \in \mathbb{C}$. These eigenvalues are

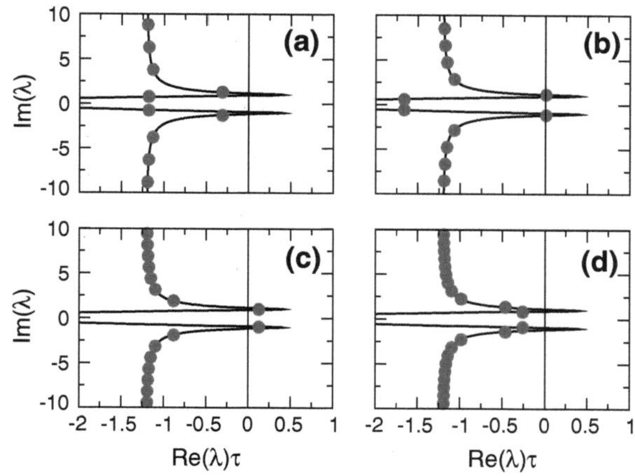

Fig. 5.3 Eigenvalue spectrum of the characteristic equation (5.2.5) for different time delays: **a** $\tau = 2.5$, **b** $\tau = 3.32$, **c** $\tau = 5$, and **d** $\tau = 7$. Parameter values are: $\zeta = 0.1$, $p = 0.3$. The *solid lines* show the asymptotic pseudocontinuous spectrum given by Eq. 5.3.7a of Sect. 5.3

given by the complex roots of the characteristic equation. Using an exponential ansatz $z(t) \sim \exp(\lambda\, t)$ yields the following equation

$$\lambda^2 + 2\zeta\lambda + 1 + p\lambda^2 e^{-\lambda\tau} = 0. \qquad (5.2.5)$$

The solutions of this transcendental equation in terms of λ cannot be derived analytically, for instance, by the Lambert function due to the mixed occurrence of exponential and polynomial terms [14, 15]. Therefore Eq. 5.2.5 must be solved numerically.

Figure 5.3 depicts the eigenvalue spectrum in the complex plane for different time delays τ. Panels (a), (b), (c), and (d) correspond to $\tau = 2.5$, 3.32, 5, and 7, respectively. It can be observed that for small time delays as in Fig. 5.3a the steady state is stable, as all the eigenvalues are in the left half-plane. As time delay increases, a pair of complex conjugate eigenvalues crosses the imaginary axis, as demonstrated in Fig. 5.3b, leading to an instability shown in Fig. 5.3c. For further increase of the time delay, the unstable eigenvalues return to the left half-plane, thus restoring the stability. Note that all eigenvalues lie on a set of two curves. The formula for this curve will be derived later as asymptotic pseudocontinuous spectrum.

At this point, let me discuss the domain of stability in the (τ, p) plane. Asymptotic properties for large delays will be investigated in Sect. 5.3.

To find an exact analytical expression for the stability boundary, one has to consider a vanishing real part of the eigenvalue λ and thus substitute $\lambda = i\omega$ into the characteristic equation (5.2.5). After separating real and imaginary parts, this gives

$$1 - \omega^2 - p\omega^2 \cos(\omega\tau) = 0 \qquad (5.2.6a)$$

$$2\zeta\omega + p\omega^2 \sin(\omega\tau) = 0. \qquad (5.2.6b)$$

Squaring and adding these equations gives

$$\left(1-\omega^2\right)^2 + 4\zeta^2\omega^2 = p^2\omega^4 \tag{5.2.7a}$$

$$\Leftrightarrow \quad (1-p^2)\omega^4 + (4\zeta^2-2)\omega^2 + 1 = 0. \tag{5.2.7b}$$

The last equation can be solved as

$$0 = \omega^4 + 2\frac{2\zeta^2-1}{1-p^2}\omega^2 + \frac{1}{1-p^2} \tag{5.2.8a}$$

$$\Rightarrow \quad \omega_{1,2}^2 = \frac{1}{1-p^2}\left[1-2\zeta^2 \pm \sqrt{\left(1-2\zeta^2\right)^2 - 1 + p^2}\right]. \tag{5.2.8b}$$

In fact, Eq. 5.2.7b provides an expression for stability boundary value of p as parameterized by the Hopf frequency ω:

$$0 = (1-p^2)\omega^4 + (4\zeta^2-2)\omega^2 + 1 \tag{5.2.9a}$$

$$\Leftrightarrow \quad p(\omega) = \frac{1}{\omega^2}\sqrt{\omega^4 + 2\omega^2(2\zeta^2-1) + 1}. \tag{5.2.9b}$$

The corresponding value of the time delay at the stability boundary can be derived from Eqs. 5.2.6a and b

$$1-\omega^2 = p\omega^2\cos(\omega\tau), \quad 2\zeta\omega = p\omega^2\sin(\omega\tau) \tag{5.2.10a}$$

$$\Leftrightarrow \quad \frac{2\zeta\omega}{1-\omega^2} = \tan(\omega\tau) \tag{5.2.10b}$$

$$\Rightarrow \quad \tau(\omega) = \frac{1}{\omega}\left[\arctan\frac{2\zeta\omega}{\omega^2-1} \pm \pi k\right], \tag{5.2.10c}$$

where $k = 0, 1, 2, \ldots$

To summarize previous derivations, Eqs. 5.2.9b and 5.2.10c are parametric formulas for the boundary of stability in dependence on the Hopf frequency ω. Figure 5.4 illustrates this dependence of critical mass ratio p on the time delay τ which ensures the stability of the steady state below the curves in the yellow shaded area. It is noteworthy that if $|p| > 1$, the steady state is unstable for any positive time delay τ; on the other hand, if $|p| < 1$ and $\zeta > 1/\sqrt{2}$, then the steady state is asymptotically stable for any positive time delay τ. For $\zeta < 1/\sqrt{2}$, there is a lower bound on the value of $p_{\min} = 2\zeta\sqrt{1-\zeta^2}$, so that an asymptotic stability is guaranteed for all $\tau > 0$ provided $p < p_{\min}$ [12]. This lower bound will be derived in Sect. 5.3.

If one includes also a delayed viscous damping term to the hybrid system depicted in Fig. 5.2 of Sect. 5.1, one can study the influence of velocity feedback on the stability of neutral delay differential equations. The modified system's equations read

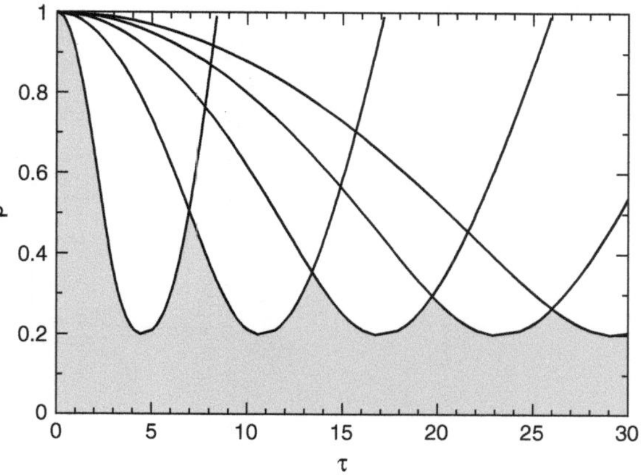

Fig. 5.4 Exact stability boundary of the characteristic equation (5.2.5) in the (τ, p) parameter plane for $\zeta = 0.1$. The steady state is stable in the *yellow* (*shaded*) *area*. The curves correspond to the parametric formulas of Eqs. 5.2.9b and 5.2.10c for $p(\omega)$ and $\tau(\omega)$, respectively

$$\frac{d^2z(t)}{dt^2} + 2\zeta_1\frac{dz(t)}{dt} + z(t) + p\frac{d^2z(t-\tau)}{dt^2} + 2\zeta_2\frac{dz(t-\tau)}{dt} = 0, \tag{5.2.11}$$

where ζ_1 and ζ_2 denote the damping rates of the undelayed and delayed velocities, respectively. This equation was introduced in Ref. [12], where it was shown that depending on the difference between two damping parameters ζ_1 and ζ_2, the stability domain may shrink and even split into separate stability regions in the parameter plane which are called death islands since the oscillations decay.

The stability of the steady state $z^* = 0$ of the extended system (5.2.11) can be inferred by the solution of the characteristic equation which reads

$$\lambda^2 + 2\zeta_1\lambda + 1 + p\lambda^2 e^{-\lambda\tau} + 2\zeta_2\lambda e^{-\lambda\tau} = 0. \tag{5.2.12}$$

Figure 5.5 depicts the computed eigenvalue spectrum according to Eq. 5.2.12 for a time delay of $\tau = 2$, 2.725, 3.5, and 5 in panels (a), (b), (c), and (d), respectively. The parameters are chosen as $\zeta_1 = 0.25$, $\zeta_2 = 0.24$, and $p = 0.3$. From this figure it follows that similar to the situation without velocity feedback as shown in Fig. 5.3, the system undergoes successive stability switches as the time delay is varied. For small τ the fixed point at the origin is stable. See Fig. 5.5a where all eigenvalues have negative real part. As the time delay increases, the fixed point looses eventually its stability as can be seen in Fig. 5.5b and c. For even larger values of τ the steady state can regain its stability as all the eigenvalues lie again in the left half-plane, i.e., $\mathrm{Re}(\lambda) < 0$ as indicated in Fig. 5.5d. Compare also the case without velocity feedback of Fig. 5.3. Similar to this case, on can calculate the domain of stability analytically from the characteristic equation (5.2.12).

At the points, where the stability changes, the eigenvalues are purely imaginary, i.e., $\lambda = i\omega$, and the real and imaginary part of the characteristic equation can written as

$$1 - \omega^2 - p\omega^2 \cos\omega\tau + 2\zeta_2\omega \sin\omega\tau = 0 \tag{5.2.13a}$$

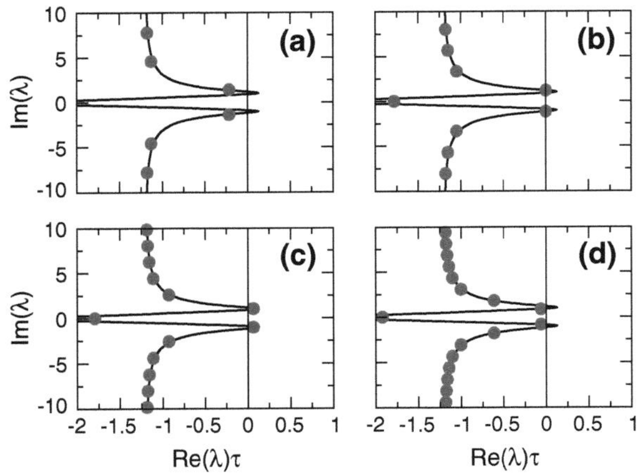

Fig. 5.5 Eigenvalue spectrum of the characteristic equation (5.2.12) for different time delays: **a** $\tau = 2$, **b** $\tau = 2.725$, **c** $\tau = 3.5$, and **d** $\tau = 5$. Parameter values are: $\zeta_1 = 0.25$, $\zeta_2 = 0.24$, $p = 0.3$. The *solid lines* show the asymptotic pseudocontinuous spectrum given by Eq. 5.3.19 of Sect. 5.3

$$2\omega\zeta_1 + p\omega^2 \sin\omega\tau + 2\omega\zeta_2 \cos\omega\tau = 0. \tag{5.2.13b}$$

Squaring these two equations gives

$$\left(1 - \omega^2\right)^2 = \left(p\omega^2 \cos\omega\tau - 2\zeta_2\omega \sin\omega\tau\right)^2 \tag{5.2.14a}$$

$$\left(2\zeta_1\omega\right)^2 = \left(-p\omega^2 \sin\omega\tau - 2\zeta_2\omega\cos\omega\tau\right)^2. \tag{5.2.14b}$$

Adding the squared equations of the real and imaginary parts yields

$$\left(1 - \omega^2\right)^2 + 4\zeta_1^2\omega^2 = p^2\omega^4 + 4\zeta_2^2\omega^2 \tag{5.2.15a}$$

$$\Leftrightarrow \quad \left(1 - p^2\right)\omega^4 + \left(4\zeta_1^2 - 4\zeta_2^2 - 2\right)\omega^2 + 1 = 0 \tag{5.2.15b}$$

$$\Leftrightarrow \quad \omega^4 + 2\frac{\left(2\zeta_1^2 - 2\zeta_2^2 - 1\right)}{1 - p^2}\omega^2 + \frac{1}{1 - p^2} = 0 \tag{5.2.15c}$$

such that the following parameterization of p the Hopf frequency in dependence on ω holds:

$$\omega_{1,2}^2 = \frac{1}{1 - p^2}\left[1 - 2\zeta_1^2 + 2\zeta_2^2 \pm \sqrt{\left(1 - 2\zeta_1^2 + 2\zeta_2^2\right)^2 - 1 + p^2}\right]. \tag{5.2.16}$$

Similar to the previous case, one can derive parametric expressions for $p(\omega)$ and $\tau(\omega)$ from Eqs. 5.2.13a, b:

$$0 = \left(1 - p^2\right)\omega^4 + \left(4\zeta_1^2 - 4\zeta_2^2 - 2\right)\omega^2 + 1 \tag{5.2.17a}$$

$$= -p^2\omega^4 + \omega^4 - 2\omega^2 + 1 + 4\left(\zeta_1^2 - \zeta_2^2\right)\omega^2 \tag{5.2.17b}$$

$$\Rightarrow \quad p(\omega) = \frac{1}{\omega^2}\sqrt{\left(\omega^2 - 1\right)^2 + 4\left(\zeta_1^2 - \zeta_2^2\right)\omega^2}. \tag{5.2.17c}$$

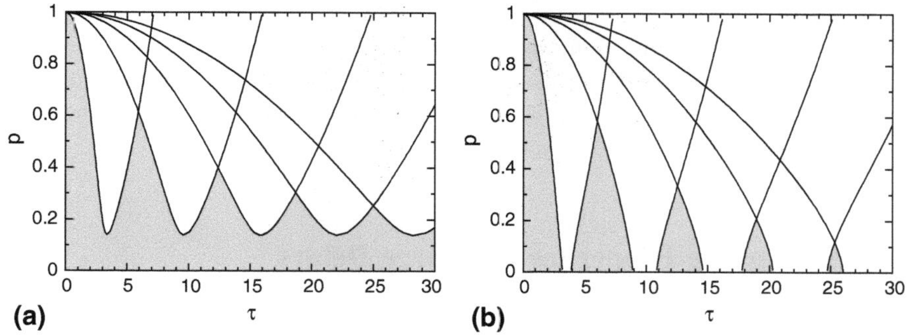

Fig. 5.6 Exact stability boundary of the characteristic equation (5.2.12) in the (τ, p) parameter plane. The steady state is stable in the *yellow* (*shaded*) *area*. Parameter values are: **a** $\zeta_1 = 0.25$ and $\zeta_2 = 0.24$, **b** $\zeta_1 = 0.23$ and $\zeta_2 = 0.25$

The corresponding value of the time delay at the stability boundary can be derived similar to Eq. 5.2.10c. Starting from the real part of the characteristic equation (5.2.13a) leads to the following expression

$$\sin \omega\tau = \frac{1}{2\zeta_2\omega}\left(\omega^2 - 1 + p\omega^2 \cos \omega\tau\right) \qquad (5.2.18)$$

which can be inserted into the imaginary part of the characteristic equation given by Eq. 5.2.13b

$$2\zeta_1 + \frac{p\omega}{2\zeta_2\omega}\left(\omega^2 - 1 + p\omega^2 \cos \omega\tau\right) + 2\zeta_2 \cos \omega\tau = 0 \qquad (5.2.19a)$$

$$\Leftrightarrow \quad 4\zeta_1\zeta_2 + p\left(\omega^2 - 1 + p\omega^2 \cos \omega\tau\right) + 4\zeta_2^2 \cos \omega\tau = 0 \qquad (5.2.19b)$$

$$\Leftrightarrow \quad 4\zeta_1\zeta_2 + p\left(\omega^2 - 1\right) + \cos \omega\tau(4\zeta_2^2 + p^2\omega^2) = 0. \qquad (5.2.19c)$$

Solving for τ yields the value of the time delay at the stability boundary:

$$\tau(\omega) = \frac{1}{\omega}\left[2\pi n - \arccos \frac{p(1-\omega^2) - 4\zeta_1\zeta_2}{p^2\omega^2 + 4\zeta_2^2}\right], \quad n = 1, 2, 3, \ldots \qquad (5.2.20)$$

Figure 5.6 depicts the parametric dependence of the critical mass ration p and the time delay τ on the Hopf frequency ω according to Eqs. 5.2.17c and 5.2.20 for two different combinations of ζ_1 and ζ_2. The steady state is stable below the curves in the yellow shaded areas. Note how the stability boundary which is given as yellow area is affected by the relation between ζ_1 and ζ_2. In particular, it can be stated that when $\zeta_1 = \zeta_2$ the stability boundary touches the τ axis ($p = 0$), and for $\zeta_2 > \zeta_1$, the stability area consists of non-overlapping death islands, inside which the oscillations are damped, and the steady state is stable.

To understand the dynamics of the system in the neighborhood of these stability changes, one can use the framework of pseudocontinuous spectrum proposed by Yanchuk et al. [16] for the analysis of scaling behavior of eigenvalues for large time delays in ordinary differential subject to time-delayed feedback control. This will be elaborated in the next section.

5.3 Asymptotic Properties for Large Delays

Following the approach of Sect. 3.6 and Ref. [16] for the case of ordinary differential equations with time-delayed feedback, one can express the asymptotic approximation of the eigenvalues for large τ as

$$\lambda = \frac{1}{\tau}\gamma + i\left(\Omega + \frac{1}{\tau}\phi\right) + \mathcal{O}\left(\frac{1}{\tau^2}\right), \tag{5.3.1}$$

where γ, Ω, and ϕ are real-valued quantities, which are associated with the real and imaginary part of the eigenvalue λ, respectively. Substituting this representation into the characteristic equation (5.2.5), yields

$$0 = \lambda^2 + 2\zeta\lambda + 1 + p\lambda^2 e^{-\lambda\tau} \tag{5.3.2a}$$

$$\begin{aligned} &= \left[\frac{1}{\tau}\gamma + i\left(\Omega + \frac{1}{\tau}\phi\right)\right]^2 + 2\zeta\left[\frac{1}{\tau}\gamma + i\left(\Omega + \frac{1}{\tau}\phi\right)\right] + 1 \\ &+ p\left[\frac{1}{\tau}\gamma + i\left(\Omega + \frac{1}{\tau}\phi\right)\right]^2 \exp\left[\frac{1}{\tau}\gamma + i\left(\Omega + \frac{1}{\tau}\phi\right)\right] + \mathcal{O}\left(\frac{1}{\tau^2}\right). \end{aligned} \tag{5.3.2b}$$

This gives to the leading order in $\mathcal{O}(1/\tau)$

$$1 - \Omega^2 + 2i\zeta\Omega - p\Omega^2 e^{-\gamma} e^{-i\phi} = 0 \tag{5.3.3}$$

with an additional constraint $\Omega = \Omega^{(n)} = 2\pi n/\tau$, $n = \pm 1, \pm 2, \pm 3, \ldots$ To derive an expression for the real part γ of the eigenvalue as a function of the Hopf frequency Ω it is helpful to separate Eq. 5.3.3 into real and imaginary part

$$0 = 1 - \Omega^2 - p\omega^2 e^{-\gamma} \cos\phi \tag{5.3.4a}$$

$$0 = 2\zeta\Omega + p\Omega^2 e^{-\gamma} \sin\phi. \tag{5.3.4b}$$

Squaring and adding these equations yields

$$(1 - \Omega^2)^2 = p^2\omega^4 e^{-2\gamma} \cos^2\phi \tag{5.3.5a}$$

$$4\zeta^2\Omega^2 = p^2\Omega^4 e^{-2\gamma} \sin^2\phi \tag{5.3.5b}$$

$$\Rightarrow \quad (1 - \Omega^2)^2 + 4\zeta^2\Omega^2 = p^2\Omega^4 e^{-2\gamma}. \tag{5.3.5c}$$

Solving this equation for real part γ of the eigenvalue leads to a formula $\gamma(\Omega)$ in dependence on the Hopf frequency Ω

$$\gamma(\Omega) = -\frac{1}{2}\ln\frac{(1-\Omega^2)^2 + 4\zeta^2\Omega^2}{p^2\Omega^4} \tag{5.3.6a}$$

$$= -\frac{1}{2}\ln\frac{1}{p^2}\left[1 + \frac{4\zeta^2 - 2}{\Omega^2} + \frac{1}{\Omega^4}\right]. \tag{5.3.6b}$$

A steady state can lose its stability via a Hopf bifurcation, at which point the tip of curve $\gamma(\Omega)$ will cross the imaginary axis. This instability can be prevented, provided the interval of unstable frequencies $\Omega_1 < \Omega < \Omega_2$ lies inside the interval $\left[\Omega^{n_0}, \Omega^{n_0+1}\right]$ for some n_0. Here, $\Omega_{1,2}$ are two positive roots of the equation $\gamma(\Omega) = 0$, which can be found from Eq. 5.3.6a as

$$\gamma(\Omega) = 0 \Leftrightarrow \frac{1}{p^2}\left[1 + \frac{4\zeta^2 - 2}{\Omega^2} + \frac{1}{\Omega^4}\right] = 1 \tag{5.3.7a}$$

$$\Rightarrow \quad p^2\Omega^4 = \Omega^4 + (4\zeta^2 - 2)\Omega^2 + 1 \tag{5.3.7b}$$

$$\Leftrightarrow \quad 0 = \Omega^4 + 2\frac{2\zeta^2 - 1}{1 - p^2}\Omega^2 + \frac{1}{1 - p^2} \tag{5.3.7c}$$

$$\Leftrightarrow \quad \Omega_{1,2}^2 = \frac{1}{1 - p^2}\left[1 - 2\zeta^2 \pm \sqrt{(1 - 2\zeta^2)^2 - 1 + p^2}\right]. \tag{5.3.7d}$$

For further analytical progress, one can expand this expression for small values of ζ, which gives

$$\Delta\Omega = \Omega_1 - \Omega_2 = \frac{1}{\sqrt{1-p}} - \frac{1}{\sqrt{1+p}} - \frac{\zeta^2}{p}\left(\frac{1}{\sqrt{1+p}} + \frac{1}{\sqrt{1-p}}\right). \tag{5.3.8}$$

Since the actual values of the frequencies are $\Omega^{(n)} = 2\pi\, n/\tau$ for any integer n, the distance between any two successive frequencies is $2\pi/\tau$, and hence the necessary condition for stability $\Delta\Omega < 2\pi/\tau$ can be written as

$$\frac{1}{\sqrt{1-p}} - \frac{1}{\sqrt{1+p}} - \frac{\zeta^2}{p}\left(\frac{1}{\sqrt{1+p}} + \frac{1}{\sqrt{1-p}}\right) < 2\pi/\tau \tag{5.3.9}$$

or equivalently

$$\left(1 - \frac{\zeta^2}{p}\right)\frac{1}{\sqrt{1-p}} - \left(1 + \frac{\zeta^2}{p}\right)\frac{1}{\sqrt{1+p}} < 2\pi/\tau. \tag{5.3.10}$$

In the limit of $\tau \to 0$, p approaches unity. Thus the first term of Eq. 5.3.9 dominates and one obtains

$$\frac{1}{\sqrt{1-p}} \approx 2\pi/\tau \tag{5.3.11a}$$

$$\Rightarrow \quad p \approx 1 - \tau^2/4\pi^2 \tag{5.3.11b}$$

which is an approximation for the critical mass ratio.

For large enough time delay τ, p asymptotically approaches a lower bound of stability which corresponds to $\Delta\Omega = 0$. It can be obtained from Eq. 5.3.7a by using

$$0 = (\Omega_1 - \Omega_2)(\Omega_1 + \Omega_2) = \Omega_1^2 - \Omega_2^2 \tag{5.3.12a}$$

$$= \frac{2}{1-p^2}\sqrt{(1-2\zeta^2)^2 - 1 + p^2} \tag{5.3.12b}$$

which yields

$$p = 2\zeta\sqrt{1-\zeta^2} \approx 2\zeta. \tag{5.3.13}$$

Alternatively, this result can be derived from Eq. 5.2.9b by calculation of the minima in Fig. 5.4

$$\frac{dp(\omega)}{d\omega} = \frac{d}{d\omega}\left(\frac{1}{\omega^2}\sqrt{\omega^4 + 2\omega^2(2\zeta^2-1)+1}\right) \tag{5.3.14a}$$

$$= \frac{4\omega^2(2\zeta^2-1)+4}{2\omega^3\sqrt{\omega^4+2\omega^2(2\zeta^2-1)+1}} = 0 \tag{5.3.14b}$$

$$\Rightarrow \quad \omega_{\min}^2 = \frac{1}{1-2\zeta^2}. \tag{5.3.14c}$$

The corresponding value p recovers the result of Eq. 5.3.13

$$p(\omega_{\min}) = \frac{1}{\omega_{\min}^2}\sqrt{\omega_{\min}^4 + 2\omega_{\min}^2(2\zeta^2-1)+1} \tag{5.3.15a}$$

$$= \sqrt{1 + 2\frac{2\zeta^2-1}{\omega_{\min}^2} + \frac{1}{\omega_{\min}^4}} \tag{5.3.15b}$$

$$= \sqrt{1 - 2(1-2\zeta^2)^2 + (1-2\zeta^2)^2} \tag{5.3.15c}$$

$$= 2\zeta\sqrt{1-\zeta^2}. \tag{5.3.15d}$$

Figure 5.7 shows the plot of the approximate stability boundary according to (5.3.9) as a function of time delay τ for a given small value of the damping ζ.

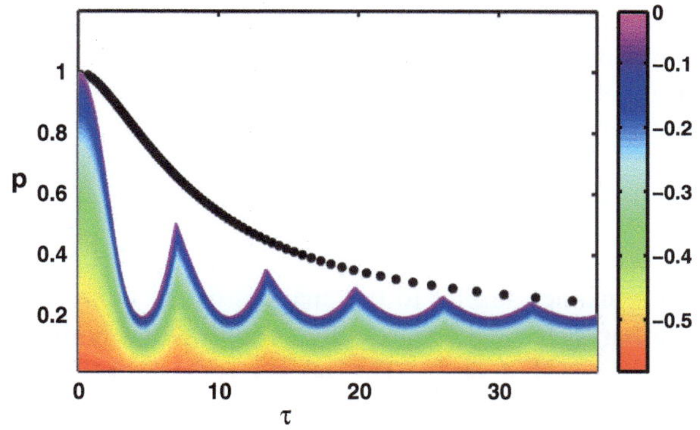

Fig. 5.7 Comparison of the approximate upper bound of stability according to Eq. 5.3.9 (*dotted line*) with an exact stability boundary for $\zeta = 0.1$ in the (τ, p) plane. The *color code* encodes the value of the largest real part of the complex eigenvalues λ

The color code in this figure indicates the value of the largest real part of the eigenvalues in the spectrum of the characteristic equation (5.2.5) for each value of p and τ. As it follows from Fig. 5.7, the analytically derived formula (5.3.9) for the maxima on the stability boundary provides a good approximation for large time delay τ. While it deviates from the exact stability peaks, which correspond to codimension-two Hopf bifurcation, for small delays, it still correctly approaches the asymptotic value of $p = 1$ as $\tau \to 0$.

As in the previous Section, one can include effects of delayed velocity feedback also in the treatment of the asymptotic properties. Assuming in Eq. 5.2.12 the same asymptotic behavior (5.3.1) of the eigenvalues for large time delay, i.e., the real part of the eigenvalue scales as $1/\tau$, gives to the leading order

$$1 - \Omega^2 + 2i\zeta_1\Omega - p\Omega^2 e^{-\gamma} e^{-i\phi} + 2i\zeta_2\Omega e^{-\gamma} e^{-i\phi} = 0 \tag{5.3.16}$$

with the constraint $\Omega = \Omega^{(n)} = 2\pi\ n/\tau$, $n = \pm 1, \pm 2, \pm 3, \ldots$ Similar to the derivation of Eq. 5.3.7a, separating Eq. 5.3.16 into real and imaginary parts yields

$$-(1 - \Omega^2) = -p\Omega^2 e^{-\gamma} \cos\phi + 2\zeta_2\Omega e^{-\gamma} \sin\phi \tag{5.3.17a}$$

$$-2\zeta_1\Omega = p\Omega^2 e^{-\gamma} \sin\phi + 2\zeta_2\Omega e^{-\gamma} \cos\phi. \tag{5.3.17b}$$

Squaring and adding these equations leads to the following expression after some algebraic manipulations such as $\cos^2(\cdot) + \sin^2(\cdot) = 1$

$$(1 - \Omega^2)^2 + 4\zeta_1^2\Omega^2 = \left(p^2\Omega^4 + 4\zeta_2^2\Omega^2\right)e^{-2\gamma}. \tag{5.3.18}$$

Solving this equation for γ gives in dependence on the Hopf frequency Ω at the Hopf bifurcation

$$\gamma(\Omega) = -\frac{1}{2}\ln\frac{\left(1 - \Omega^2\right)^2 + 4\zeta_1^2\Omega^2}{p^2\Omega^4 + 4\zeta_2^2\Omega^2}. \tag{5.3.19}$$

Note that vanishing ζ_2 recovers Eq. 5.3.6a without delayed velocity feedback. Transition to instability occurs when $\gamma(\Omega) = 0$, which gives the expression for instability frequencies

$$\Omega_{1,2}^2 = \frac{1}{1-p^2}\left[1 + 2\zeta_2^2 - 2\zeta_1^2 \pm \sqrt{\left(1 + 2\zeta_2^2 - 2\zeta_1^2\right)^2 - 1 + p^2}\right]. \tag{5.3.20}$$

In a manner similar to the analysis of the delayed force feedback, one can make further analytical progress by assuming that both damping coefficients are small: $|\zeta_1| \ll 1$, $|\zeta_2| \ll 1$. The necessary stability condition $\Delta\Omega = \Omega_1 - \Omega_2 < 2\pi/\tau$ gives the following asymptotic approximation for the maxima of the stability boundary:

$$\frac{1}{\sqrt{1-p}} - \frac{1}{\sqrt{1+p}} - \frac{\zeta_1^2 - \zeta_2^2}{p}\left(\frac{1}{\sqrt{1+p}} + \frac{1}{\sqrt{1-p}}\right) < 2\pi/\tau. \tag{5.3.21}$$

The expression (5.3.21) can be further simplified for small time delay τ to give $p \approx 1 - \tau^2/4\pi^2$ which is identical to the limit of p for small τ in the case without delayed velocity feedback. For large time delay the asymptotic behavior can be obtained in a manner similar to Eq. 5.3.13. Setting the difference ${\Omega^2}_1 - {\Omega^2}_2$ to zero yields

$$0 = \Omega_1^2 - \Omega_2^2 \tag{5.3.22a}$$

$$= \frac{2\sqrt{\left(1 + 2\zeta_2^2 - 2\zeta_1^2\right)^2 - 1 + p^2}}{1-p^2} \tag{5.3.22b}$$

$$\Leftrightarrow \quad p^2 = \left(1 + 2\zeta_2^2 - 2\zeta_1^2\right)^2 - 1 \tag{5.3.22c}$$

$$\Leftrightarrow \quad = 2\left(2\zeta_2^2 - 2\zeta_1^2\right) + \left(2\zeta_2^2 - 2\zeta_1^2\right)^2 \tag{5.3.22d}$$

which gives the asymptotic value

$$p = 2\sqrt{(\zeta_1^2 - \zeta_2^2)(1 - \zeta_1^2 + \zeta_2^2)} \approx 2\sqrt{\zeta_1^2 - \zeta_2^2}. \tag{5.3.23}$$

It is noteworthy that the inequality (5.3.21) provides a good approximation for the stability boundary even when actual values of damping coefficients ζ_1 and ζ_2 are large, as long as the difference $(\zeta_1^2 - \zeta_2^2)$ is small by the absolute value. Figure 5.8 shows an excellent agreement between the asymptotic approximation (5.3.21) and the exact stability boundary.

After the discussion of asymptotic behavior for large time delays, the next Section is devoted to the application of time-delayed feedback to the system without velocity feedback. The goal is to enlarge the parameter range for which the

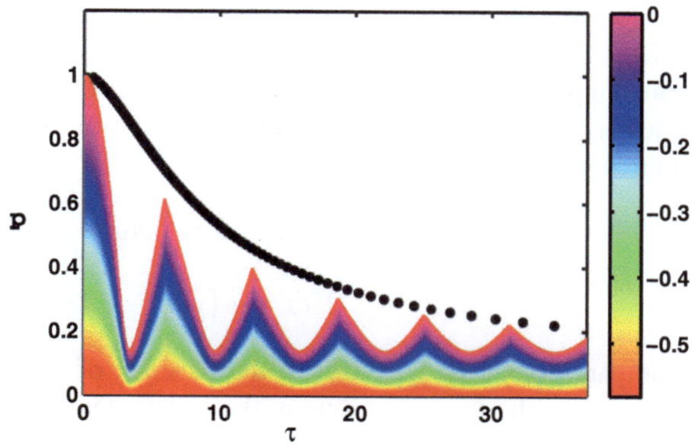

Fig. 5.8 Comparison of the approximate upper bound of stability according to Eq. 5.3.21 (*dotted line*) with an exact stability boundary for $\zeta_1 = 0.25$ and $\zeta_2 = 0.24$ in the (τ, p) plane. The *color code* refers to the value of the largest real part of the complex eigenvalues λ

steady state is stable. It will be shown that the relation between the system's delay and the control delay becomes crucial.

5.4 Control by Time-Delayed Feedback

The previous Sections introduced a model system of a neutral delay differential equation and discussed its asymptotic behavior for large time delays. It was shown that there are regions in the parameter space for which the steady state at the origin $z^* = 0$ is unstable. The purpose is this Section is to influence the stability of this steady state by means of time-delayed feedback control and thus increase the parameter range with stable steady state. To this end, the original system (5.2.1) is replaced by its modified version [11]

$$\ddot{z}(t) + 2\zeta\dot{z}(t) + z(t) + p\ddot{z}(t - \tau_1) = K[z(t) - z(t - \tau_2)], \tag{5.4.1}$$

where $K > 0$ is the strength of the control force, and $\tau_2 > 0$ is the time delay of the control. This modified system has the same steady state as the original system for any time delay τ_2, i.e., $z^* = 0$, but now the stability of this equilibrium is determined by the roots of the modified characteristic equation:

$$\lambda^2 + 2\zeta\lambda + 1 + p\lambda^2 e^{-\lambda\tau_1} = K\left(1 - e^{-\lambda\tau_2}\right). \tag{5.4.2}$$

Note the additional exponential on the right-hand side compared to Eq. 5.2.5. Vanishing feedback strength $K = 0$ or control delay $\tau_2 = 0$ recovers the original uncontrolled model with an unstable steady state of Sect. 5.2. The goal is now to find a relation between K and τ_2 in terms of the original system parameters p, ζ, and τ_1, such that in the absence of control there is a pair of unstable complex conjugate eigenvalues, while in the presence of control all the eigenvalues are stable.

The strategy is the following: At first, I will investigate the case where the control delay τ_2 matches the system's delay τ_1. Afterwards I will also take different time delays $\tau_1 \neq \tau_2$ into account.

Consider the case when the time delay of the control coincides with the time delay of the original system $\tau_2 = \tau_1 = \tau$. The stability boundary is determined by vanishing real part of Λ in the characteristic equation (5.4.2), i.e., $\mathrm{Re}(\lambda) = 0$. Looking for solutions of Eq. 5.4.2 in the form $\lambda = i\omega$ and separating real and imaginary parts yields

$$-\omega^2 + 1 - K = \left(p\omega^2 - K\right)\cos\omega\tau \tag{5.4.3a}$$

$$2\zeta\omega = -\left(p\omega^2 - K\right)\sin\omega\tau. \tag{5.4.3b}$$

Compare also Eqs. 5.2.6a, b in the absence of time-delayed feedback. Squaring and adding the two equations in the last system gives

$$\left(-\omega^2 + 1 - K\right)^2 + 4\zeta^2\omega^2 = \left(p\omega^2 - K\right)^2. \tag{5.4.4}$$

The last equation can be solved for p as follows:

$$p(\omega) = \frac{1}{\omega^2}\left(\sqrt{\left(-\omega^2 + 1 - K\right)^2 + 4\zeta^2\omega^2} + K\right) \tag{5.4.5}$$

which is a parametric expression of p in dependence on the Hopf frequency ω.

Also, dividing the equation of the imaginary part (5.4.3b) by the corresponding equation of the real part (5.4.3a), the critical time delay can be expressed as

$$\tau(\omega) = \frac{1}{\omega}\left[\arctan\frac{2\zeta\omega}{\omega^2 - 1 + K} \pm \pi k\right], k = 0, 1, 2, \ldots \tag{5.4.6}$$

Now, for each fixed value of the control strength K and the damping ζ, the stability boundary can be parameterized in the (τ, p) plane according to Eqs. 5.4.5 and 5.4.6. Figure 5.9a shows that it is possible to raise the stability boundary by increasing the control strength K. Compare the solid, dashed, and dotted curves which correspond to a value of $K = 0$ (no control), 0.1, and 0.2, respectively. This improve overall stability of the system. This also means that those points which were unstable in the uncontrolled system, are now stable as they are moved by the control action to the area below the new stability boundary. It is worth mentioning that if $K < 0$, this actually lowers the stability boundary, and hence reduces the stability.

So far, the control strength K was fixed and the stability boundary was studied as a curve in the (τ, p) plane. At the same time, one can consider p and τ fixed. This corresponds to a single unstable fixed point of the uncontrolled system and the system (5.4.3a, b) can be solved for a tuple (K, ω). This would give a full picture of controllability of the system by providing the value of the minimum control strength K required to stabilize the steady state for the given values of p and τ. The results of this computation are shown in Fig. 5.9b, where the steady state is stable above the surface. This plot suggests that the more unstable is a steady state, i.e., the higher it is above its stability boundary in (τ, p) plane, the higher should be the control strength K for stabilization of this steady state.

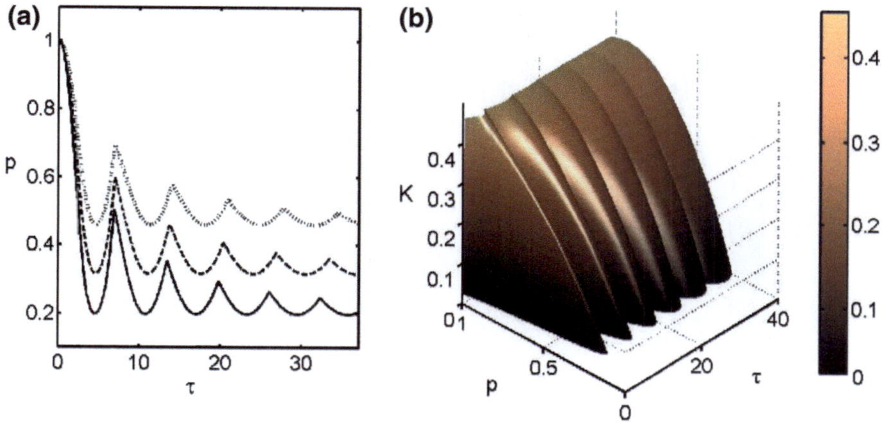

Fig. 5.9 **a** Stability boundary of the controlled system for $\tau_1 = \tau_2 = \tau$, $\zeta = 0.1$ and different control strengths: $K = 0$ (*solid*), $K = 0.1$ (*dashed*) and $K = 0.2$ (*dotted*). The steady state is stable below the curves. **b** Minimum stabilizing control strength K as a function of p and $\tau_1 = \tau_2 = \tau$

It is noteworthy that time-delayed feedback control scheme fails to achieve controllability of the steady state if $|p| > 1$ independently on the intrinsic time delay and control parameters. This highlights a major difference between stabilizing unstable steady states in neutral and non-neutral time-delayed systems.

When $\tau_2 \neq \tau_1$, the characteristic equation (5.4.2) has two time delays simultaneously present, which significantly complicates a stability analysis. Several approaches have been recently put forward to study stability of equations with multiple time delays [17, 18]. Beretta and Kuang have developed geometric stability switch criteria for equations with delay-dependent parameters, and this method can also be used to analyze systems with two time delays [19]. Sipahi and Olgac have suggested in Ref. [20] a systematic procedure of finding eigenvalues of characteristic equations for systems with multiple time delays by means of a substitution, which replaces an original transcendental characteristic equation with a polynomial. In the present case, it is convenient to use a parameterization of the stability boundary in the parameter plane of two time delays, as outlined in Ref. [21]. The idea is to rewrite the characteristic equation (5.4.2) in an equivalent form

$$1 + a_1(\lambda)e^{-\lambda\tau_1} + a_2 e^{-\lambda\tau_2} = 0 \tag{5.4.7}$$

with the following abbreviations

$$a_1(\lambda) = \frac{p\lambda^2}{1 - K + 2\zeta\lambda + \lambda^2}, \quad a_2(\lambda) = \frac{K}{1 - K + 2\zeta\lambda + \lambda^2}. \tag{5.4.8}$$

The stability border can now be parameterized by a Hopf frequency $\omega \in \Omega$, where $\Omega = \bigcup_{i=1}^{k} \Omega_i$ consists of a finite number of intervals of finite length. The critical time delays at the stability boundary in the (τ_1, τ_2) plane are then given by

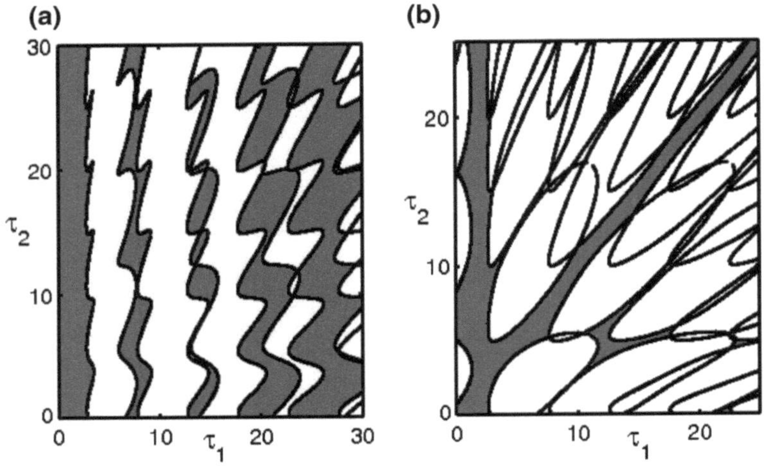

Fig. 5.10 Stability boundaries of the characteristic equation (5.4.2) for $p = 0.4$ and $\zeta = 0.1$. **a** $K = 0.1$, **b** $K = 0.3$. The steady state is stable inside the *shaded area*

$$\tau_1 = \frac{\arg[a_1(i\omega)] + (2u - 1)\pi \pm \theta_1}{\omega} \tag{5.4.9a}$$

$$u = u_0^{\pm}, u_0^{\pm} + 1, u_0^{\pm} + 2, \ldots \tag{5.4.9b}$$

$$\tau_2 = \frac{\arg[a_2(i\omega)] + (2v - 1)\pi \pm \theta_1}{\omega} \tag{5.4.9c}$$

$$v = v_0^{\pm}, v_0^{\pm} + 1, v_0^{\pm} + 2, \ldots, \tag{5.4.9d}$$

where θ_1, $\theta_2 \in [0, \pi]$ are calculated as

$$\theta_1 = \arccos\left(\frac{1 + |a_1(i\omega)|^2 - |a_2(i\omega)|^2}{2|a_1(i\omega)|}\right) \tag{5.4.10a}$$

$$\theta_2 = \arccos\left(\frac{1 + |a_2(i\omega)|^2 - |a_1(i\omega)|^2}{2|a_2(i\omega)|}\right) \tag{5.4.10b}$$

and $u_0^{\pm}$ and $v_0^{\pm}$ are determined as the smallest positive integers such that the corresponding $\tau_1^{u_0^+}$, $\tau_1^{u_0^-}$, $\tau_2^{v_0^+}$, $\tau_2^{v_0^-}$ are all non-negative.

Figure 5.10 shows the stability boundaries of the controlled system in the parameter space of the time delays τ_1 and τ_2, as given by the formulas (5.4.9a, b) for different values of $K = 0.1$ and 0.3 in panels (a) and (b), respectively. Note that when $K = 0.1$, the stability boundaries are represented by zig-zag curves with vertical axes. For larger values of the control strength K, the stability boundaries become closed curves. The steady state is stable inside the shaded areas.

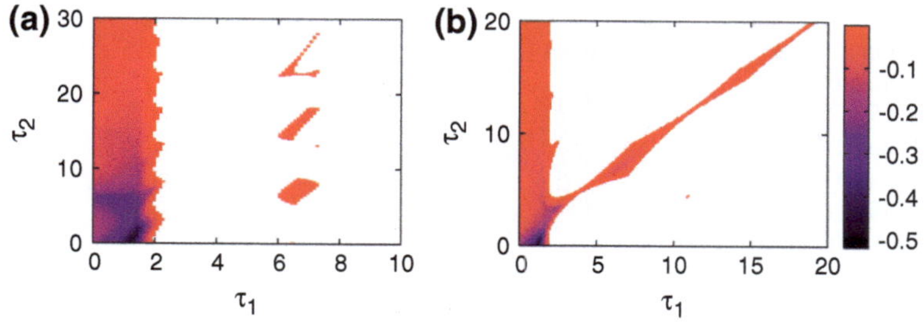

Fig. 5.11 Real part of the leading eigenvalue of the characteristic equation (5.4.2) versus delay times τ_1 and τ_2 for $p = 0.4$ and $\zeta = 0.1$. (a) $K = 0.1$. (b) $K = 0.3$. The *color code* denotes the value of $\mathrm{Re}(\lambda)$; only those parts are shown where it is negative. The *colored* areas in both plots correspond to the regions where the steady state is stable

To deepen the understanding of the stability changes depending on the relation between the two time delays, one can compute the maximum of the real part of the characteristic eigenvalues, which is shown in Fig. 5.11 wherever it is negative. To calculate the real part of the leading eigenvalue of the characteristic equation (5.4.2) pseudospectral differentiation methods can be used. For details see Refs. [22–24]. Note some of the very small stable domains with $\mathrm{Re}(\lambda)$ very close to zero shown in Fig. 5.10 are not fully resolved numerically in Fig. 5.11.

The control strength K appears as a parameter in a_1 and a_2, and thus determines the values of τ_1 and τ_2 at the stability threshold. At the same time, in order to determine the controllability of the system it is more convenient to fix particular values of the parameters τ_1, p, and ζ of the original system, and then study the stability border in the (τ_2, K) space. Figure 5.12a shows how the real part of the leading characteristic eigenvalue depends on K and τ_2, and in Fig. 5.12b for illustrative purposes I present three possible regimes for different values of the control strength K. One can observe that while for small K the steady state is unstable, it becomes stable starting with some K, and the larger K is, the more negative the real part of the leading eigenvalue becomes. It is clear from Fig. 5.12a that the stability islands in (τ_2, K) parameter plane are finite, and hence if the control strength is very large, the steady state will be unstable again.

If the parameters are taken closer to the stability boundary of the uncontrolled system, i.e., for smaller values of p, a much smaller control strength K is required to stabilize an unstable equilibrium. In this case all stability boundaries are closed curves shown in Fig. 5.13 for fixed $K = 0.1$ and $K = 0.3$, and the steady state is stable in the shaded area outside those curves. Note that the parameter p is fixed and $p = 0.25$, i.e., smaller than in Figs. 5.10–5.12. The corresponding plot of the real part of the leading eigenvalue, shown in Fig. 5.14, reveals the existence of several regions of values of τ_2, for which the steady state can be stabilized for the same value of the control strength K and the original time delay τ_1. If I fix τ_1 and consider the plane of control parameters K and τ_2, the dependence of the leading eigenvalue of the characteristic equation on these parameters is qualitatively the same as the one for $p = 0.4$, whereas the stability regions are more pronounced,

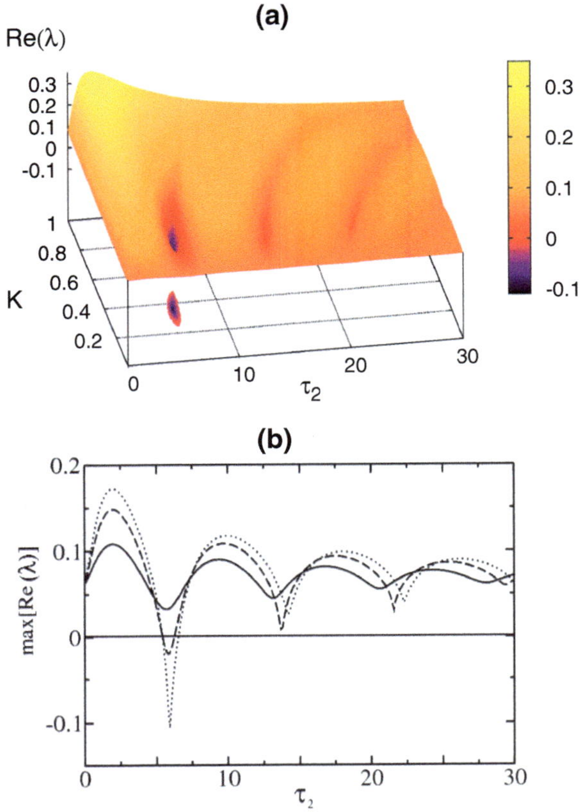

Fig. 5.12 **a** Real part of the leading eigenvalue of the characteristic equation (5.4.2) for $p = 0.4$, $\zeta = 0.1$, and $\tau_1 = 5$. **b** Dependence of the real part of the leading eigenvalue on the control time delay τ_2 for $K = 0.15$ (*solid*), $K = 0.28$ (*dashed*), and $K = 0.36$ (*dotted*) with fixed $\tau = 5$

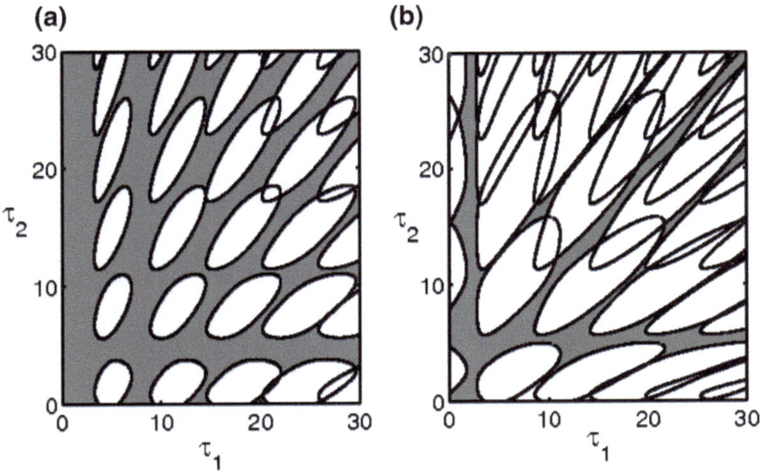

Fig. 5.13 Stability boundaries of the characteristic equation (5.4.2) for $p = 0.25$ and $\zeta = 0.1$. The feedback strength K is chosen as $K = 0.1$ and $K = 0.3$ in **a** and **b**, respectively. The *shaded areas* correspond to a stable steady state

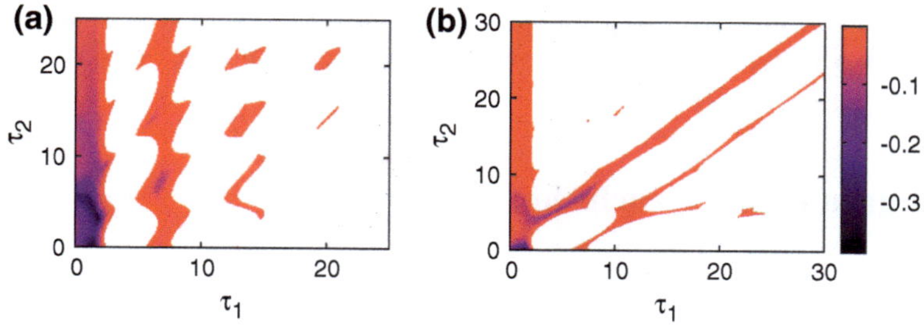

Fig. 5.14 Real part of the leading eigenvalue of the characteristic equation (5.4.2) versus delay times τ_1 and τ_2 for $p = 0.25$ and $\zeta = 0.1$. The feedback strength K is chosen as $K = 0.1$ and $K = 0.3$ in **a** and **b**, respectively. The *color code* denotes the value of Re(λ); only those parts are shown where it is negative. The *colored areas* in both plots correspond to the regions where the steady state is stable

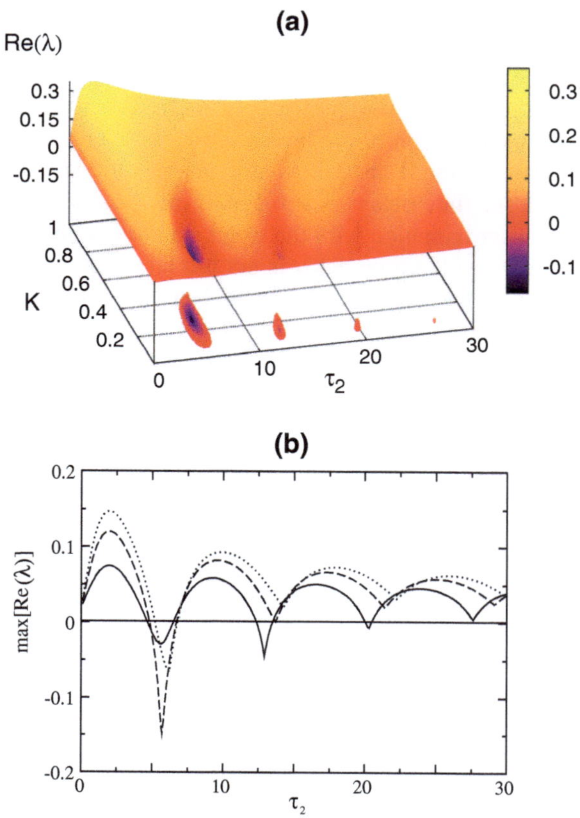

Fig. 5.15 **a** Real part of the leading eigenvalue of the characteristic equation (5.4.2) for $p = 0.25$, $\zeta = 0.1$, and $\tau_1 = 5$. **b** Dependence of the real part of the leading eigenvalue on the control time delay τ_2 for $K = 0.15$ (*solid*), $K = 0.28$ (*dashed*), and $K = 0.36$ (*dotted*)

see Fig. 5.15. In comparison to the case of $p = 0.4$, there is more than one interval of time delay τ_2, for which stabilization of the steady state is achieved for a fixed value of the control strength K. Note that similar behavior as in Figs. 5.13 and 5.14

was found for an electronic oscillator where two time delays were added in the form of two Pyragas-type time-delayed feedback control terms [18].

5.5 Intermediate Conclusion

In this chapter, I have discussed systems where the occurrence of a time delay is due to the intrinsic structure of the model. This intrinsic delay arises from various sources such as propagation delays, calculation times for numerical computation etc. I have investigated the dynamics of two models which belong to the class of neutral delay differential equations since the delayed variable enters in the highest derivative. The analysis of steady state solutions has shown that the systems exhibit stability switches, where stability is lost and regained depending on the time delay.

In the case when delayed velocity feedback is present, the interplay between the time delay and the two damping coefficients produced different stability regimes in the parameter plane. For some parameter values the stability area collapsed into separate islands. On the basis of a characteristic equations, I found an analytical parameterization of the boundary of stability.

In addition, I have derived an asymptotic approximation of the stability peaks for large time delays where I used universal scaling arguments. I have compared this approximation with the exact stability boundary. The results agree quite well even when the time delay is not too large, and give excellent agreement for large delays. The results were obtained by numerical simulations of the characteristic spectrum.

Next to systems with intrinsic time delays, there are those where the time delay is introduced externally in order to stabilize unstable periodic orbits and steady states. To this end, time-delayed feedback control as introduced in Chap. 2 has been demonstrated to be an efficient tool for stabilization of these states. Since the intrinsic delay eventually destabilized the steady state of the neutral differential delay equation, the application of time-delayed feedback led to a larger parameter range for which this fixed point was stable. The interplay between the intrinsic time delay of the original system and the time delay of the control force was crucial.

For a given value of the control strength, it was possible to find an analytical parameterization of stability boundaries in terms of the system's delay and the control delay. I found the following result: The more unstable was a steady state, as determined by the real part of the leading eigenvalue of the corresponding characteristic equation, the higher should be the control strength required to stabilize it for a fixed value of the control time delay. At the same time, if the steady state is close to a Hopf bifurcation, it was easier to control it. In the latter case there was a larger number of islands in a parameter space where stabilization was successfully achieved. This results held even in the case where the two time delays did not match.

Until this point I have only considered single systems subject to time-delayed feedback. These were steady states in systems with and without intrinsic time delays as well as periodic orbits. In the following chapter, I will also apply this control method to coupled systems. These systems will be realized as excitable units which rest in a fixed point unless they are kicked above a certain threshold. This excitation can be caused by random fluctuations as well as delayed input from the coupled systems. The goal will be to investigate the underlying dynamics and to characterize the relevant timescales.

References

1. Engelborghs K, Luzyanina T, Samaey G (2001) Dde-biftool v. 2.00: a matlab package for bifurcation analysis of delay differential equations. Tech. Rep. TW-330, Department of Computer Science, K.U.Leuven, Belgium
2. Szalai R (2005) PDDE-CONT: a continuation and bifurcation software for delay-differential equations. http://seis.bris.ac.uk/~rs1909/pdde
3. Balanov AG, Janson NB, McClintock PVE, Tucker RW, Wang CHT (2003) Bifurcation analysis of a neutral delay differential equation modelling the torsional motion of a driven drill-string. Chaos Solitons Fractals 15:381
4. Blakely JN, Corron NJ (2004) Experimental observation of delay-induced radio frequency chaos in a transmission line oscillator. Chaos 14:1035
5. Bocharov G, Hadeler KP (2000) Structured population models, conservation laws, and delay equations. J Differ Equ 168:212
6. Hadeler KP, Bocharov G (2003) Delays in population models and where to put them, in particular in the neutral case. Can Appl Math Quart 11:159
7. Hadeler KP (2008) Neutral delay equations from and for population dynamics. Proc 8th Coll QTDE, Electron J Qual Theory Differ Equ 11:1
8. Kyrychko YN, Hogan SJ, Gonzalez-Buelga A, Wagg DJ (2007) Modelling real-time dynamic substructuring using partial delay differential equations. Proc R Soc A 463:1509
9. Gonzalez-Buelga A, Wagg D, Neild S (2007) Parametric variation of a coupled pendulum-oscillator system using real-time dynamic substructuring. Struct Control Health Monit 14:991
10. Kyrychko YN, Blyuss KB, Hövel P, Schöll E (2009) Asymptotic properties of the spectrum of neutral delay differential equations. Dyn Syst 24:361
11. Blyuss KB, Kyrychko YN, Hövel P, Schöll E (2008) Control of unstable steady states in neutral time-delayed systems. Eur Phys J B 65:571
12. Kyrychko YN, Blyuss KB, Gonzalez-Buelga A, Hogan SJ, Wagg DJ (2006) Real-time dynamic substructuring in a coupled oscillator–pendulum system. Proc R Soc A 462:1271
13. Kolmanovskii VB, Myshkis A (1992) Applied theory of functional differential equations. Kluwer Academic Publishers
14. Hale JK (1977) Theory of functional differential equations. Springer, New York
15. Hale JK, Verduyn Lunel SM (1993) Introduction to functional differential equations. Springer, New York
16. Yanchuk S, Wolfrum M, Hövel P, Schöll E (2006) Control of unstable steady states by long delay feedback. Phys Rev E 74:026201
17. Ahlborn A, Parlitz U (2004) Stabilizing unstable steady states using multiple delay feedback control. Phys Rev Lett 93:264101
18. Ahlborn A, Parlitz U (2005) Controlling dynamical systems using multiple delay feedback control. Phys Rev E 72:016206
19. Beretta E, Kuang Y (2002) Geometric stability switch criteria in delay differential systems with delay dependent parameters. SIAM J Math Anal 33:1144

20. Sipahi R, Olgac N (2005) Complete stability robustness of third-order LTI multiple time-delay systems. Automatica 41:1413
21. Gu K, Niculescu S-I, Chen J (2005) On stability crossing curves for general systems with two delays. J Math Anal Appl 311:231
22. Engelborghs K, Roose D (2002) On stability of LMS methods and characteristic roots of delay differential equations. SIAM J Math Anal 40:629
23. Breda D (2006) Solution operator appro
ximations for characteristic roots of delay differential equations. Appl Numer Math 56:305
24. Breda D, Maset S, Vermiglioa R (2006) Pseudospectral approximation of eigenvalues of derivative operators with non-local boundary conditions. Appl Numer Math 56:318

Chapter 6
Neural Systems

Delay is preferable to error.

(Thomas Jefferson: Letter to George Washington)

Synchronization of neurons or neural populations can have multiple biological reasons. Groups of neurons can operate in a synchronized manner to enhance data transmission for processing of biological information [1–3]. In the latter case, synchronization in the brain is discussed in terms of binding-by-synchrony which addresses the question how information processing in distant regions of the brain is executed in parallel and how the relatedness of the results is encoded for a coherent perception. The temporal synchrony is realized with biologically fast precision on the millisecond scale. For a review on this matter see Ref. [4]. In this context, synchronized behavior can be a desirable, advantageous pattern. Synchronization of neural spiking can also cause regular action potentials which are associated with pathological rhythmic brain activity and symptoms in Parkinson's disease, essential tremor as well as epilepsy [5–8]. Modern concepts of time-delayed feedback control have recently been applied to suppress this undesired synchrony [9–15].

Since the network of neurons in the brain exhibits a subtle balance of dynamic chaos and selforganized order, understanding the spontaneous emergence of neural synchrony is of crucial importance for clinical applications. The goal is to gain insight into the underlying mechanisms which then can lead to the development of new techniques in order to manipulate and influence the neural dynamics [12, 16–18].

This chapter is devoted to the application of time-delayed feedback control to excitable neural systems. I will mainly focus on the effects concerning a single neuron or two coupled neural elements. In the context of larger networks, this can be seen as a first step towards network motifs which build up the network and consist of only a few neural oscillators.

The chapter is organized as follows: in Sect. 6.1, I will introduce the model which will be used throughout this chapter. The model consists of a FitzHugh–Nagumo system [19, 20] which is a two-dimensional simplification of the

P. Hövel, *Control of Complex Nonlinear Systems with Delay*,
Springer Theses, DOI: 10.1007/978-3-642-14110-2_6,

Hogkin–Huxley model [21]. After the discussion of the system's equations, I will present tools for the characterization of the dynamics. Two coupled FitzHugh–Nagumo systems will be the subject of Sect. 6.2 where I will develop measures for the detection of synchronization. Sections 6.3–6.5 will deal with the effects of time-delayed feedback control. In Sect. 6.3, the control scheme will be applied to a single excitable neural oscillator. The subject of Sect. 6.4 will be the influence of time-delayed feedback in its extended version by Socolar [22] on the cooperative dynamics of two coupled neural elements where special emphasis lies on the role of the memory parameter. In Sect. 6.5, I will discuss effects of different coupling schemes to the compound system of two FitzHugh–Nagumo models. The different coupling possibilities arise from the construction and application of the delayed feedback. Section 6.6 will focus on potential applications and analysis of time-delayed feedback in the context of networks, where I will discuss a technique called master stability function and present some preliminary results. Finally, Sect. 6.7 concludes this chapter in an intermediate summary.

6.1 Single FitzHugh–Nagumo System

Before investigating the effects of time-delayed feedback control in Sects. 6.3–6.6, I will first discuss the dynamics of the uncontrolled system. This section is devoted to the analysis of a single neural system and Sect. 6.2 deals with the interaction of two coupled systems.

As a paradigmatic model for neural systems, I consider the FitzHugh–Nagumo system throughout this chapter [19, 20]. This two-dimensional system is a simplification of the four-dimensional Hodgkin–Huxley model [21] and can be written as

$$\varepsilon \frac{du(t)}{dt} = u(t) - \frac{u(t)^3}{3} - v(t) \tag{6.1.1a}$$

$$\frac{dv(t)}{dt} = u(t) + a + D\xi(t). \tag{6.1.1b}$$

The variable u is related to the transmembrane voltage and v corresponds to various quantities connected to the electrical conductance of the relevant ion currents. The neuron is driven by Gaussian white noise $\xi(t)$ with zero mean and unity variance. The noise intensity is denoted by D. The parameter ε is the timescale ratio between the activator u and the inhibitor v.

Figure 6.1 depicts the vector field of the system (6.1.1a, b) in the (u, v) phase space. Panels (a) and (b) correspond to two different timescale ratios $\varepsilon = 0.1$ and 0.005, respectively. The parameter a is fixed at $a = 1.05$. The red (dotted) and blue (dashed) curves show the nullclines of Eqs. 6.1.1a and 6.1.1b, respectively. These curves are determined by $du/dt = 0$ and $dv/dt = 0$ which yields

$$v(t) = u(t) - \frac{u(t)^3}{3} \quad \text{and} \quad u(t) = -a. \tag{6.1.2}$$

The fixed point (u^*, v^*) of the neural system is given by the intersection of the u- and v-nullclines

$$(u^*, v^*) = \left(-a, -a + \frac{a^3}{3}\right). \tag{6.1.3}$$

Since the time derivative of u and v vanishes on the respective nullcline, the vector field is horizontal and vertical at these coordinates, respectively. The difference of the timescale ratio ε results in a faster motion in the u-direction for smaller values of ε.

A linearization of Eqs. 6.1.1 around the fixed point (u^*, v^*) yields for $D = 0$

$$\begin{pmatrix} \frac{du(t)}{dt} \\ \frac{dv(t)}{dt} \end{pmatrix} = \underbrace{\begin{pmatrix} \frac{1-a^2}{\varepsilon} & -\frac{1}{\varepsilon} \\ 1 & 0 \end{pmatrix}}_{=\mathbf{J}} \begin{pmatrix} u(t) \\ v(t) \end{pmatrix}. \tag{6.1.4}$$

The eigenvalues λ of the Jacobian matrix $\mathbf{J}$ determine the stability of the fixed point. They are given as roots of the characteristic equations

$$0 = \det\begin{pmatrix} \frac{1-a^2}{\varepsilon} - \lambda & -\frac{1}{\varepsilon} \\ 1 & -\lambda \end{pmatrix} \tag{6.1.5a}$$

$$= \lambda^2 - \mathrm{tr}(\mathbf{J})\lambda + \det(\mathbf{J}) \tag{6.1.5b}$$

$$= \lambda^2 - \frac{1-a^2}{\varepsilon}\lambda + \frac{1}{\varepsilon} \tag{6.1.5c}$$

$$\Rightarrow \quad \lambda = \frac{1 - a^2 \pm \sqrt{(1-a^2)^2 - 4\varepsilon}}{2\varepsilon}. \tag{6.1.5d}$$

According to the characterization of two-dimensional systems elaborated in Sect. 3.1, one can conclude the following: Since ε is a positive constant, the determinant of $\mathbf{J}$ is also positive, i.e., $\det(\mathbf{J}) > 0$. Thus, the system falls into the right half-plane of Fig. 6.1. The trace of the Jacobian matrix $\mathrm{tr}(\mathbf{J}) = (1 - a^2)/\varepsilon$ is positive if the absolute value of the parameter a is smaller than unity and negative for $|a| > 1$. Therefore, the fixed point is stable for $|a| > 1$, whereas for $|a| < 1$ the intersection of the nullclines occurs in the range of positive slope of the u-nullcline. Then the fixed point is unstable. In the nonlinear system, this change of stability happens by a Hopf bifurcation above which periodic oscillations exist.

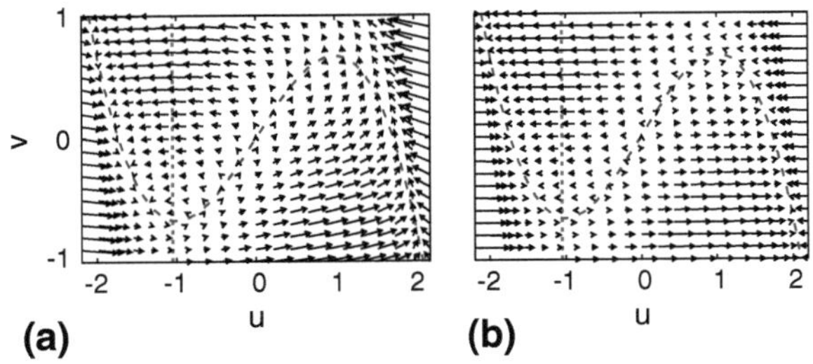

Fig. 6.1 Vector field of the FitzHugh–Nagumo system with u- and v-nullclines as blue (*dashed*) and red (*dotted*) curves, respectively. The timescale ratio ε is fixed at $\varepsilon = 0.1$ and $\varepsilon = 0.005$ in panels (**a**) and (**b**), respectively. Other parameters: $a = 1.05$ (excitable regime) and $D = 0$

Fig. 6.2 Statistics of the noise generator for 10^6 numbers compared to the exact Gaussian distribution

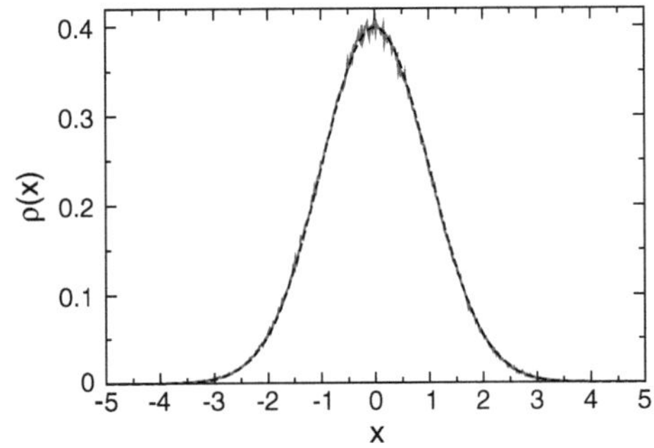

In the case of a stable fixed point, the system reaches the rest state (u^*, v^*) after an initial transient.

In order to create an excursion in phase space, it is necessary to overcome an excitability threshold. This can be done by application of an external force. In the case of Eqs. 6.1.1, the excitation is triggered by random fluctuations $D\xi(t)$. Since this noise term is realized as Gaussian white noise, the system will eventually overcome the threshold of excitation.

Figure 6.2 provides an exemplary realization of statistics of the random number generator used in this chapter. The red curve show the distribution of 10^6 random numbers. The black dashed curve refers to the analytic Gaussian distribution with zero mean and unity variance given by

$$\rho(x) = \frac{1}{\sqrt{2\pi}} \exp\left(-\frac{x^2}{2}\right) \tag{6.1.6}$$

and is added as a references.

Figure 6.3 displays the temporal dynamics of system (6.1.1) for a noise intensity of $D = 0.02$. Panel (a) shows the trajectory in the (u, v) phase space, where the nullclines are added for better orientation. Panel (b) depicts the time series of the activator and inhibitor as red (solid) and green (dashed) curves, respectively. An excitation cycle exhibits the following stages: starting at the fixed

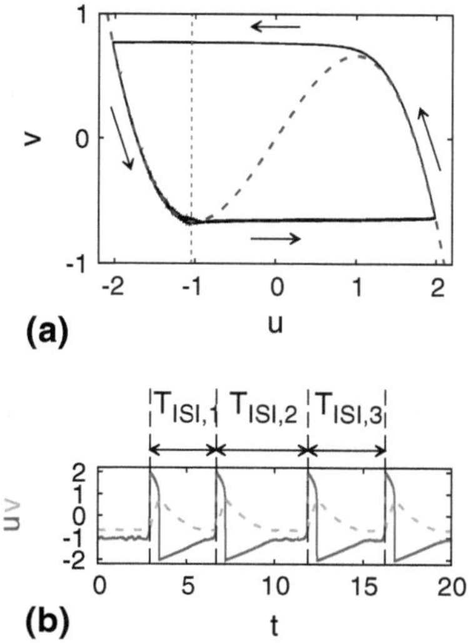

Fig. 6.3 Panel (**a**): phase portrait with nullclines of the single FitzHugh–Nagumo system (1). The activator u-nullcline is shown as blue (*dashed*) curve and the red (*dotted*) line depicts the inhibitor v-nullcline. The arrows indicate the propagation direction of the trajectory. Panel (**b**) Time series of the activator u and inhibitor v as red (solid) and green (*dashed*) curves, respectively. The vertical dashed lines mark the times of the first four spikes. Parameters: $\varepsilon = 0.005$, $a = 1.05$, and $D = 0.02$

point, the system performs subthreshold oscillations and is eventually lifted above the threshold, where the activation time is determined by the stochastic input. Then the trajectory jumps horizontally to the distant branch of the u-nullcline. This rapid change of the activator variable u is due to the small timescale ratio ε. Next, the trajectory slowly follows the nullcline to its local maximum from which it returns to the left branch of the u-nullcline. There the system undergoes a refractory period in which it relaxes to the fixed point and is susceptible for another excitation. Note that the excursion time is given by the deterministic dynamics of the system. After the discussion of the time series, I will introduce some tools for the analysis of this temporal dynamics in the following.

The vertical dashed lines in panel (b) mark the times of the first four spikes calculated from the activator variable u. The interspike intervals are labeled by $T_{\mathrm{ISI},j}$ with $j = 1, 2, 3$. Since the temporal dynamics is noise-induced, the interspike intervals are not identical. In fact, the activation time depends on the stochastic input $D\xi(t)$.

Note that one could also simplify Eqs. 6.1.1a, b by replacing the cubic dynamical function $f(u, v) = u - u^3/3 - v$ by a piecewise linear function

$$f(u, v) = \begin{cases} -u - v, & u < -1 \\ u - v, & -1 \leq u < 1 \\ -u - v, & 1 \leq u \end{cases}. \tag{6.1.7}$$

Then, the dynamics is analytically solvable in each interval, where the function f is linear, and the complete solution can be derived via proper matching conditions at $u = \pm 1$ [23].

There are various measures to characterize the dynamics and its underlying timescales in order to investigate the spiking behavior. One of these measures that will be used repeatedly in this chapter is the average interspike interval

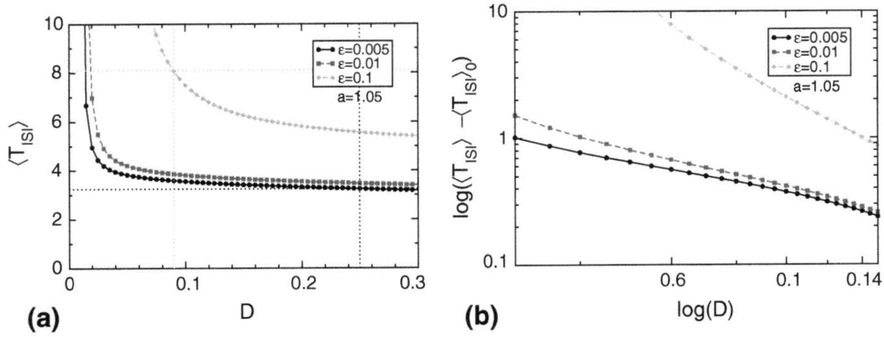

Fig. 6.4 Average interspike interval for $\varepsilon = 0.005$, 0.01, and 0.1 in dependence on the noise intensity D. The parameter a is fixed at $a = 1.05$. The intersections of the *dotted lines* in panel (**a**) indicate the values for the average value $\langle T_{\text{ISI}} \rangle$ of the respective interspike interval distribution in Fig. 6.5. Panel (**b**) displays the same data as panel (**a**) in double-logarithmic scale, where $\langle T_{\text{ISI}} \rangle_0$ refers to the mean value at $D = 0.3$ shown in panel (**a**)

$\langle T_{\text{ISI}} \rangle$. Another measure is the correlation time which will be addressed at the end of this section. In the case of coupled systems, an additional phase shift between the time series of the subsystems needs to be taken into account. Different measures to quantify this phase synchronization will be introduced in Sect. 6.2. In the following, I will analyze the average average interspike interval $\langle T_{\text{ISI}} \rangle$.

Figure 6.4 depicts this average duration between two consecutive spikes which is calculated from the activator variable u, in dependence on the noise intensity D. The green, red, and black curves correspond to different timescale ratios $\varepsilon = 0.1, 0.01$, and 0.005, respectively. One can see that the mean interspike interval decreases for larger D. Since the system is prepared in the excitable regime, excitations are due to the stochastic input of the random fluctuations $D\xi(t)$. Thus, the activation time until a spike is emitted corresponds to the stochastic part of $\langle T_{\text{ISI}} \rangle$. In fact, for small noise intensities D the time between two spikes becomes very long. Note that there is a lower bound of $\langle T_{\text{ISI}} \rangle$ in panel (a) which depends on the choice of ε. This value of $\langle T_{\text{ISI}} \rangle$ corresponds to the excursion time which is dominated by the deterministic dynamics of the system [24]. During an excursion the system is not receptive to external input in order to emit a next spike. The black and green dotted lines mark noise intensities D_1 and the respective average interspike intervals of the the distributions shown in Fig. 6.5. Panel (b) displays the same data as panel (a) in double-logarithmic scale, where the offset $\langle T_{\text{ISI}} \rangle_0$ which refers to the mean value at $D = 0.3$ shown in panel (a) is subtracted. For small noise intensities, there is a curves show a straight line which corresponds to a power-law dependence between D and $\langle T_{\text{ISI}} \rangle$ [25].

Detailed information of the timescales present in the neural system can be inferred by the interspike interval distribution, whereas in Sect. 6.2, the average interspike interval is a valuable tool to check, for instance, for frequency synchronization.

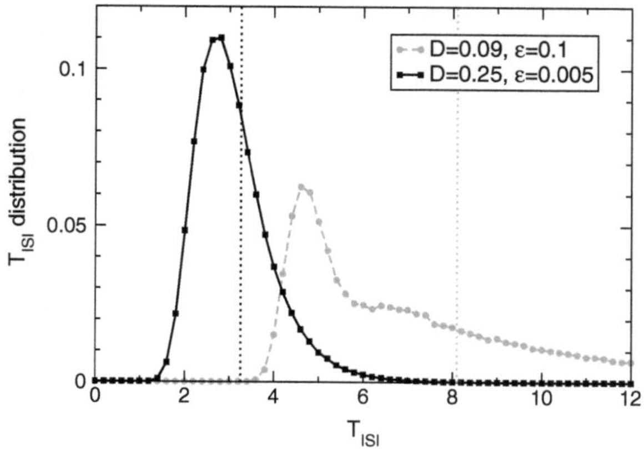

Fig. 6.5 Interspike interval distribution for $\varepsilon = 0.005$, $D = 0.25$ (*black*) and $\varepsilon = 0.1$, $D = 0.09$ (*green dashed*). The parameter a is fixed at $a = 1.05$. The *dotted lines* denote the average interspike interval given in Fig. 6.4

Figure 6.5 shows this distribution for two specific combinations of ε and D in the range of $T_{\text{ISI}} \in [0, 12]$. The black and green curves correspond to $\varepsilon = 0.005, D = 0.25$ and $\varepsilon = 0.1$, $D = 0.09$, respectively. These combinations are given as dotted lines in Fig. 6.4. The dotted lines in Fig. 6.5 indicates the value the mean interspike interval, i.e., $\langle T_{\text{ISI}} \rangle = 8.1$ for $\varepsilon = 0.1$, $D = 0.09$ and $\langle T_{\text{ISI}} \rangle = 3.25$ for $\varepsilon = 0.005, D = 0.25$ [26] which is calculated according to the definition of the mean value of T_{ISI} given by

$$\langle T_{\text{ISI}} \rangle = \int_0^{\infty} p(T_{\text{ISI}}) T_{\text{ISI}} dT_{\text{ISI}} \qquad (6.1.8)$$

with the interspike interval distribution $p(T_{\text{ISI}})$. Note that the maximum peak in the distributions does not match with the respective mean value. Especially when the distribution is asymmetric it exhibits a long tail shown, for instance, in the green curve the sole information of the mean value can be misleading for the discussion of the system's timescale.

Figure 6.6 depicts the interspike interval distribution as color code in dependence on the noise intensity D for different timescale ratios $\varepsilon = 0.005$, 0.01, and 0.1 in panels (a), (b), and (c), respectively. Thus, the curves in Fig. 6.5 are horizontal cuts through the respective panel of Fig. 6.6. One can see that the peak in the distribution becomes sharper with a less pronounced tail as the value of ε decreases.

Similar information about the timescales in the system can be extracted from the power spectrum S of a time series $x(t)$ which is defined as

$$S(\omega) = |F(\omega)|^2 \qquad (6.1.9)$$

with the Fourier spectrum

$$F(\omega) = \mathcal{F}(x) = \frac{1}{2\pi} \int_{-\infty}^{\infty} x(t) e^{-i\omega t} dt. \qquad (6.1.10)$$

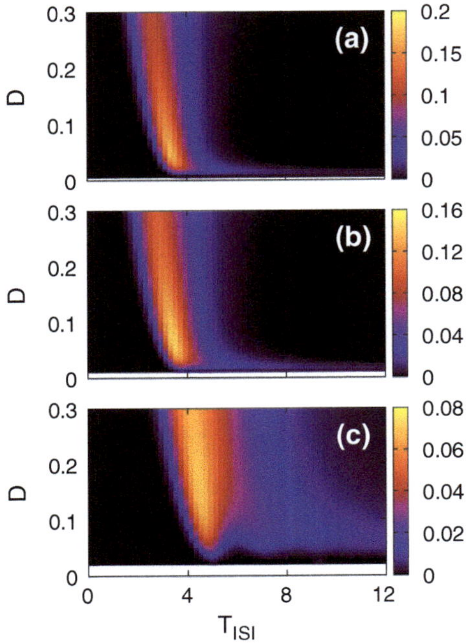

Fig. 6.6 Interspike interval distribution in dependence on the noise intensity D for $\varepsilon = 0.005$, 0.01, and 0.1 in panels (**a**), (**b**), and (**c**), respectively. The parameter a is fixed at $a = 1.05$

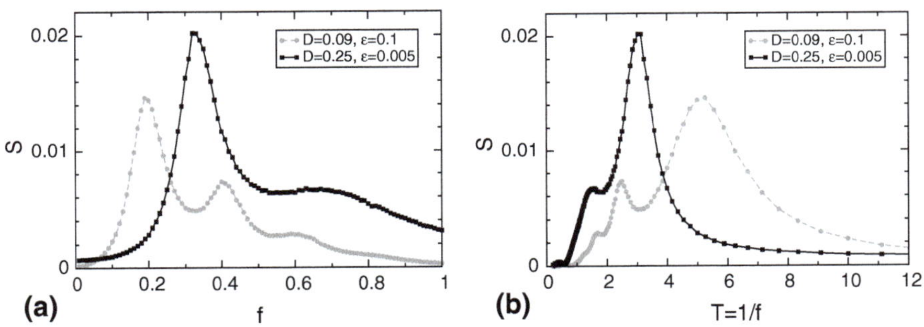

Fig. 6.7 Power spectrum S for $\varepsilon = 0.005$, $D = 0.25$ (*black*) and $\varepsilon = 0.1$, $D = 0.09$ (*green dashed*). Panels (**a**) and (**b**) show the power spectrum in dependence on the frequency f and the period $T = 1/f$, respectively. The parameter a is fixed at $a = 1.05$

Figure 6.7 displays the power spectrum for the same combinations of ε and D as in Fig. 6.5, i.e., $\varepsilon = 0.005$, $D = 0.25$ and $\varepsilon = 0.1$, $D = 0.09$ for the black and green curves. Panels (a) and (b) show the same power spectra in dependence on the frequency f and the period $T = 1/f$, respectively. Note that the period of the mean peak in panel (b) coincides with the peak in the interspike interval distribution.

Similar to Fig. 6.6, Fig. 6.8 shows the power spectrum as color code in dependence on the noise intensity D for $\varepsilon = 0.005$, 0.01, and 0.1 in panels (a), (b), and (c), respectively. One can observe as in Fig. 6.6 that the peak becomes sharper

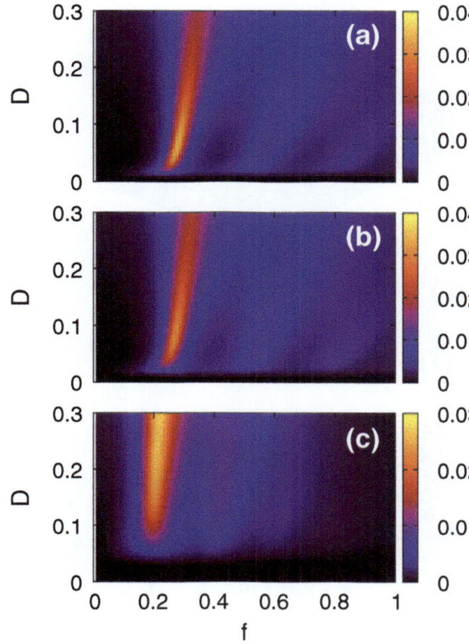

Fig. 6.8 Power spectrum in dependence on the noise intensity D for $\varepsilon = 0.005$, 0.01, and 0.1 in panels (**a**), (**b**), and (**c**), respectively. The parameter a is fixed at $a = 1.05$

for decreasing ε. Note that higher harmonics are also visible in the power spectra, but not in the interspike interval distribution of Fig. 6.6.

Figure 6.9 reproduces the power spectrum for $\varepsilon = 0.005$ in dependence on both the frequency f in panels (a) and (b) and on the period T in panels (c) and (d) as two- and three-dimensional plots. Let me stress again the correspondence to the respective panel of interspike interval distribution given in Fig. 6.6. Compare in particular the power spectrum versus $T = f^{-1}$ in Fig. 6.9d with Fig. 6.6a. Therefore, the analysis of the power spectrum provides a helpful method to analyze the dynamics due to coupling effects and time-delayed feedback in the subsequent sections.

Before discussing two coupled neural systems in the next section, I will consider another quantity which is frequently used in the analysis of systems with stochastic input, namely the correlation time t_{cor}.

Starting from the definition of the autocorrelation function $\Psi(s)$ for a time series $x(t)$ [27–29]

$$\Psi(s) = \frac{\langle [x(t) - \langle x \rangle][x(t-s) - \langle x \rangle] \rangle_t}{\langle [x(t) - \langle x \rangle]^2 \rangle_t}, \tag{6.1.11}$$

the power spectrum is related to $\Psi(s)$ via the Wiener–Khinchin theorem by a Fourier transform as follows:

$$S(\omega) = \frac{1}{\pi} \int_0^\infty \Psi(s) \cos(\omega s) ds \tag{6.1.12a}$$

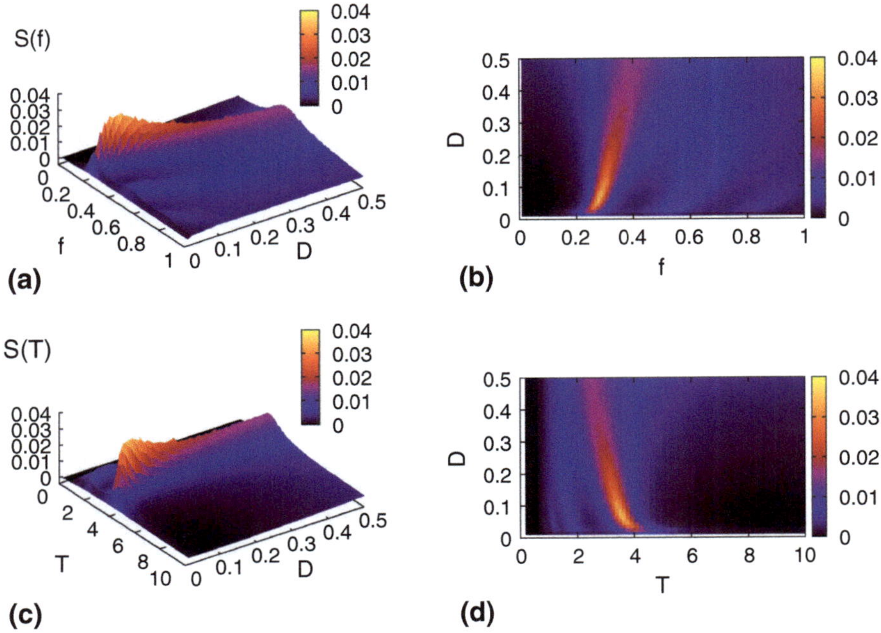

Fig. 6.9 Power spectrum S calculated from the u-variable in dependence on the noise intensity. Panels (**a**) and (**b**): S in dependence on the frequency f. Panels (**c**) and (**d**): S in dependence on the period $T = f^{-1}$. The parameters are chosen as $a = 1.05$ and $\varepsilon = 0.005$

$$\Psi(s) = \int_0^\infty S(\omega)\cos(\omega s)d\omega \tag{6.1.12b}$$

which results from the following expression by symmetry arguments concerning $\Psi(s)$

$$S(\omega) = \frac{1}{2\pi}\int_{-\infty}^{\infty} \Psi(s)e^{-i\omega s}ds \tag{6.1.13a}$$

$$\Psi(s) = \int_{-\infty}^{\infty} S(\omega)e^{i\omega s}d\omega. \tag{6.1.13b}$$

For linear stochastic systems the power spectrum has Lorentzian shape

$$S(\omega) = \frac{\alpha}{\alpha^2 + (\omega - \omega_0)^2}, \tag{6.1.14}$$

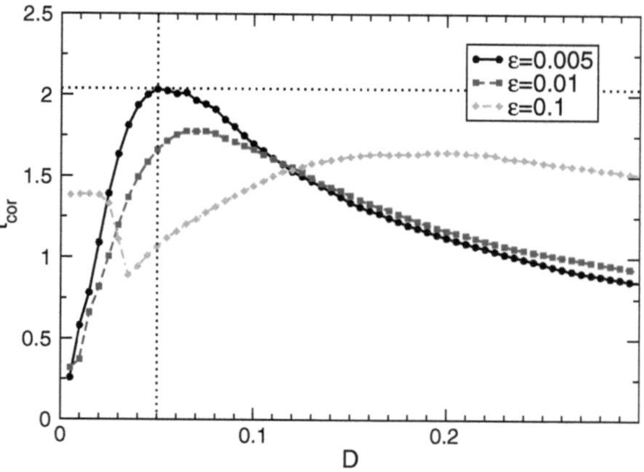

Fig. 6.10 Correlation time calculated from the time series of the activator u in Eq. 6.1.1 for $\varepsilon = 0.005, 0.01$, and 0.1 in dependence on the noise intensity D as *black* (*dots*), *red* (*squares*), and *green* (*diamonds*) curves, respectively. The parameter a is fixed at $a = 1.05$

where the height of the peak is given by $1/\alpha$ and ω_0 denotes the center frequency. The half-width σ relates the power spectrum of Eq. 6.1.14 to the autocorrelation function as follows:

$$\Psi(s) = \exp(-\sigma s)\cos(\omega_0 s). \tag{6.1.15}$$

Thus the autocorrelation function consists of an oscillating function whose amplitude decays exponentially. Furthermore, this exponential decay is related to the correlation time by [30, 31]

$$t_{\text{cor}} = \frac{2}{\pi\sigma}. \tag{6.1.16}$$

If the stochastic process is not linear, this relation does not hold anymore, but the correlation time can be calculated via the integral of the absolute value of the normalized autocorrelation function

$$t_{\text{cor}} = \frac{1}{\Psi(0)}\int_0^{\infty} |\Psi(s)|ds. \tag{6.1.17}$$

For linear stochastic processes with $\omega_0 \ll 1/t_{\text{cor}}$, Eqs. 6.1.16 and 6.1.17 coincide.

Figure 6.10 depicts the correlation time in dependence on the noise intensity D and shows the phenomenon of coherence resonance [32, 33] for $\varepsilon = 0.01$ and 0.005, i.e., the correlation is maximum at some finite noise intensity.

An exemplary power spectrum of the single FitzHugh–Nagumo system (6.1.1) for a noise intensity $D = 0.05$. is shown in Fig. 6.11a. One can see that the shape of the power spectrum is not Lorentzian, but has more structures such as secondary peaks. Accordingly, the autocorrelation function depicted in Fig. 6.11b differs from the formula (6.1.15). This can be seen from the exponential function included

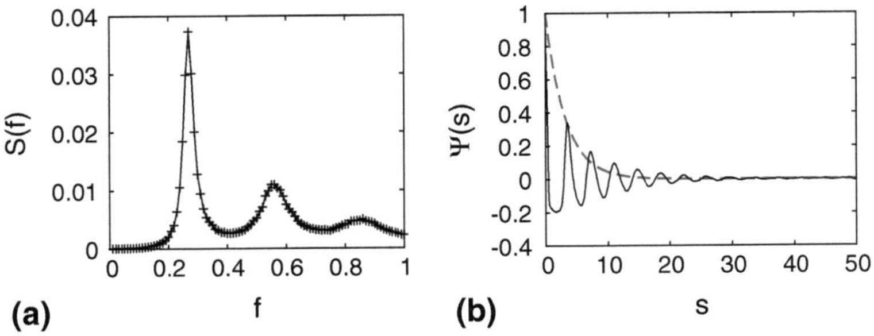

Fig. 6.11 Panel (**a**): power spectrum S of noise-induced oscillations in the FitzHugh–Nagumo system (6.1.1). Panel (**b**): autocorrelation function $\Psi(s)$ calculated from the u-variable. The exponent of the decay shown as *red curve* is taken from the *dotted lines* in Fig. 6.10. The parameters are chosen as $a = 1.05$, $\varepsilon = 0.005$, and $D = 0.05$

as red curve which shows the exponential decay in Eq. 6.1.15 for a correlation time of $t_{\mathrm{cor}} = 2.04$. This value is calculated according to Eq. 6.1.17. See also the dotted lines in Fig. 6.10 which display the correlation time for different timescale ratios $\varepsilon = 0.005, 0.01$, and 0.1 as black, red, and green curves in dependence on the noise intensity D. The plateau for small D in the green curve, i.e., for $\varepsilon = 0.1$, is due to subthreshold oscillations. In this regime, the random fluctuations are not large enough to tripper spikes. I will show in Sect. 6.4 that the interpretation of the correlation time can be intriguing and difficult in the presence of time-delayed feedback. See also Refs. [30, 31, 34].

After the introduction of the FitzHugh–Nagumo system and the investigation of a single unit, the following section deals with two instantaneously coupled neural systems of this type.

6.2 Two Coupled FitzHugh–Nagumo Systems

The topic of this section is the investigation of two instantaneously coupled neural systems [26]. As in the Sect. 6.1, I consider activator–inhibitor equations of FitzHugh–Nagumo type which are prepared in the excitable regime and are subject to random fluctuations realized as Gaussian white noise. For details on the noise source see Fig. 6.2 in Sect. 6.1. Including diffusive coupling between the two subsystems, the model can be written as

$$\begin{aligned} \varepsilon_1 \frac{du_1(t)}{dt} &= u_1(t) - \frac{u_1^3(t)}{3} - v_1(t) + C[u_2(t) - u_1(t)] \\ \frac{dv_1(t)}{dt} &= u_1(t) + a + D_1\xi_1(t) \end{aligned} \tag{6.2.1a}$$

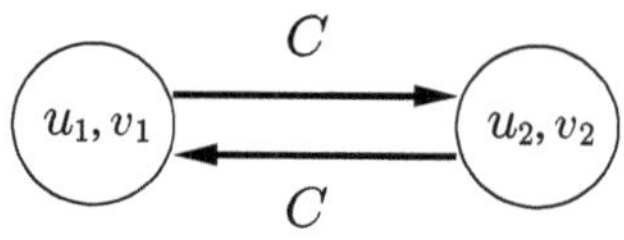

Fig. 6.12 Schematic diagram of two coupled FitzHugh–Nagumo systems of Eqs. 6.2.1

$$\begin{aligned} \varepsilon_2 \frac{du_2(t)}{dt} &= u_2(t) - \frac{u_2^3(t)}{3} - v_2(t) + C[u_1(t) - u_2(t)] \\ \frac{dv_2(t)}{dt} &= u_2(t) + a + D_2 \xi_2(t), \end{aligned} \tag{6.2.1b}$$

where the parameter C denotes the coupling strength. Figure 6.12 depicts a schematic diagram of this configuration. For reason of comparison with results when time-delayed feedback is applied, I will consider in this section the case where no control is applied to the system. The effects of time-delayed feedback will then be discussed in subsequent sections.

One could also consider a time delay in the coupling which accounts for propagation delays in the interneural connections. The effects of this delay will be discussed in detail in Sect. 6.3 [35–38]. It is also possible to analyze periodic solutions in the presence of delayed coupling in the context of a leading-order problem [39].

Note that the timescale ratios ε between the activator u_i and the inhibitor v_i ($i = 1, 2$) can be chosen independently. This will lead to substantially different intrinsic timescales of each subsystem in the uncoupled case. See also Sect. 6.1, e.g., Fig. 6.5. In addition, the two subsystems are subject to independent noise sources $D_i \xi_i(t)$ whose noise intensities D_i ($i = 1, 2$) can also be varied independently. Without this external input, the two neural oscillators relax to their stable fixed points since they are prepared in the excitable regime. Random fluctuations lead to spiking [40]. The threshold parameter a is fixed at $a = 1.05$ throughout this section. Although a is called threshold parameter, the threshold for excitation differs in the two subsystems due to different timescales ε_1 and ε_2. This can be seen from the vector field depicted in Fig. 6.13 which is an enlargement of Fig. 6.1 in the region of the fixed point. For smaller ε and in panel (b), the arrows are almost horizontal which leads to a rapid change in the u-direction and a smaller threshold.

In order to investigate the dynamics of a large number of coupled neurons in a large network, one can group several neurons into a single effective element which shows excitable dynamics and which is coupled to other elements [9, 11]. The simplest network motif is given by two coupled elements. Thus, Eqs. 6.2.1 can be interpreted as two coupled neural oscillators or neural populations. The goal is to investigate the cooperative dynamics of this simple motif. This basic study has important consequences and applications in biology and medicine as mentioned in the introduction of this chapter [2, 6, 7].

Figure 6.14 displays the time series of two coupled FitzHugh–Nagumo systems as given by Eqs. 6.2.1 for different noise intensities $D_1 = 0, 0.05$, and 1. The noise intensity D_2 is fixed a small value $D_2 = 0.09$ such that even for a vanishing noise

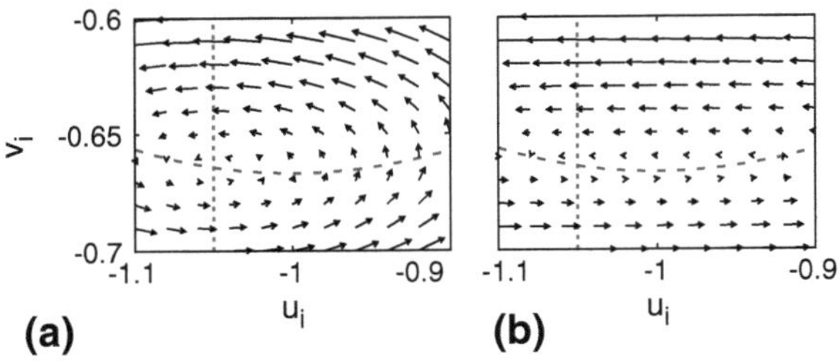

Fig. 6.13 Enlargement of the vector field of a single FitzHugh–Nagumo system around the fixed point with u_i- and v_i-nullclines with $i = 1, 2$ as blue (*dashed*) and red (*dotted*) curves, respectively. The second subsystems is shown in panel (**a**) for a timescale ratio $\varepsilon = 0.1$ and the first subsystems is displayed in panel (**b**) for $\varepsilon = 0.005$. Other parameters: $a = 1.05$, $C = 0$, and $D_1 = D_2 = 0$

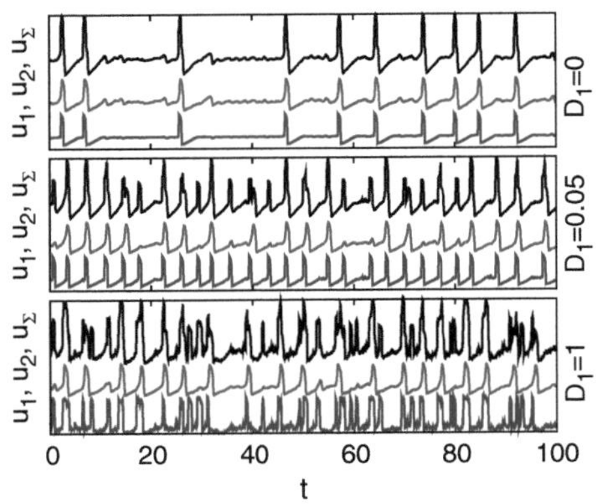

Fig. 6.14 Time series of two coupled FitzHugh–Nagumo systems: the *blue*, *red*, and *black curves* correspond to u_1, u_2, and their sum $u_\Sigma = u_1 + u_2$, respectively. The noise intensities are chosen as $D_1 = 0$, 0.05, and 1 in the respective panels. Other parameters: $\varepsilon_1 = 0.005$, $\varepsilon_2 = 0.1$, $a = 1.05$, $C = 0.07$, and $D_2 = 0.09$

amplitude in the first subsystem spikes occur occasionally. The blue, red, and black curves refer to the time series of the activator variable of the first subsystem, i.e., u_1, of the second system u_2, and the summarized signal $u_\Sigma = u_1 + u_2$, respectively. The timescale ratios are chosen as $\varepsilon_1 = 0.005$ and $\varepsilon_2 = 0.1$ throughout this section. The two neural oscillators are only weakly coupled since the coupling strength C is small, i.e., $C = 0.07$. For proper choices of the noise intensities and coupling strength, the two subsystems exhibit cooperative dynamics.

For $D_1 = 0$ the two subsystems are prefect synchrony since system 1 exhibits no independent dynamics, but it is forced by subsystem 2 via the coupling. Every emission of a spike in u_2 triggers a spike in the first system and thus, the variables u_1 and u_2 show excellent synchrony. Note that the notion of synchronization cannot be applied for vanishing noise intensity D_1 since there are no oscillations in the first subsystems in the uncoupled case. If the noise intensity D_1 increases, however, there are also noise-induced spikes in the first subsystem which are transmitted to system 2. This is not the case for all spiking events since there are also occasions in which only subsystem 1 spikes. In general the number of spikes increases since the excitations occur more frequently.

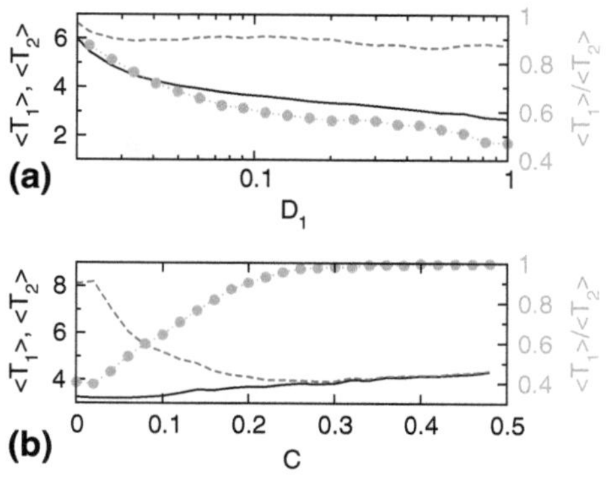

Fig. 6.15 Average interspike intervals calculated from the activator variable of each subsystem ($\langle T_1 \rangle$ as *black solid curve* and $\langle T_2 \rangle$ as *red dashed curve*) and their ratio ($\langle T_1 \rangle / \langle T_2 \rangle$ as *green dotted curve*) in dependence on the noise intensity D_1, the coupling strength C in panels (**a**) for fixed $C = 0.07$ and (**b**) for $D_1 = 0.25$, respectively. Other parameters as in Fig. 6.14

For a large value of $D_1 = 1$, the time series of u_1 is noise-dominated, but as the coupling strength is small, the second neural oscillator is not excited by every spike and its frequency of spikes is similar to the previous case of $D_1 = 0.05$.

In order to quantify the cooperative dynamics of the coupled neural oscillators, it is helpful to consider the mean interspike interval of each subsystem. The ratio of these average values provides information about the synchronization of the systems in Eqs. 6.2.1. To be precise, the ratio of the average interspike intervals is a measure to investigate frequency synchronization. If the ratio has the value 1, both systems show on average the same spiking frequency. Note that this quantity is only one measure of synchronization [41]. One has to check also for phase synchronization which will be quantified by a phase synchronization index and average phase synchronization intervals towards the end of this section.

Figure 6.15 shows the mean interspike intervals $\langle T_1 \rangle$ and $\langle T_2 \rangle$ calculated from the activator variable of the respective subsystem as black and red curves, respectively, as well as their ratio $\langle T_1 \rangle / \langle T_2 \rangle$. Panel (a) displays the dependence on the noise intensity D_1 for fixed coupling strength $C = 0.07$ and panel (b) shows the dependence on C while D_1 is constant at $D_1 = 0.25$. The noise intensity D_2 is small as in Fig. 6.14, i.e., $D_2 = 0.09$. As the noise intensity D_1 increases in Fig. 6.15a the ratio $\langle T_1 \rangle / \langle T_2 \rangle$ becomes smaller and thus, the two subsystems are less synchronized. Figure 6.15b shows the opposite tendency: as the coupling strength C becomes larger the two neurons are better synchronized because the value $\langle T_1 \rangle / \langle T_2 \rangle$ tends to 1. The two neural oscillators are in 1:1 synchronization. Note that $C = 0$ recovers the mean interspike interval of the uncoupled configuration of Sect. 6.1 since Eqs. 6.2.1 decouple. See also the respective dotted lines of Fig. 6.5.

Figure 6.15a and b is one-dimensional projections of the parameter space. Figure 6.16 depicts a two-dimensional cut through this parameter space in dependence on both the coupling strength C and the noise intensity D_1.

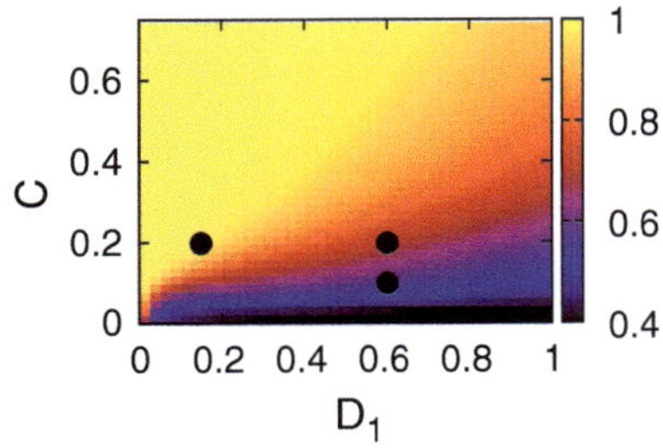

Fig. 6.16 Ratio of interspike intervals $\langle T_1\rangle/\langle T_2\rangle$ of the two subsystems as color code in dependence on the coupling strength C and noise intensity D_1. No control is applied to the system. The *dots* mark the parameter choice for different synchronization regimes used in the following. Other parameters as in Fig. 6.14

Calculating the ratio of the average interspike interval of the two neurons $\langle T_1\rangle$ and $\langle T_2\rangle$, i.e., $\langle T_1\rangle/\langle T_2\rangle$ which is given by the color code, one can see in Fig. 6.16 how the frequency synchronization changes in dependence on these two parameters. For a small value of D_1 and large coupling strength, the two subsystems display synchronized behavior, $\langle T_1\rangle/\langle T_2\rangle \approx 1$. The timescales in the interacting systems are adjusted to 1:1 synchronization. On average, they show the same number of spikes indicated by yellow color. The two subsystems are less synchronized in the blue region. The black dots define three different cases: moderately, weakly, and strongly synchronized systems which are realized by the specific choices of the coupling strength C and noise intensity D_1 in the first subsystem. Strong synchronization can be found for small noise intensity D_1 and large C, e.g., $D_1 = 0.15$ and $C = 0.2$. Moderate synchronization is given for a choice of $D_1 = 0.6$ and $C = 0.2$, and weak synchronization can be realized by $D_1 = 0.6$ and $C = 0.1$ [40].

These different cases will be used as configurations in the following and in Sects. 6.3–6.5, when time-delayed feedback will be applied to these moderately, weakly, and strongly synchronized systems.

In terms of the time series, the three configurations of moderate, weak, and strong synchronization are displayed in Fig. 6.17 as panels (a), (b), and (c), respectively. In all panels, the black, red, and blue curves refer to the global u_Σ-variable, the u_2-, and u_1-variable, respectively, where u_Σ is given by $u_\Sigma = u_1 + u_2$ [40].

Moderately synchronized systems perform mostly synchronized spiking as depicted in Fig. 6.17a. However, there are certain events where only one system oscillates. In the case of weak synchronization, the spikes of the two subsystems coincide less as can be seen from the time series of the summarized signal u_Σ in panel (b) of Fig. 6.17. For strongly synchronized subsystems (see Fig. 6.17c), the time series of the u_1- and u_2-variable exhibit spiking at the same time [40].

Before discussing the results with respect to phase synchronization at the end of this section, I will consider timescales present in the coupled system (6.2.1) by

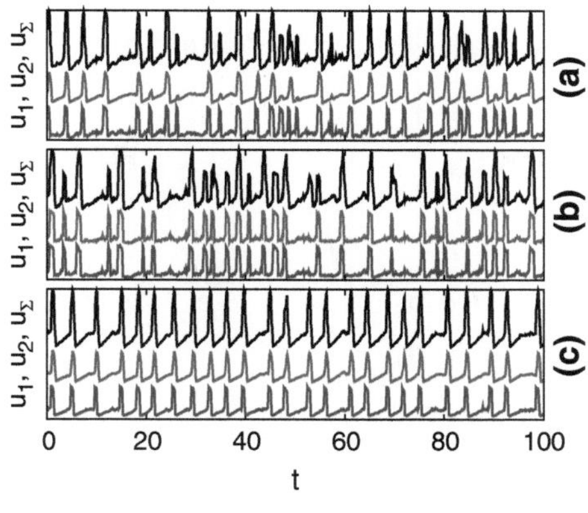

Fig. 6.17 Time series of two coupled FitzHugh–Nagumo systems (2). Panels (**a**), (**b**), and (**c**) correspond to moderately ($C = 0.2$, $D_1 = 0.6$), weakly ($C = 0.1$, $D_1 = 0.6$), and strongly ($C = 0.2$, $D_1 = 0.15$) synchronized systems, respectively. In all panels, the *black*, *red*, and *blue curves* refer to the summarized variable $u_\Sigma = u_1 + u_2$, the u_2-, and the u_1-variable, respectively. Other parameters as in Fig. 6.14

analysis of the power spectrum S in the following. For a definition of S see Eqs. 6.1.9 and 6.1.10 in Sect. 6.1. The power spectrum provides information which timescales are predominant in the coupled subsystems.

Figure 6.18 depicts the power spectrum of the two coupled FitzHugh–Nagumo systems (6.2.1). The red and blue curves correspond to the power spectrum, of the u_1- and u_2-variable, respectively. Panels (a), (b), and (c) refer to the case of moderate, weak, and strong synchronization, respectively. This change of frequency synchronization is reached by choosing different coupling strengths C and noise intensities D_1. In the case of moderate synchronization ($C = 0.2, D_1 = 0.6$) the peaks of the power spectrum partly overlap. Weak synchronization ($C = 0.1$, $D_1 = 0.6$) shows a smaller overlap, whereas for strong synchronization ($C = 0.2$, $D_1 = 0.15$) the spectra almost coincide [40].

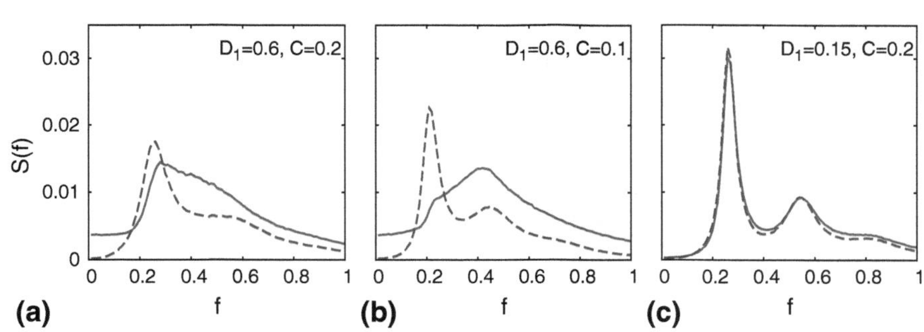

Fig. 6.18 Power spectrum of the activator variables u_1 and u_2 of the two coupled subsystems (6.2.1) as *red solid* and *blue dashed* curves, respectively. Panels (**a**), (**b**), and (**c**) correspond to the case of moderate, weak, and strong synchronization, respectively. Other parameters as in Fig. 6.14

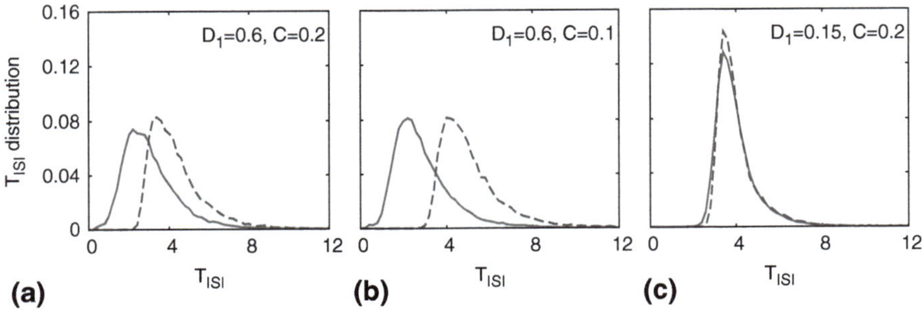

Fig. 6.19 Interspike interval distribution of the activator variables u_1 and u_2 of the two coupled subsystems (6.2.1) as *red solid* and *blue dashed* curves, respectively. Panels (**a**), (**b**), and (**c**) correspond to the case of moderate, weak, and strong synchronization, respectively. Other parameters as in Fig. 6.14

Similar information can be obtained from the interspike interval distribution of the three different combinations of D_1 and C. Figure 6.19 shows this distribution with the same color coding as in Fig. 6.18. Panels (a), (b), and (c) corresponds to moderate, weak, and strong synchronization. For moderate synchronization, one can see that the peaks of u_1 and u_2 are separated, but they overlap to a large extent. This overlap is reduced since centers of the distribution are further apart. For strong synchronization, the interspike interval distributions of the individual subsystem match. Note that the interspike interval corresponding to the center of the peaks relates to the frequency of the respective panel in Fig. 6.18 by its inverse value.

In order to gather more information about the combined system of two neural oscillators, one can discuss the frequencies present in the time series of the summarized activator signal $u_\Sigma = u_1 + u_2$. Figure 6.20 depicts the power spectrum of the summarized variable as well as the power spectrum of the individual

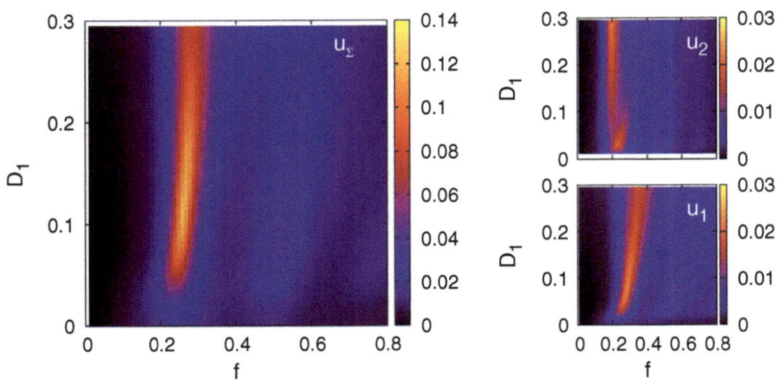

Fig. 6.20 Power spectrum S of noise-induced oscillations in color code for the summarized variable $u_\Sigma = u_1 + u_2$ and the individual activator variables u_1 and u_2 of the coupled FitzHugh–Nagumo systems (6.2.1) in dependence on the noise intensity D_1 for a fixed coupling strength $C = 0.2$. Other parameters as in Fig. 6.14

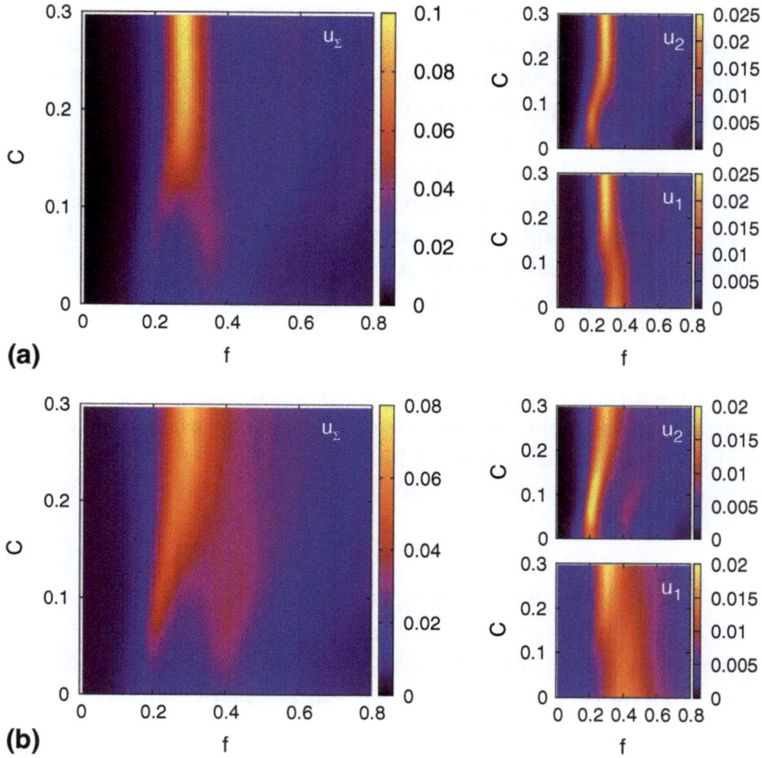

Fig. 6.21 Power spectrum S of noise-induced oscillations for the summarizes variable $u_\Sigma = u_1 + u_2$ and the individual activator variables u_1 and u_2 of the coupled FitzHugh–Nagumo systems (6.2.1) in dependence on the coupling strength C. The noise intensity D_1 is set to $D_1 = 0.25$ and 0.5 in panels (**a**) and (**b**), respectively. Other parameters as in Fig. 6.14

oscillators in dependence on the noise intensity D_1 for fixed coupling strength $C = 0.2$ as color code. Note that there is a distinct maximum in the power spectrum indicated by yellow color. For increasing D_1 this main peak is slightly shifted towards larger frequencies.

Figure 6.21 displays the power spectrum of u_Σ and the individual activator variables u_1 and u_2 for varying coupling strength C. The noise intensity D_1 is fixed at $D_1 = 0.2$ and 0.5 in panels (a) and (b), respectively. For small C one can distinguish in the power spectrum of the summarized signal two peaks which arise from the main frequencies of the uncoupled individual subsystems. Compare with Fig. 6.7 in Sect. 6.1 when the case of a single FitzHugh–Nagumo system is discussed.

There are at least two ways which lead to synchronization: frequency locking and the suppression natural dynamics. This observation holds for periodic systems as well as for noise-induced and chaotic oscillations [42]. Depending on the specific choice of the system's parameters such as the timescale separation ε, both scenarios can be realized for neural oscillators [26] and are displayed in Fig. 6.21a and b.

The mechanism of frequency locking can be seen in Fig. 6.21a where the peaks of the individual activator variables u_1 and u_2 are clearly distinguishable and change smoothly as the coupling strength C increases. They meet in a single peak for large values of C. Thus, the mutually coupled subsystems operate on the same timescale. Figure 6.21b shows the case when the natural dynamics of the first system is suppressed. This can be inferred from the observation that in the power spectrum of u_1 the peak originating from the uncoupled configuration, i.e., $C = 0$, vanishes as C becomes larger. At the same time, a second spectral peak emerges at the frequency of u_2 indicated by the yellow color.

The measures for cooperative dynamics considered so far, i.e., the ratio of average interspike intervals and the power spectrum, are insensitive to phase relations. They are a valuable tool the detection of frequency synchronization, but a phase shift between the subsystems cannot be detected. A measure of phase synchronization can be obtained by introducing phase variables for each subsystems, and monitoring their difference [40].

For definition of such a phase variable, one can generate a phase from the time series of spikes as follows

$$\varphi(t) = 2\pi\frac{t - t_{i-1}}{t_i - t_{i-1}} + 2\pi(i - 1), \tag{6.2.2}$$

where t_i denotes the time of the ith spike. With this definition the phase increases by a value of 2π for each spike [26, 41, 43, 44]. The phase difference $\Delta\varphi$ between two subsystems can be defined for general $n{:}m$ synchronization as follows

$$\Delta\varphi_{n,m}(t) = \left|\varphi_1(t) - \frac{m}{n}\varphi_2(t)\right|, \tag{6.2.3}$$

where $\varphi_1(t)$ and $\varphi_2(t)$ denote the phases of the respective subunits according to Eqs. 6.1.1. In the following, I will restrict the discussion to 1:1-synchronization and skip the indices of $\Delta\varphi_{1,1}(t)$ [40].

Figure 6.22 displays the time series of the phase difference $\Delta\varphi$. The black, red, and green curves correspond to the case of moderate ($C = 0.2$, $D_1 = 0.6$), weak ($C = 0.1$, $D_1 = 0.6$), and strong ($C = 0.2$, $D_1 = 0.15$) synchronization, respectively. The inset depicts an enlargement for moderate synchronization. One can see that for better synchronization, the slope of $\Delta\varphi$ becomes flatter as in the green curve which corresponds to strongly synchronized neural oscillators. Note that in principle the curves do not need to be monotonicly increasing if the second subsystem performs a spike while the first one remains subthreshold. For the chosen parameters, in particular the different timescale ratios $\varepsilon_1 = 0.005$ and $\varepsilon_2 = 0.1$ and small noise intensity $D_2 = 0.09$, the first subsystem is more excitable and exhibits a larger number of spikes. Thus, the phase φ_1 increases faster than φ_2.

In Refs. [26, 45], a measure for the phase synchronization is derived from the definition of the phase difference (6.2.2) which is called phase synchronization index γ. This quantity is defined using the phase difference $\Delta\varphi$ as follows

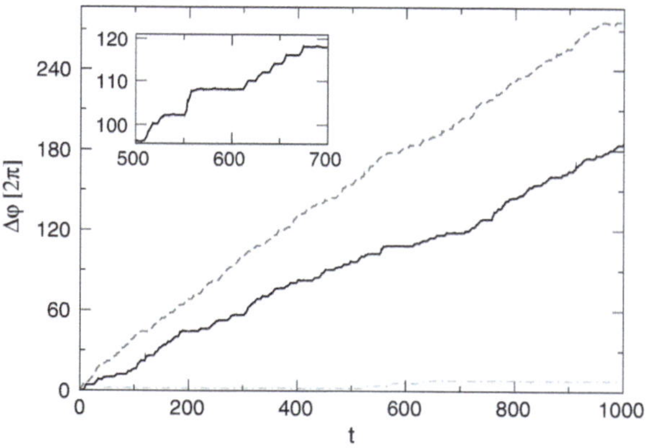

Fig. 6.22 Phase difference between the activator signals u_1 and u_2 of Eqs. 6.1.1 in units of 2π. The *black solid*, *red dashed*, and *green dash-dotted* curves correspond to the case of moderate ($C = 0.2$, $D_1 = 0.6$), weak ($C = 0.1$, $D_1 = 0.6$), and strong ($C = 0.2$, $D_1 = 0.15$) synchronization, respectively. The *inset* shows an enlargement for moderate synchronization. Other parameters as in Fig. 6.14

$$\gamma = \sqrt{\langle \cos \Delta\varphi(t)\rangle^2 + \langle \sin \Delta\varphi(t)\rangle^2}. \tag{6.2.4}$$

It varies between 0 which corresponds to no synchronization and 1 associated with perfect synchronization [26, 40].

Figure 6.23 depicts the synchronization index γ in dependence on the coupling strength C and noise intensity D_1. As shown in Ref. [26], the dependence of the γ resembles the dependence of the ratio $\langle T_1\rangle/\langle T_2\rangle$ used for the detection of frequency synchronization. This can be seen by compare of Fig. 6.23 to Fig. 6.16. Therefore, phase and frequency synchronization occur simultaneously in the coupled FitzHugh–Nagumo systems without time-delayed feedback. Note that the definition of γ involves time averages and the range of γ is bounded between 0 and 1. Thus, one cannot infer from this scalar a timescale during which the two subsystems are in phase synchrony. This can be done using a different quantity namely the average phase synchronization interval δ. This quantity will be introduced in the following.

If the two subsystems are, for instance, already strongly synchronized, it is helpful to consider the time intervals during which the phase difference stays in a

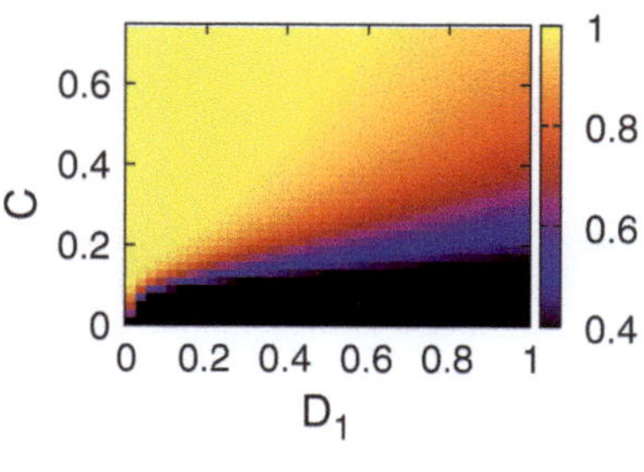

Fig. 6.23 Phase synchronization index γ as color code in dependence on the coupling strength C and noise intensity D_1. No control is applied to the system. The *dots* mark the parameter choice for different synchronization regimes used in the following. Other parameters as in Fig. 6.14

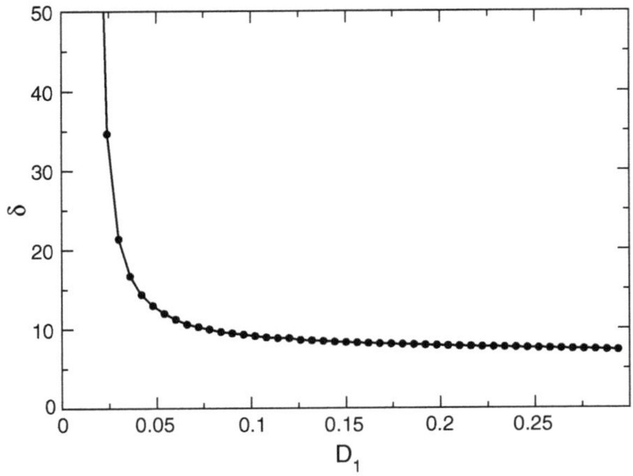

Fig. 6.24 Average phase synchronization interval in dependence on the noise intensity D_1 for fixed coupling strength $C = 0.07$. Other parameters as in Fig. 6.14

2π-phase range [46–48]. In these synchronization intervals, the subsystems exhibit the same number of spikes. From the inset of Fig. 6.22, one can clearly see the plateaus in between phase jumps of 2π. At these jumps, only one subsystem shows a spike whereas the other one remains subthreshold. In the case of strong synchronization, for instance, $\Delta\varphi$ shows only a few phase jumps and remains in a 2π-range for large time intervals. The quantity δ measures the average lengths of these intervals [40].

Figure 6.24 shows the dependence of the average phase synchronization interval δ on the noise intensity D_1 for fixed coupling strength $C = 0.07$. One can see that δ decreases monotone as D_1 becomes larger. Note that for small noise intensities the average phase synchronization interval can be arbitrarily large as the length of the phase synchronization interval is only determined by the length of the time series. Compare also to the time series for $D_1 = 0$ shown in Fig. 6.14.

Figure 6.25 depicts the dependence of δ on the coupling strength C while the noise intensity D_1 is fixed at $D_1 = 0.25$ and 0.5 in the black and red curves,

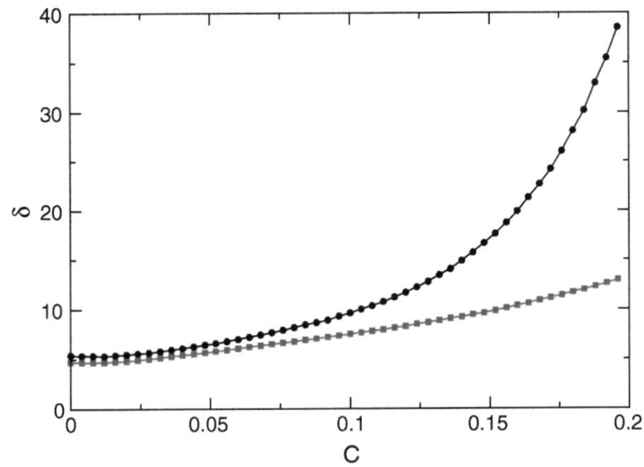

Fig. 6.25 Average phase synchronization interval in dependence on the coupling strength C for fixed $D_1 = 0.25$ (*black dots*) and 0.5 (*red squares*). Other parameters as in Fig. 6.14

respectively. One can see that the two subsystems exhibit stronger phase synchronization for larger coupling strengths.

In summary, this section demonstrated that two instantaneously coupled neural oscillators can exhibit cooperative dynamics depending on the coupling parameter and the individual noise intensities. Various tools for the characterization of both frequency synchronization and phase synchronization were introduced: interspike interval distribution, ratio of the average interspike intervals, power spectrum, phase synchronization index, and average phase synchronization interval. These measures will be used in the following sections when time-delayed feedback is added in different configurations in order to influence the neural dynamics. In the next Sect. 6.3, time-delayed feedback control will be applied to a single excitable neuron modeled as FitzHugh–Nagumo systems. I will discuss a single as well as two feedback terms in the equations. The latter case be be compared to two delay-coupled neurons with additional time-delayed feedback. Section 6.4 is devoted to a detailed discussion of the influence of the control scheme on the dynamics of two coupled neural elements. Especially, I will consider effects due to an additional control parameter, namely the memory parameter of extended time-delayed feedback control [22].

6.3 Single FitzHugh–Nagumo System and Time-Delayed Feedback

The previous sections introduced the FitzHugh–Nagumo system as a paradigmatic model for excitable neural dynamics and presented various measures to characterize the cooperative dynamics of coupled neural elements. The focus of this section is the effects of time-delayed feedback control on neural systems, where I will restrict the investigation at this point to the case of a single, isolated neural element. Effects concerning the synchronization of two coupled FitzHugh–Nagumo systems will be elaborated in Sect. 6.4.

A single FitzHugh–Nagumo system with activator u and activator v is given by the following equation including time-delayed feedback

$$\varepsilon\frac{du(t)}{dt} = u(t) - \frac{u(t)^3}{3} - v(t) \tag{6.3.1a}$$

$$\frac{dv(t)}{dt} = u(t) + a + D\xi(t) + K[v(t-\tau) - v(t)]. \tag{6.3.1b}$$

As in the previous Sects. 6.1 and 6.2, the system is prepared in the excitable regime with $a = 1.05$ and the activator u operates on a faster timescale than the inhibitor variable v which is taken into account by the small timescale ratio $\varepsilon = 0.005$ unless stated otherwise. By this choice, the timescale of Eqs. 6.3.1

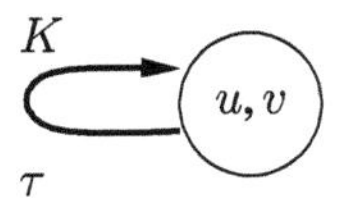

Fig. 6.26 Schematic diagram of a single FitzHugh–Nagumo system including time-delayed feedback according to Eqs. 6.3.1

corresponds to the natural timescale of the inhibitor variable v. This variable is the dominating contribution to the interspike interval because it determines the duration of the refractory period.

The system is subject to a noise term with noise intensity D and time-delayed feedback. Since the focus of this section is the control scheme, D is fixed at $D = 0.09$. The control method of Pyragas type is determined by the feedback gain K and time delay τ [49]. For details on this control scheme see Chap. 2.

Figure 6.26 depicts a schematic diagram of the configuration given in Eq. 6.3.1. In order to investigate the effects of time-delayed feedback control, I will discuss the interspike interval distribution and the power spectrum. Measures for synchronization as introduced in Sect. 6.2 are obsolete for a single system, but will come into play again in Sect. 6.4.

At first, I will consider histograms of the interspike interval distribution T_{ISI} in dependence on the time delay τ, where the feedback gain K is fixed at $K = 0.2$. Figures 6.27 and 6.28 depict these histograms in linear and half-logarithmic scale, respectively. In both figures, panels (a)–(l) correspond to a value of $\tau = 0, 1, 2, \ldots, 11$ and the black curve shows the case without feedback, i.e., $\tau = 0$ which is added for better comparison as reference case.

In the absence of control shown in panel (a), the histogram exhibits a single peak at $T_{\text{ISI}} \approx 4$ which is due to noise-induced oscillations. For small time delays as in panels (b) to (c), the peak is shifted towards larger interspike intervals. If the time delay is in the range of the uncontrolled interspike intervals as displayed in panels (d)–(h) for $\tau = 3, 4, 5, 6,$ and 7, one can see a reduced number of intervals in the histogram at $T_{\text{ISI}} = \tau + 1$. This is due to a suppression of timescales which match with the time delay in the controller. For even larger delays as in panels (i)–(l), the main peak returns to its initial position, but the total number of spikes is reduced. The half-logarithmic plot shows the same effects, of course. In addition, one can see that the dependence of the number above the peak in panels (a)–(c) for $\tau = 0, 1,$ and 2 and beyond the minimum in all other panels is linear in half-logarithmic scale, i.e., there is an exponential distribution of interspike intervals on top of a fixed excursion time as excepted from theory [24].

After these first results on the dynamics of a single FitzHugh–Nagumo system with time-delayed feedback control, Fig. 6.29 depicts the interspike interval distribution as color code in dependence on the time delay τ for a fixed feedback gain $K = 0.2$ and 0.5 in panels (a) and (b), respectively. The yellow lines are guides to the eye for $T_{\text{ISI}} = n\tau$ with $n = 1, 2, 3$. A detailed analysis of the characteristic timescales in the FitzHugh–Nagumo system subject to constant and time-delayed feedback can be found in Ref. [24].

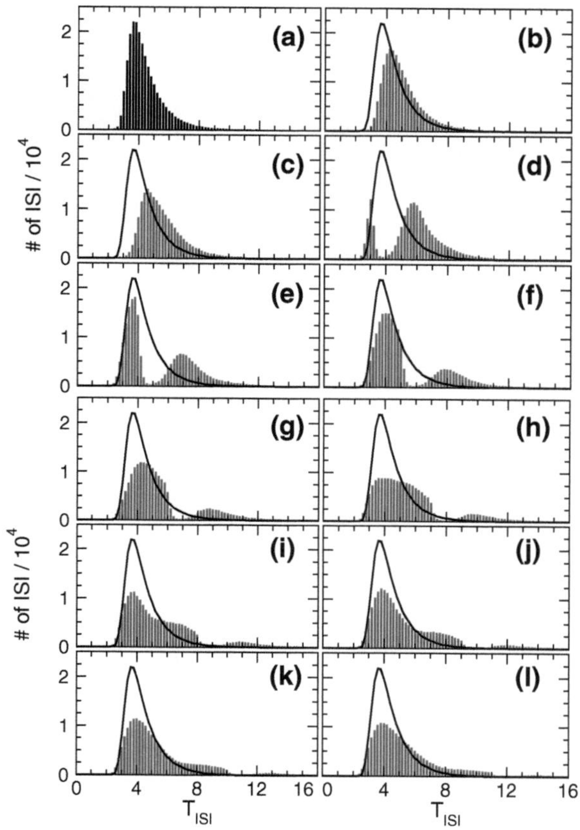

Fig. 6.27 Histogram of the interspike intervals T_{ISI} of a single FitzHugh–Nagumo system including time-delayed feedback for different time delays τ. Panels (**a**)–(**h**) correspond to $\tau = 0, 1, 2,\ldots, 11$, respectively. The *black curve* refers to the case when no control force is applied. The feedback is applied in the inhibitor variable v according to Eqs. 6.3.1 with a feedback gain $K = 0.2$. Other parameters: $a = 1.05$, $\varepsilon = 0.005$, and $D = 0.09$

Similar to the findings in the histograms, one can see a shift on the maximum of the distribution in the range of small time delays τ. Furthermore, the distribution is reduced for T_{ISI} slightly above τ as indicated by the black diagonal stripe to the right of the line $T_{\mathrm{ISI}} = \tau$. For $K = 0.5$ in panel (b), one an see a second valley in the distribution slightly above $T_{\mathrm{ISI}} = 2\tau$. At the same time, the peak for large τ is less pronounced in panel (b).

After the investigation of the interspike interval distribution, I will discuss the power spectrum for the same parameters in the following. Figure 6.30 displays the power spectrum as color code for varying τ and fixed feedback gain $K = 0.2$ and 0.5 in panels (a), (b) and (c), (d), respectively. The corresponding interspike interval distribution is given in Fig. 6.29a and b. Note that both columns depict the same data, but the power spectra are presented in dependence on the frequency f in panels (a), (c) and on the period $T = f^{-1}$ in panels (b), (d).

One can see that time-delayed feedback control leads to a modulation of the power spectrum. Recall that the uncontrolled power spectrum has only a single peak. Compare Fig. 6.9 of Sect. 6.1. Higher harmonics can clearly be observed in

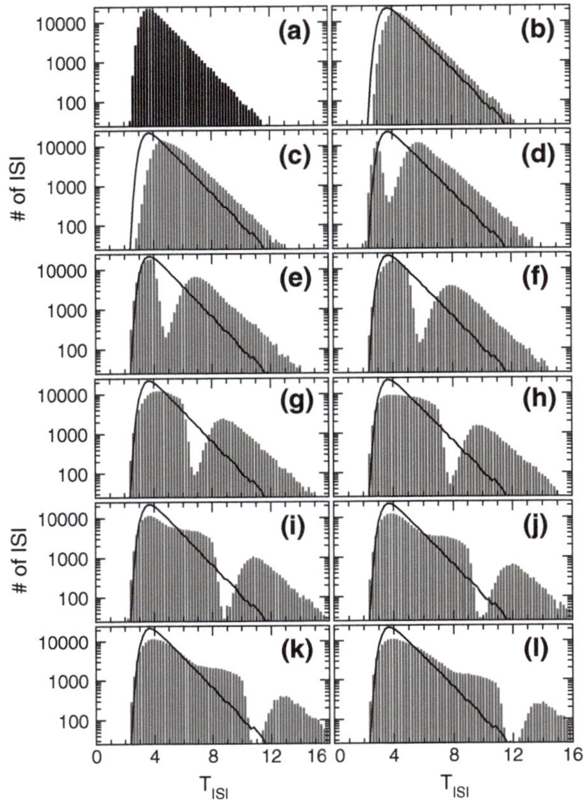

Fig. 6.28 Histogram of the interspike intervals T_{ISI} of a single FitzHugh–Nagumo system including time-delayed feedback for different time delays τ as Fig. 6.27 in half-logarithmic scale. Parameters as in Fig. 6.27

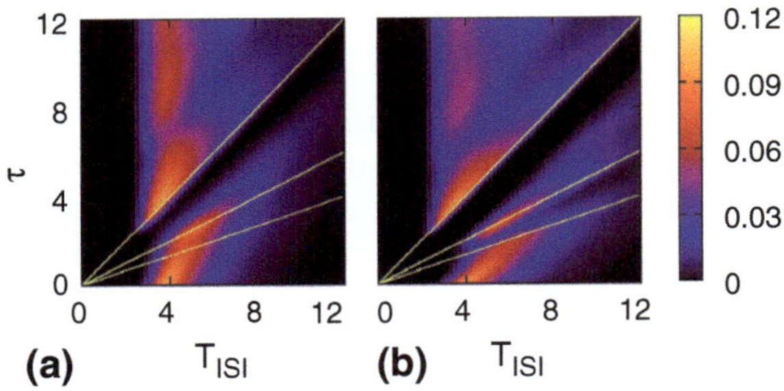

Fig. 6.29 Interspike interval distribution of a single FitzHugh–Nagumo system including time-delayed feedback according to Eqs. 6.3.1 as color code in dependence on the time delays τ. The feedback gain is fixed at $K = 0.2$ and 0.5 in panels (**a**) and (**b**), respectively. The *yellow lines* correspond to $T_{\mathrm{ISI}} = n\tau$ with $n = 1, 2, 3$ and are added as guide to the eye. Other parameters as in Fig. 6.27

all panels. In the power spectrum corresponding to a larger feedback of $K = 0.5$ in panels (c), (d), the peaks are sharper as their widths is decreased compared to the case of $K = 0.2$ shown in panels (a), (b).

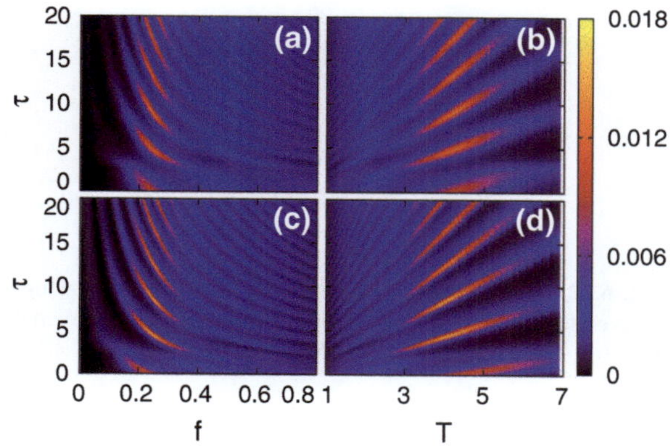

Fig. 6.30 Power spectrum S of a single FitzHugh–Nagumo system with time-delayed feedback as color code in dependence on the time delays τ. Panels (**a**), (**b**) and (**c**), (**d**) correspond to a fixed feedback gain $K = 0.2$ and 0.5, respectively, and show the power spectrum vs the frequency f and period $T = f^{-1}$. Parameters as in Fig. 6.27

Before considering extensions to the present case shown schematically in Fig. 6.26, I will briefly discuss the influence of time-delayed feedback on the correlation time.

At the end of Sect. 6.1, the correlation time t_{cor} was defined as the integral of the absolute value of the autocorrelation function. See Eq. 6.1.17. Figure 6.31 depicts t_{cor} in dependence on the time delay τ where the green (dashed) and red (solid) curve refer to the two above chosen feedback gains $K = 0.2$ and 0.5, respectively. Similar to the power spectrum, one can see a modulation of the correlation time as τ increases. Note that the larger value of K yields a larger correlation time. As presented in Fig. 6.10 of Sect. 6.1, the correlation time is proportional to the exponent of envelope of the autocorrelation function for linear stochastic systems. A similar linear connection can be derived relating t_{cor} to the width of the peak in the power spectrum. See Eqs. 6.1.14–6.1.16. In the case of time-delayed feedback these findings do not hold anymore. This can be seen, for instance, in Fig. 6.30, where the application to time-delayed feedback control leads to multiple peaks in the power spectrum.

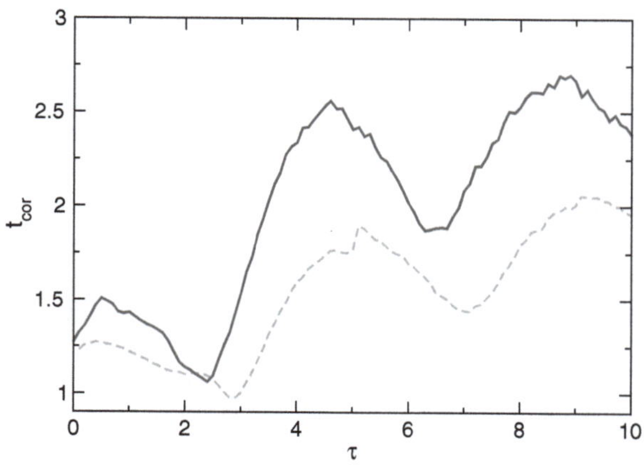

Fig. 6.31 Correlation time of a single FitzHugh–Nagumo system with time-delayed feedback vs the time delays τ. The green (*dashed*) and red (*solid*) curves correspond to a fixed feedback gain of $K = 0.2$ and 0.5, respectively. Parameters as in Fig. 6.27

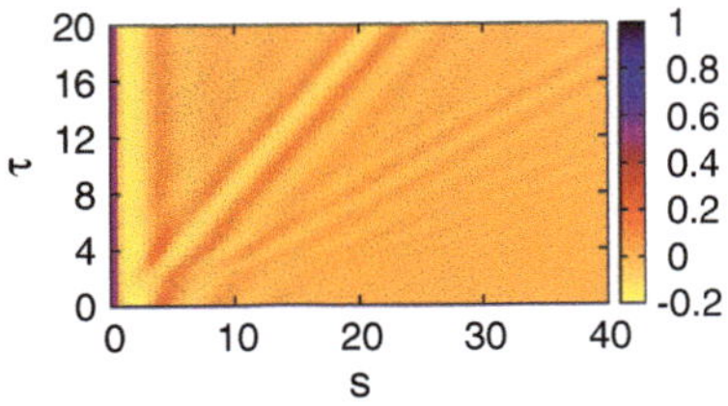

Fig. 6.32 Autocorrelation function Ψ of a single FitzHugh–Nagumo system including time-delayed feedback as color code in dependence on the time delays τ. The feedback gain is fixed at $K = 0.2$. Parameters as in Fig. 6.27

Figures 6.30 and 6.31 are in agreement with Refs. [50, 51]. In these works, a single FitzHugh–Nagumo system was investigated with the following parameters $\varepsilon = 0.01$ and $a = 1.1$. As in Eqs. 6.3.1, the neural model was subject to time-delayed feedback control of Pyragas type applied to the equation of inhibitor v. Analytical approximations were developed in Refs. [24, 52]. In addition, Refs. [50, 51] present a similar investigation of the Van-der-Pol system which is equivalent to the FitzHugh–Nagumo system for $a = 0$ as will be shown in the following: Starting from the activator equation (6.3.1a), a time derivative leads to:

$$\varepsilon \frac{d^2u(t)}{dt^2} = \frac{du(t)}{dt} - u(t)^2 \frac{du(t)}{dt} - \frac{dv(t)}{dt}. \tag{6.3.2}$$

Replacing the last term by the inhibitor equation (6.3.1b) yields

$$\varepsilon \frac{d^2u(t)}{dt^2} = \frac{du(t)}{dt} - u(t)^2 \frac{du(t)}{dt} - u(t) + a + D\xi(t). \tag{6.3.3}$$

For $\varepsilon = 1$, i.e., similar timescales of $u(t)$ and $v = du(t)/dt$, this equation is equivalent to a Van-der-Pol equation with constant input a and Gaussian white noise. In Refs. [50, 51], time-delayed feedback is applied to the Van-der-Pol system in the activator variable. This is a coupling scheme different from the present case of Eq. 6.3.1a. I will investigate the effects of different coupling schemes in Sect. 6.5. Time-delayed feedback was also considered for the Van-der-Pol system in Refs. [30, 31] and in Ref. [34] the extended form of the control method [22] was applied.

Figure 6.32 shows the autocorrelation function $\Psi(s)$ as color code for a feedback gain $K = 0.2$ in dependence on the time delay as color code. Note that especially for small, fixed values of τ the autocorrelation function decreases rapidly in the beginning indicated by the small dark stripe for small s. This rapid decay is then followed by an additional slow decay for large time shifts s.

Next, I will briefly consider an extension of the original Pyragas scheme which involves integer multiples of a basic time delay τ [22]. This extended time-delayed feedback control was introduced in Sect. 2.2 and will be investigated in detail in the context of two coupled FitzHugh–Nagumo systems in Sect. 6.4.

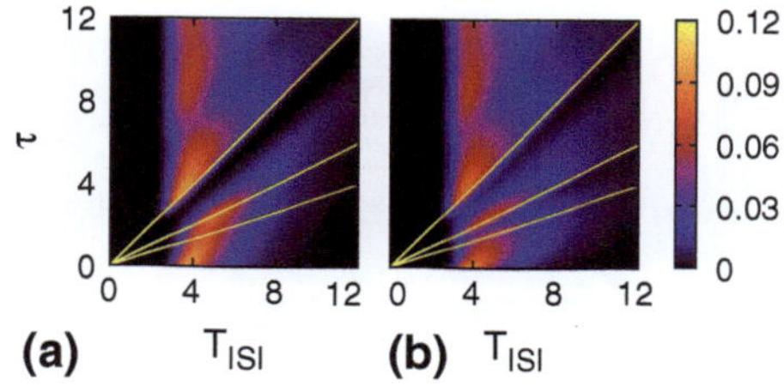

Fig. 6.33 Interspike interval distribution of a single FitzHugh–Nagumo system including extended time-delayed feedback according to Eqs. 6.3.4 in dependence on the time delay τ. Panels (**a**) and (**b**) correspond to a memory parameter $R = 0$ and 0.5, respectively. The feedback gain is fixed at $K = 0.2$. Parameters as in Fig. 6.27

Including extended time-delayed feedback, the feedback term in the Eqs. 6.3.1 needs to be modified such that the equations become

$$\varepsilon \frac{du(t)}{dt} = u(t) - \frac{u(t)^3}{3} - v(t) \tag{6.3.4a}$$

$$\frac{dv(t)}{dt} = u(t) + a + D\xi(t) + K \sum_{n=0}^{\infty} R^n [v(t - (n+1)\tau) - y(t - n\tau)], \tag{6.3.4b}$$

where R denotes the additional control parameter called *memory parameter*. The absolute value of R is smaller than unity and it serves as a weight of information further in the past. Note that Eqs. 6.3.1 is recovered for vanishing R.

In the following, Figs. 6.33 and 6.34 depict the interspike interval distribution and the power spectrum in dependence on the time delay τ for fixed feedback gain $K = 0.2$, but different memory parameters R.

In Fig. 6.33, the yellow lines are added as guides to the eye for $T_{\mathrm{ISI}} = n\tau$ with $n = 1, 2, 3$ as in Fig. 6.29. Panels (a) and (b) refer to a memory parameter $R = 0$ and 0.5, respectively. It can be seen that non-zero R leads to a reduced distribution for multiples of the time delay τ indicated by the dark stripe slightly to the right of the line at $T_{\mathrm{ISI}} = 2\tau$. Thus, $R > 0$ has a similar effect as increasing feedback gain K. Compare Fig. 6.33b and Fig. 6.29b. Note that opposed to the increase of K, non-zero R has only a minor effect on the distribution for large τ.

Comparing the power spectra depicted in Fig. 6.34 for different values of $R = 0$ and 0.5 in panels (a) and (b), respectively one can see that the peaks become sharper, but at the same time, the background of the power spectrum broadens.

As another modification of the a single FitzHugh–Nagumo system subject to time-delayed feedback as given by Eqs. 6.3.1 or 6.3.4, I will return to the original Pyragas control, i.e., $R = 0$, but add a second control loop whose time delay and feedback gain can be varied independently. Using the notation of Refs. [37, 53], the system is given for this configuration by the following equations

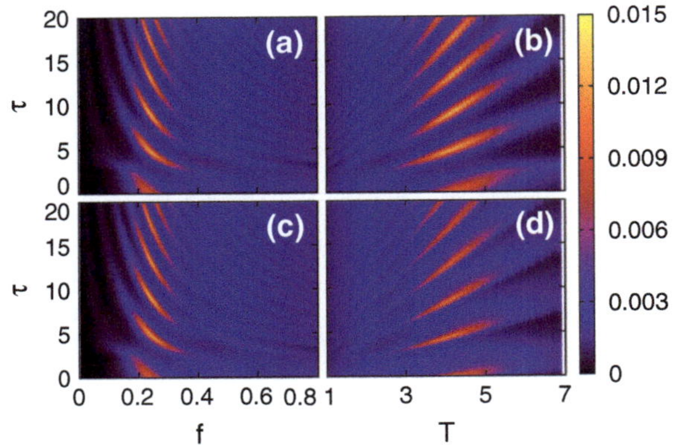

Fig. 6.34 Power spectrum of a single FitzHugh–Nagumo system including extended time-delayed feedback as given by Eqs. 6.3.4 in dependence on the time delay τ. Panels (**a**), (**b**) and (**c**), (**d**) correspond to a memory parameter $R = 0$ and 0.5, respectively. The feedback gain is fixed at $K = 0.2$. Panel (**a**)/(**c**) displays the same data as (**b**)/(**d**), but the power spectrum is shown in dependence on the frequency f instead of the period $T = f^{-1}$. Parameters as in Fig. 6.27

$$\varepsilon \frac{du(t)}{dt} = u(t) - \frac{u(t)^3}{3} - v(t) + C[u(t-\tau) - u(t)] + K[u(t-\tau_K) - u(t)] \tag{6.3.5a}$$

$$\frac{dv(t)}{dt} = u(t) + a \tag{6.3.5b}$$

and Fig. 6.35 depicts a schematic diagram. Note that opposed to Eqs. 6.3.1 the control force is applied to the equation of the activator u and consists of the current and delayed version of the same component. This is done for better comparison with Refs. [37, 53] and the Master's thesis [35]. This comparison can be found at the end of this section. A detailed investigation of different coupling schemes will be the subject of Sect. 6.5. For the same reason of comparison, the threshold parameter a is chosen as $a = 1.3$ and the timescale ratio is fixed at $\varepsilon = 0.01$. The control parameters in one of the control loops remain constant at $C = 0.5$ and $\tau = 6$. Note that there is no noise term in Eqs. 6.3.5. A first spike is triggered by the initial conditions. These are implemented as follows: the system is located at the fixed $(u, v) = (-a, -a + a^3/3)$ for $t < 0$. At $t = 0$, a rectangular pulse is applied for a period $t = 0.05$. This is enough to start a first pulse in the system which triggers for appropriate time delays and feedback gains the spiking behavior.

Figure 6.36 depicts the timeseries for six different scenarios. The control parameters K and τ_K are given in Table 6.1. The different scenarios are described in the following: panel (a) shows oscillations with a period of $\tau = 6$. Between two

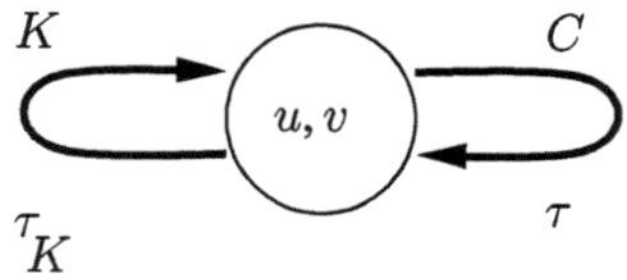

Fig. 6.35 Schematic diagram of a single FitzHugh–Nagumo system including time-delayed feedback with two control loops according to Eqs. 6.3.5

consecutive spikes, there are small, subthreshold oscillations after a time $\tau_K = 3$ induced by the (τ_K, K)-control loop. Due to the small value of the feedback gain $K = 0.05$ the respective feedback is not amplified by an amount which could cause the next excitation. If the value of K is larger as in panels (b)–(d), periodic spiking occurs with a period of $T = \tau_K$. Note that the ratio of τ and τ_K in these panels is equal to 3:1, 2:1, and 4:1, respectively. Then, the feedback by the (τ, C)-control loop coincides with multiple runs of the (τ_K, K)-loop. For other ratios, oscillators death occurs as in panel (e) for $\tau_K = 0.9$ and $K = 0.9$. The reason is that the feedback is applied to the system while it is still in the refractory phase. For $\tau_K = 3.2$ and $K = 0.5$, bursting behavior is found as displayed in panel (f).

Similar scenarios arise in the system schematically depicted by Fig. 6.37. The corresponding equations are given by

$$\begin{aligned}\varepsilon\frac{du_1(t)}{dt} &= u_1(t) - \frac{u_1(t)^3}{3} - v_1(t) + C[u_2(t-\tau) - u_1(t)] + K[u_1(t-\tau_K) - u_1(t)]\\ \frac{dv_1(t)}{dt} &= u_1(t) + a\end{aligned} \tag{6.3.6a}$$

$$\begin{aligned}\varepsilon\frac{du_2(t)}{dt} &= u_2(t) - \frac{u_2(t)^3}{3} - v_2(t) + C[u_1(t-\tau) - u_2(t)]\\ \frac{dv_2(t)}{dt} &= u_2(t) + a.\end{aligned} \tag{6.3.6b}$$

These equations describe two delay-coupled FitzHugh–Nagumo systems where the first subsystem is subject to self-feedback in the activator variables u_1.

One can see that the second element u_2, v_2 in Eqs. 6.3.7 acts as an active relay identical to the first subsystem, whereas the case depicted in Fig. 6.35 and described by Eqs. 6.3.5 corresponds to a passive relay realized by the second control loop. The discussion of active and passive relays has recently drawn much attention in the context of chaos synchronization and optical communication [54–59], where low-frequency fluctuations and bubbling corresponding to a blow-out bifurcation was investigated [60–63].

Figure 6.38 shows the timeseries of two delay-coupled FitzHugh–Nagumo systems given in Eqs. 6.3.6 for the same set of parameters as Fig. 6.36. Note that the coupling delay is set to $\tau = 3$ instead of $\tau = 6$ as used in Fig. 6.36 which

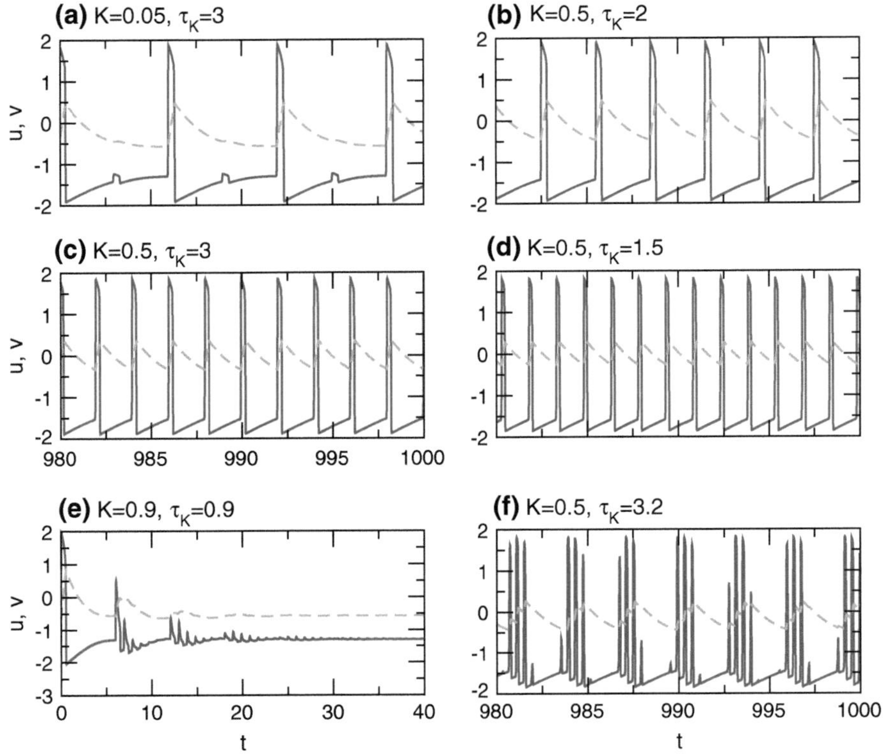

Fig. 6.36 Time series of activator u and inhibitor v of Eqs. 6.3.5 as red (*solid*) and green (*dashed*) curves, respectively. The values of K and τ_K are given in Table 6.1, Other parameters: $a = 1.3$, $\varepsilon = 0.01$, $C = 0.5$, and $\tau = 6$

Table 6.1 Parameters in Figs. 6.36, 6.38, and 6.40

Panel	a	b	c	d	e	f
K	0.05	0.5	0.5	0.5	0.9	0.5
τ_K	3	2	3	1.5	0.9	3.2, (3.25 in Fig. 6.38)

account the propagation time to the other system and back again. In addition the time delay τ_K is slightly larger than in Fig. 6.36 to compensate for the processing in the active relay u_2, v_2. See also Table 6.1 for the choices of the self-delay τ_K and

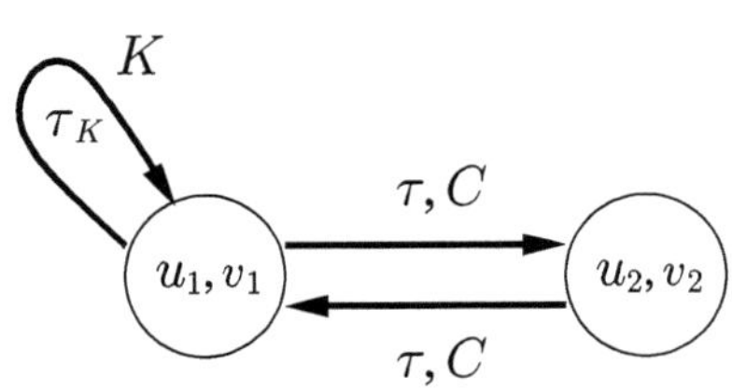

Fig. 6.37 Schematic diagram of two delay-coupled FitzHugh–Nagumo systems including time-delayed self-feedback according to Eqs. 6.3.6

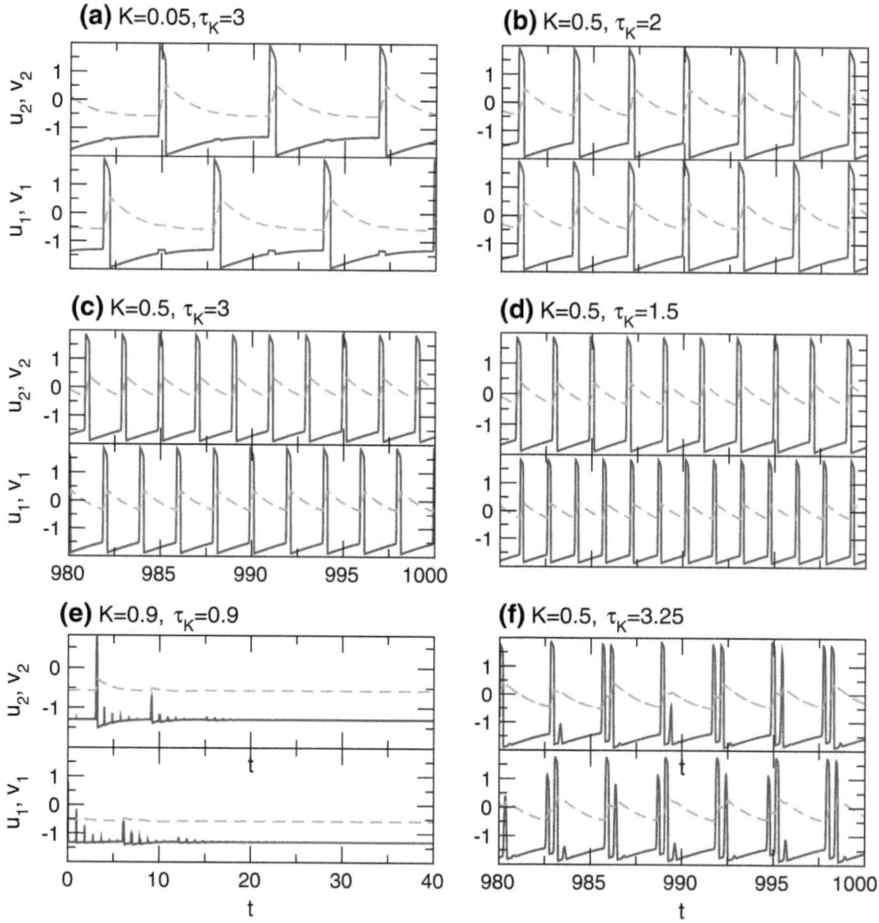

Fig. 6.38 Time series of two delay-coupled FitzHugh–Nagumo systems with time-delayed self-feedback of only one subsystem according to Eqs. 6.3.6. The red (*solid*) and green (*dashed*) curves correspond to the activator variables u_i and inhibitor variables v_i $(i = 1, 2)$, respectively. The parameters τ_K and K are given in Table 6.1. Other parameters as in Fig. 6.36

corresponding feedback gain K. A comparison between Figs. 6.38 and 6.36 yields the same dynamic scenarios described above.

Similar scenarios arise in another configuration schematically depicted by Fig. 6.39 [35, 37, 38, 53]. The corresponding equations are given by

$$\begin{aligned} \varepsilon\frac{du_1(t)}{dt} &= u_1(t) - \frac{u_1(t)^3}{3} - v_1(t) + C[u_2(t-\tau) - u_1(t)] + K[u_1(t-\tau_K) - u_1(t)] \\ \frac{dv_1(t)}{dt} &= u_1(t) + a \end{aligned} \tag{6.3.7a}$$

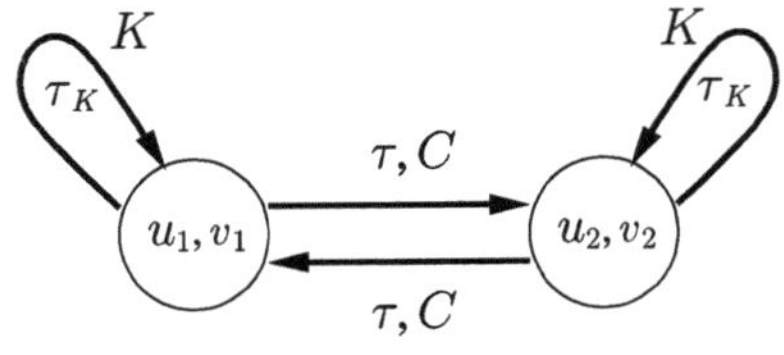

Fig. 6.39 Schematic diagram of two delay-coupled FitzHugh–Nagumo systems including time-delayed self-feedback according to Eqs. 6.3.7

$$\varepsilon \frac{du_2(t)}{dt} = u_2(t) - \frac{u_2(t)^3}{3} - v_2(t) + C[u_1(t-\tau) - u_2(t)] + K[u_2(t-\tau_K) - u_2(t)]$$
$$\frac{dv_2(t)}{dt} = u_2(t) + a. \tag{6.3.7b}$$

These equations describe two delay-coupled FitzHugh–Nagumo systems where each subsystem is subject to its own self-feedback in the activator variables u_1 and u_2, respectively. For a detailed analysis of Eqs. 6.3.7 see Ref. [37] and the Master's thesis [35].

Figure 6.40 shows the timeseries of two delay-coupled FitzHugh–Nagumo systems given in Eqs. 6.3.7 for the same set of parameters as Fig. 6.36 [37]. Note that the coupling delay is set to $\tau = 3$ instead of $\tau = 6$ as used in Fig. 6.36 which accounts for the propagation time to the other system and back again. See also Table 6.1 for the choices of the self-delay τ_K and corresponding feedback gain K. Thus, a spike which is emitted by the first subsystem returns via the second subsystem which acts as an active relay. As a consequence, an excitation is delayed twice during its return trip to the first subsystem via the relay. The initial conditions are chosen as follows: the second subsystem is at the rest state and the first subsystem is excited by a rectangular pulse in the time interval $[-0.3, 0]$ such that the activator and inhibitor is set to $u_1 = 1.8$ and $v_1 = 0$, respectively. A comparison between Fig. 6.40 and Figs. 6.38 as well as Fig. 6.36 yields the same dynamic scenarios described above.

To conclude this section, let me stress that an active and passive relay cause similar effects on the neural dynamics.

6.4 Two Coupled FitzHugh–Nagumo Systems and (Extended) Time-Delayed Feedback

This section is devoted to the effects of extended time-delayed feedback control on the dynamics of excitable FitzHugh–Nagumo systems and is mainly based on Refs. [40, 64]. This extension of the original Pyragas control scheme [49] takes informations of the system into account which are integer multiples of a basic time delay τ in the past [22]. In analogy to the Pyragas scheme, the control force is

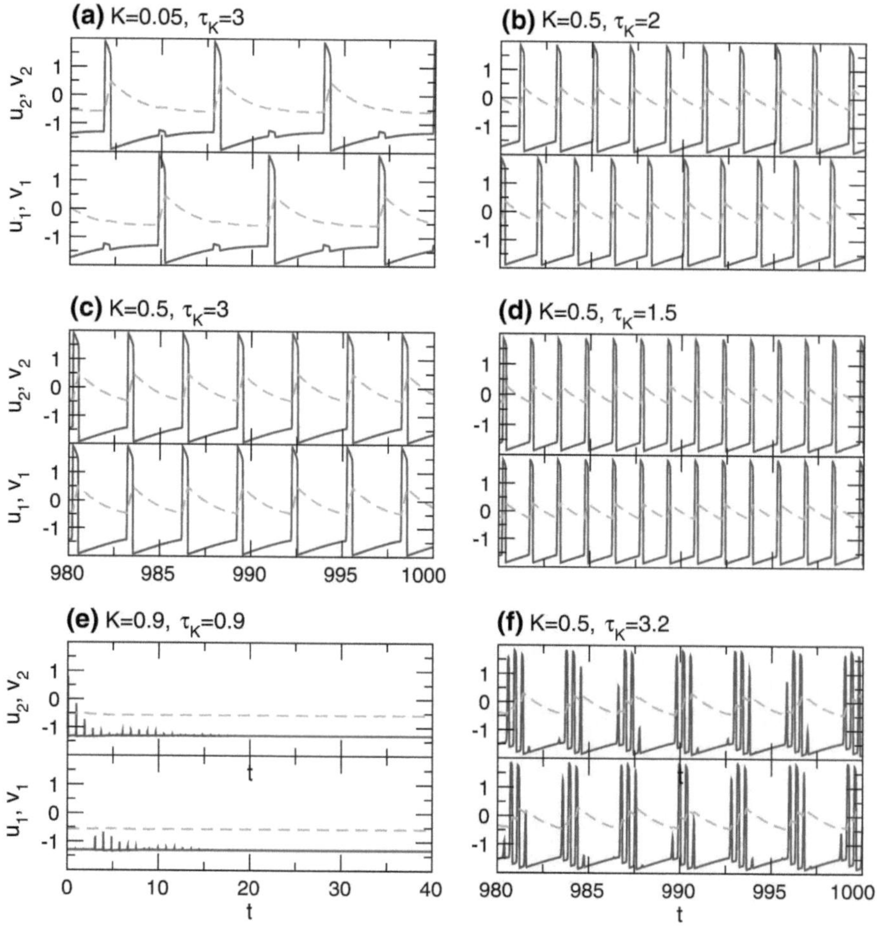

Fig. 6.40 Time series of two delay-coupled FitzHugh–Nagumo systems with time-delayed self-feedback of both subsystems according to Eqs. 6.3.7. The red (*solid*) and green (*dashed*) curves correspond to the activator variables u_i and inhibitor variables v_i ($i = 1, 2$), respectively. The parameters τ_K and K are given in Table 6.1. Other parameters as in Fig. 6.36

constructed from the difference of $\mathbf{s}(t-(n+1)\tau) - \mathbf{s}(t-n\tau)$ and a new control parameter R with $|R| < 1$ is introduced as a memory parameter which acts as a weight of states further in the past. For details concerning this extended control scheme see Sect. 2.4. This control method is known for successful stabilization at a larger range of parameters compared to the Pyragas method [65–68]. In case of noise-induced oscillations, extended feedback has been demonstrated to result in drastically improved coherence and arbitrarily long correlation times [34, 69, 70].

Including extended time-delayed feedback, the system of two instantaneously coupled FitzHugh–Nagumo systems can be rewritten as

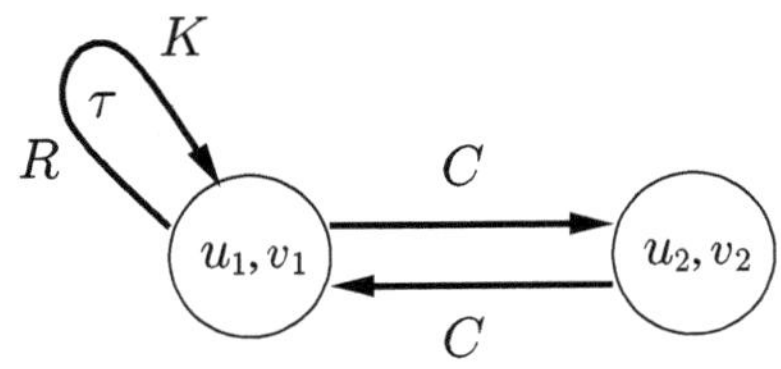

Fig. 6.41 Schematic diagram of two coupled FitzHugh–Nagumo systems with extended time-delayed feedback of one subsystem according to Eqs. 6.4.1

$$\begin{aligned}\varepsilon_1 \frac{du_1(t)}{dt} &= u_1(t) - \frac{u_1^3(t)}{3} - v_1(t) + C[u_2(t) - u_1(t)] \\ \frac{dv_1(t)}{dt} &= u_1(t) + a + D_1\xi_1(t) + K\sum_{n=0}^{\infty} R^n[v_1(t-(n+1)\tau) - y_1(t-n\tau)]\end{aligned} \tag{6.4.1a}$$

$$\begin{aligned}\varepsilon_2 \frac{du_2(t)}{dt} &= u_2(t) - \frac{u_2^3(t)}{3} - v_2(t) + C[u_1(t) - u_2(t)] \\ \frac{dv_2(t)}{dt} &= u_2(t) + a + D_2\xi_2(t),\end{aligned} \tag{6.4.1b}$$

where the control signal is generated from the inhibitor variable v_1 of the first subsystem and applied to the same component. Figure 6.41 depicts a schematic diagram of the configuration described by Eqs. 6.4.1.

Note that there exists an equivalent recursive form of the feedback $F(t)$

$$F(t) = K\sum_{n=0}^{\infty} R^n[v_1(t-(n+1)\tau) - v_1(t-n\tau)] \tag{6.4.2a}$$

$$= K[v_1(t-\tau) - v_1(t)] + RF(t-\tau). \tag{6.4.2b}$$

The equivalence was shown in Sect. 2.2. See Eqs. 2.2.2 and 2.2.3 for details. The recursive form is more amenable to experimental realization and for practical applications because the delayed feedback force $F(t-\tau)$ replaces the infinite series. This can be realized in an experiment by an additional control loop with delay and reduces the data storage in numerical simulations. See also Sect. 2.4 for further details. Note that the case of Pyragas control was investigated in Ref. [26] and the Master's thesis [44] and is included in Eqs. 6.4.1 as the limit of vanishing memory parameter R.

Before analyzing the effect of the additional control parameter R, Fig. 6.42 depicts different time series for moderate synchronization with a coupling strength $C = 0.2$ and a noise intensity $D_1 = 0.6$ in the absence of control in panel (a) compared to the cases when time-delayed is applied to the system in panels (b) and (c) for memory parameters of $R = 0$ and $R = 0.9$, respectively. The system's parameters are chosen as $a = 1.05$, $\varepsilon_1 = 0.005$, $\varepsilon_2 = 0.1$, $C = 0.07$ and

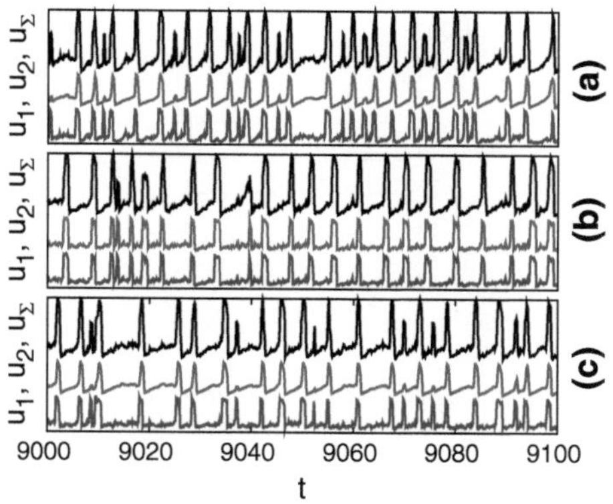

Fig. 6.42 Time series of the coupled FitzHugh–Nagumo system for moderate synchronization ($D_1 = 0.6$, $C = 0.2$). Panel (**a**) corresponds to no control. In panels (**b**) and (**c**), time-delayed feedback is applied to the system with different memory parameters $R = 0$ and $R = 0.9$, respectively. Other control parameters are fixed at $\tau = 1$ and $K = 1.5$. In all panels, the *black*, *red*, and *blue curves* refer to the summarized variable $u_\Sigma = u_1 + u_2$, the u_2-, and the u_1-variable, respectively. Other parameters: $a = 1.05$, $\varepsilon_1 = 0.005$, $\varepsilon_2 = 0.1$, $C = 0.07$, and $D_2 = 0.09$

$D_2 = 0.09$ throughout this section unless stated otherwise. This setup was already used in Sect. 6.2 for the discussion of cooperative dynamics without time-delayed feedback. The time delay and feedback gain are chosen as $\tau = 1$ and $K = 1.5$. In all panels, the black, red, and blue curves correspond to the summarized global signal $u_\Sigma = u_1 + u_2$, the u_2-, and the u_1-variable, respectively. One can see that the time-delayed feedback enhances the synchronization, i.e., events at which only one system oscillates are less frequent. In this sense the choice of $R = 0$ is more efficient compared to larger memory parameters.

After this first result on extended time-delayed feedback. I will analyze the influence of the memory parameter on the cooperative dynamics in the following by means of average interspike intervals, power spectra, and interspike interval distributions. As in Sect. 6.2, the system will be prepared in three different configurations of moderately, weakly, and strongly synchronized systems in the absence of control which is realized by combinations of the coupling strength C and noise intensity D_1 as $(C, D_1) = (0.2, 0.6)$, (0.1, 0.6), and (0.2, 0.15), respectively. These three configurations are marked by black dots in Fig. 6.16 in the absence of control. At first, I will discuss the effects of the memory parameter on the ratio of average interspike intervals.

For fixed feedback gain $K = 1.5$ and two different values of the memory parameter $R = 0$ and $R = 0.9$, Fig. 6.43 depicts the average interspike intervals of the subsystems in dependence on the time delay τ. The black, red, and dotted green curves correspond to $\langle T_1 \rangle$ and $\langle T_2 \rangle$, and their ratio $\langle T_1 \rangle / \langle T_2 \rangle$, respectively. Panels (a), (b), and (c) refer to the case of moderately, weakly, and strongly synchronized systems with combinations of C and D_1 mentioned above, respectively, and the memory parameter R is chosen as $R = 0$ and 0.9 in top and bottom panels of each subfigure.

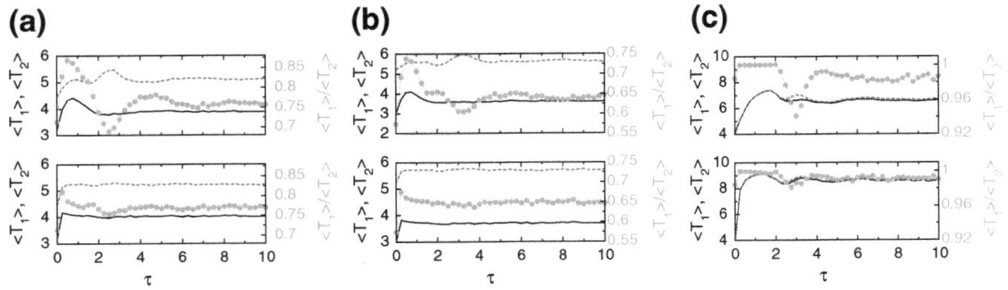

Fig. 6.43 Interspike intervals in dependence on the time delay τ. The *green dots* correspond to the ratio of the interspike intervals $\langle T_1\rangle/\langle T_2\rangle$ of the two subsystems which are also depicted by solid (*black*) and dashed (*red*) curves for $\langle T_1\rangle$ and $\langle T_2\rangle$, respectively. Panels (**a**), (**b**), and (**c**) correspond to the case of moderate, weak, and strong synchronization, respectively. The control parameters are chosen as $K = 1.5$ and the memory parameter corresponds to $R = 0$ and 0.9 in the top and bottom panels, respectively. Other parameters as in Fig. 6.42

In all cases, tuning of the time delay leads to enhanced or deteriorated synchronization for $R = 0$ depending on the value of the time delay τ. This modified synchronized behavior is given by the maxima and minima of the green curves. If more information of past states ($R \neq 0$) is included, however, the variation of the ratio $\langle T_1\rangle/\langle T_2\rangle$ is less sensitive to the specific choice of τ. The bottom panels of Fig. 6.43 do not show large deviations for the ratio of average interspike intervals. Therefore, a larger memory parameter renders the control method more robust because it is not sensitive to the specific choice of the time delay τ.

After the one-dimensional projection of the control parameter space spanned by K, τ, and R, one can also consider two-dimensional projections where the third parameter remains fixed. Parameterized by the feedback gain and time delay, Fig. 6.44 displays the ratio of $\langle T_1\rangle$ and $\langle T_2\rangle$ for moderate, weak, strong synchronization in panels (a), (b), and (c), respectively. The four panels in each subfigure correspond to a memory parameter of $R = 0$, 0.35, 0.7, and 0.9. Note that Fig. 6.43 can be understood as horizontal cuts for $K = 1.5$ in the respective diagram of Fig. 6.44.

In all three cases, one can see that an increase of the memory parameter R yields a more uniform behavior over a large range of control parameters in comparison to the case of $R = 0$. For instance, the minima at $\tau = 2$ are less pronounced as the value of R increases. The extended time-delayed feedback method is less likely to desynchronize the two subsystems. Therefore, this control method can be considered more robust in terms of a variation of the feedback gain K and time delay τ. However, not only the cases of strong desynchronization are weakened, but the realizations of enhanced cooperative dynamics loose some synchronization as well. This can be seen, for instance, in panel (c) of Fig. 6.44, where the bright regions become darker for larger R.

Similar to the one-dimensional projections displayed in Fig. 6.43, an increase of the memory parameter R at fixed K yields smaller changes of the ratio $\langle T_1\rangle/\langle T_2\rangle$. Independently on the feedback gain K, the desynchronized darker regions in Fig. 6.44 at a time delay $\tau \approx 3$ are much less pronounced. For $R = 0.9$,

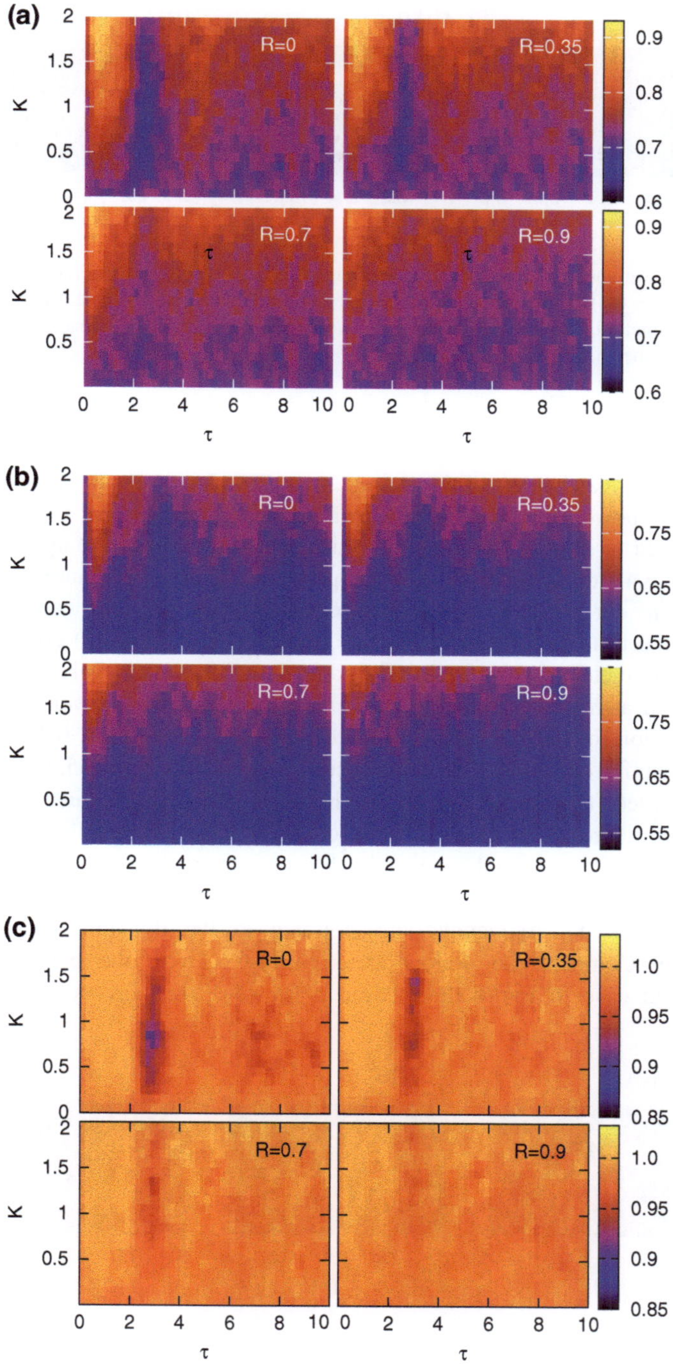

Fig. 6.44 Ratio of interspike intervals $\langle T_1\rangle/\langle T_2\rangle$ in dependence on the feedback gain K and the time delay τ for moderate ($C = 0.2$, $D_1 = 0.6$), weak ($C = 0.1$, $D_1 = 0.6$), and strong ($C = 0.2$, $D_1 = 0.15$) synchronization in panels (**a**), (**b**), and (**c**), respectively. The memory parameter are fixed at $R = 0$, 0.35, 0.7, and 0.9 in the four subfigures of all three panels. Other parameters as in Fig. 6.42

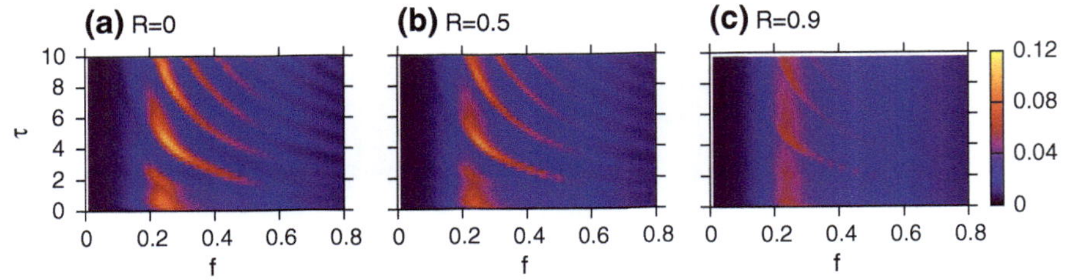

Fig. 6.45 Power spectrum of the summarized signal $u_\Sigma = u_1 + u_2$ for moderate synchronization ($C = 0.2$, $D_1 = 0.6$) in dependence on the time delay τ. Panels (**a**), (**b**), and (**c**) correspond to a memory parameter of $R = 0$, $R = 0.5$, and $R = 0.9$, respectively. The feedback strength is fixed at $K = 1.5$. Other parameters as in Fig. 6.42

the ratio of average interspike intervals is constant in a wide range of the whole control domain which reflects the robustness of this extended feedback method.

As for the investigation of the average interspike intervals and their ratio, I will consider the power spectra in the following for the same system's configurations. This will add to the information of the timescales present in the system (6.4.1).

Figures 6.45, 6.46, and 6.47 show the power spectrum of the summarized signal $u_\Sigma = u_1 + u_2$ for the three above mentioned cases of synchronization in dependence on the time delay τ. In all figures, panels (a), (b), and (c) correspond to a memory parameter of $R = 0$, $R = 0.5$, and $R = 0.9$, respectively. The feedback gain is fixed at $K = 1.5$ as in Fig. 6.43. The corresponding power spectra in the absence of control were already discussed in Sect. 6.2. See Figs. 6.18–6.21. They can also be inferred from the horizontal cut for $\tau = 0$ in Figs. 6.45, 6.46, and 6.47.

It can be seen that, depending on the choice of τ, the main frequency component is shifted which results in elongate red regions indicating large values of the power spectrum. Thus, the control scheme is able to support different timescales. Note that, for instance in the case of moderate synchronization (Fig. 6.45), the control force enhances the frequency corresponding to the dynamics of the u_1- or u_2-variable. Compare the yellow areas in Fig. 6.45 to the panel (b) of the uncontrolled case in Fig. 6.18. A time delay of $\tau \approx 3$ favors a frequency of

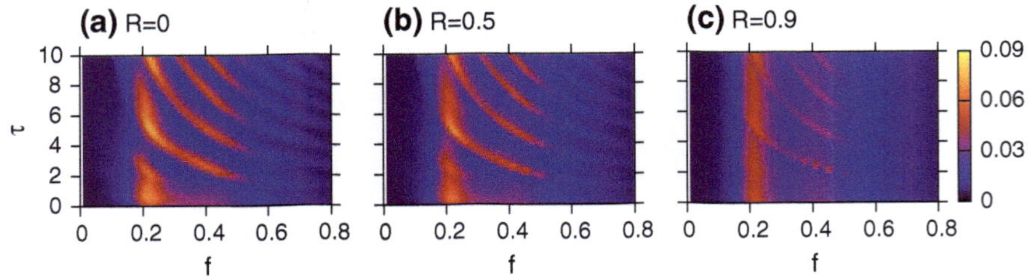

Fig. 6.46 Power spectrum of the summarized signal $u_\Sigma = u_1 + u_2$ for weak synchronization ($C = 0.1$, $D_1 = 0.6$) in dependence on the time delay τ. Panels (**a**), (**b**), and (**c**) correspond to a memory parameter of $R = 0$, $R = 0.5$, and $R = 0.9$, respectively. The feedback strength is fixed at $K = 1.5$. Other parameters as in Fig. 6.42

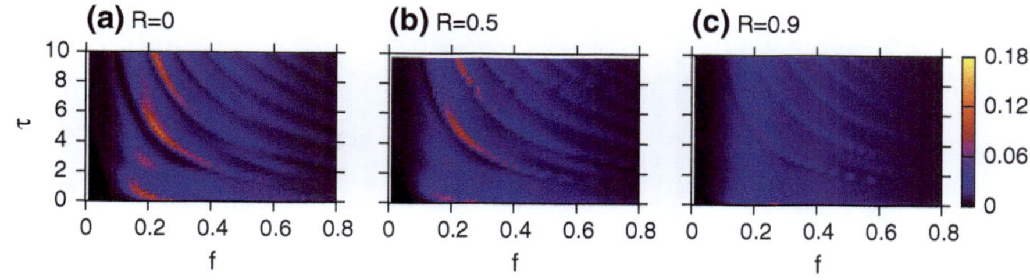

Fig. 6.47 Power spectrum of the summarized signal $u_\Sigma = u_1 + u_2$ for strong synchronization ($C = 0.2, D_1 = 0.15$) in dependence on the time delay τ. Panels (**a**), (**b**), and (**c**) correspond to a memory parameter of $R = 0$, $R = 0.5$, and $R = 0.9$, respectively. The feedback strength is fixed at $K = 1.5$. Other parameters as in Fig. 6.42

$f \approx 0.4$, i.e., the main frequency of u_1, and $\tau \approx 5$ enhances components of $f \approx 0.2$ which corresponds to the dynamics of u_2. One can also observe that the main peaks in the power spectra occur periodically as the time delay τ increases. For larger memory parameters R, the effect of a frequency shift is less pronounced. The power spectra of panel (c) in Figs. 6.45, 6.46, and 6.47 display reduced sensitivity on the specific choice of the time delay. The main peak of the spectrum is stronger localized at the frequency of the second subsystem, but the value of the power spectrum at this main frequency is much lower.

Figure 6.48 shows the power spectrum of the global summarized signal u_Σ for the three different case of synchronization and fixed time delay $\tau = 5$. In all panels, the black, red, and and green curves correspond to a memory parameter of $R = 0$, 0.5, and 0.9, respectively. The feedback gain is fixed at $K = 1.5$. Panels (a), (b), and (c) of Fig. 6.48 can be seen as horizontal cuts (at $\tau = 5$) of Figs. 6.45, 6.46, and 6.47, respectively.

Opposed to the uncontrolled case presented in Fig. 6.18, the power spectra exhibit a series of distinct peaks that are located at the harmonics of the main frequency $f \approx 0.2$. They are due to the control force which enhances not only the main frequency, but also frequency components of the higher harmonics. See also Sect. 2.6 in Chap. 2 where the mechanism of time-delayed feedback was

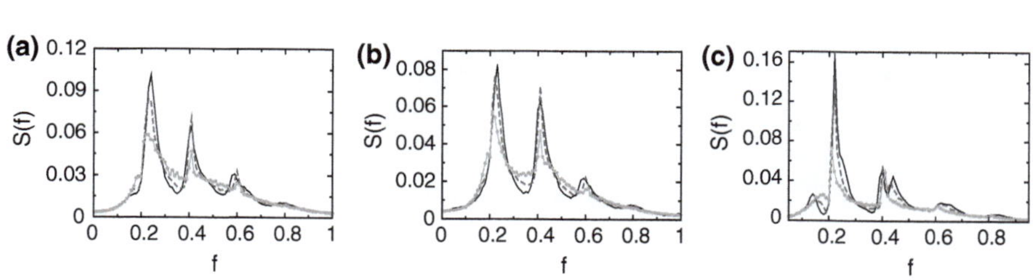

Fig. 6.48 Power spectrum of the summarized signal $u_\Sigma = u_1 + u_2$ for a fixed time delay of $\tau = 5$. Panels (**a**), (**b**), and (**c**) correspond the case of moderate ($C = 0.2$, $D_1 = 0.6$), weak ($C = 0.1$, $D_1 = 0.6$), and strong ($C = 0.2$, $D_1 = 0.15$) synchronization. The *black* (*solid*), *red* (*dashed*), and *green* (*dotted*) *curves* refer to a memory parameter of $R = 0$, 0.5, and 0.9, respectively. Other parameters as in Fig. 6.42

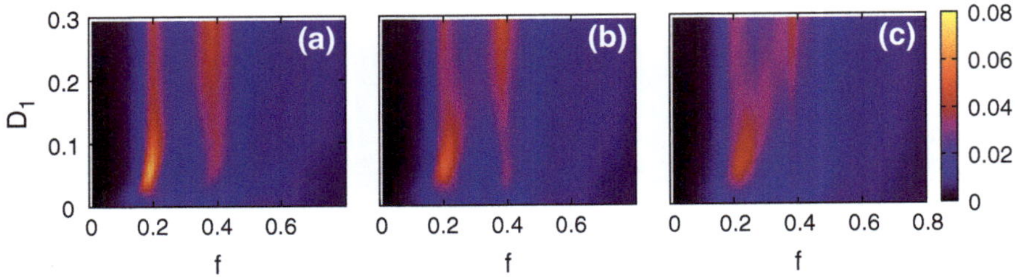

Fig. 6.49 Power spectrum S of noise-induced oscillations for the summarizes variable $u_\Sigma = u_1 + u_2$ of the coupled FitzHugh–Nagumo systems as colorcode in dependence on the noise intensity D_1 for a time delay $\tau = 2.7$. Panel (**a**), (**b**), and (**c**) correspond to a memory parameter of $R = 0, 0.5$, and $R = 0.9$, respectively. The feedback gain is fixed at $K = 0.2$. Other parameters as in Fig. 6.42

discussed in Fourier space using the formalism of transfer functions. It was shown that the transfer function of the Pyragas scheme and its extension to multiple delays yields a frequency comb with minima at multiples of a basic frequency τ^{-1}.

For increasing memory parameter R, the background becomes broader. Let me stress that a similar effect was found in the context of extended time-delayed feedback applied to the noise-induced oscillations in the Van-der-Pol system [34] and in a reaction–diffusion system [70], where, in addition, the peaks become sharper.

Before discussing the interspike interval distributions, let me consider the power spectra also in dependence on the coupling strength C and noise intensity D_1 for different memory parameters R. The reference cases in the absence of control can be found in Sect. 6.2 as Figs. 6.20 and 6.21.

Figures 6.49 and 6.50 depict the power spectrum of the summarized variable u_Σ as color code in dependence on the noise intensity D_1 of the first subsystem for different time delays $\tau = 2.7$ and 4.7. The feedback gain is fixed at $K = 0.2$ and the memory parameter is set to $R = 0, 0.5$, and 0.9 in panels (a), (b), and (c) of each figure, respectively.

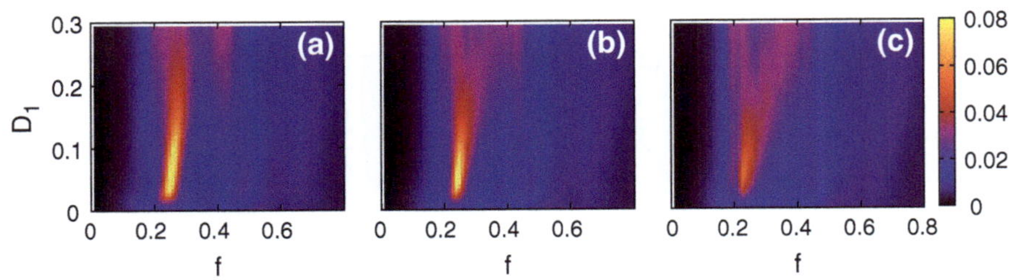

Fig. 6.50 Power spectrum S of noise-induced oscillations for the summarizes variable $u_\Sigma = u_1 + u_2$ of the coupled FitzHugh–Nagumo systems as colorcode in dependence on the noise intensity D_1 for a time delay $\tau = 4.7$. Panel (**a**), (**b**), and (**c**) correspond to a memory parameter of $R = 0, 0.5$, and $R = 0.9$, respectively. The feedback gain is fixed at $K = 0.2$. Other parameters as in Fig. 6.42

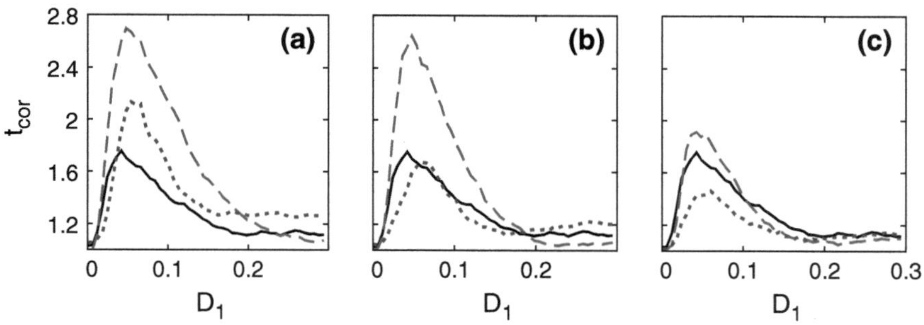

Fig. 6.51 Correlation time of noise-induced oscillations for the summarizes variable $u_\Sigma = u_1 + u_2$ of the coupled FitzHugh–Nagumo systems in dependence on the noise intensity D_1 for $R = 0$, 0.5, and 0.9 in panels (**a**), (**b**), and (**c**), respectively. The time delay is chosen as $\tau = 2.7$ and 4.7 in the blue (*dotted*) and red (*dashed*) curves, respectively, and the black (solid) curves refers to the case without control. The feedback gain is fixed at $K = 0.2$. Other parameters as in Fig. 6.42

Before further investigation of the power spectra, let me consider the choices of the time delay by a brief discussion of the correlation time t_{cor}. This quantity will be addressed in details at the end of this section. At this point, it is sufficient to note that the time delays are chosen such that the correlation time t_{cor} which is calculated from the autocorrelation function by Eq. 6.1.17 is larger for $\tau = 2.7$ and smaller for $\tau = 4.7$ in the range of large noise intensities $D_1 > 0.2$ compared to the uncontrolled case. This can be seen in Fig. 6.51 where the blue (dotted) and red (dashed) curves corresponds to the correlation time for $\tau = 2.7$ and 4.7, respectively, and the black (solid) curves refers to the reference case without control. Note that this effect is reversed in the range of small noise intensities. Panels (a), (b), and (c) correspond to a memory parameter $R = 0$, 0.5, and 0.9, respective. The effect of varying on the noise intensity persists for non-zero memory parameters.

Comparing Figs. 6.49 and 6.50, one can see maxima in all panels which correspond to the timescales of the single subsystems. Since the coupling strength is chose at a low value, i.e., $C = 0.07$, the two subsystems are only weakly coupled and they operate at different timescales.

In both figures, an increase of the memory parameter leads to a damped relief in the power spectrum and the maxima are less pronounced. For a time delay of $\tau = 2.7$ as in Fig. 6.49, the maxima persist in the whole range of displayed noise intensities D_1. In the case of $\tau = 4.7$ shown in Fig. 6.50, however, there is a distinct maximum at small D_1. For larger noise intensity, the maximum is barely visible and vanishes altogether for increases R as can be seen in panels (b) and (c). Thus, one can say that $\tau = 2.7$ is a better choice than $\tau = 4.7$ in the sense of supporting timescales already present in the uncontrolled system.

Figures 6.52 and 6.53 display the dependence of the power spectrum on the coupling strength C. As in Figs. 6.49 and 6.50, the two sets of panels correspond to time delays of $\tau = 2.7$ and 4.7, respectively. The noise intensity D_1 is fixed at

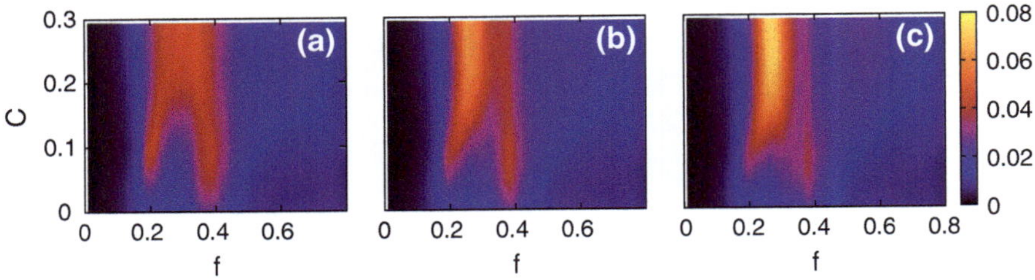

Fig. 6.52 Power spectrum S of noise-induced oscillations for the summarizes variable $u_\Sigma = u_1 + u_2$ of the coupled FitzHugh–Nagumo systems as colorcode in dependence on the coupling strength C for a time delay $\tau = 2.7$ for $R = 0, 0.5$, and 0.9 in panels (**a**), (**b**), and (**c**), respectively. The noise intensity D_1 is chosen as $D_1 = 0.25$ and the feedback gain is fixed at $K = 0.2$. Other parameters as in Fig. 6.42

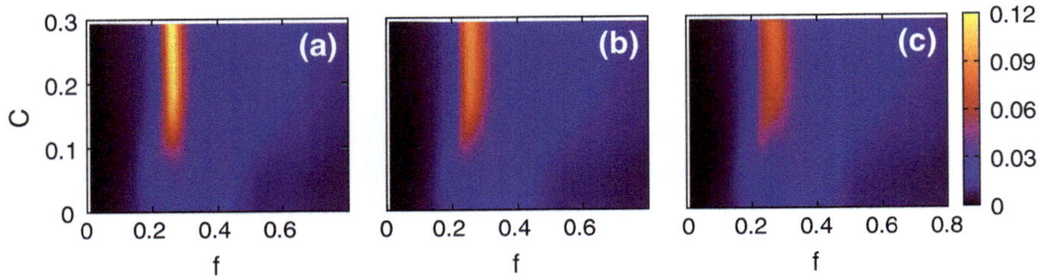

Fig. 6.53 Power spectrum S of noise-induced oscillations for the summarizes variable $u_\Sigma = u_1 + u_2$ of the coupled FitzHugh–Nagumo systems as colorcode in dependence on the coupling strength C for a time delay $\tau = 4.7$ for $R = 0, 0.5$, and 0.9 in panels (**a**), (**b**), and (**c**), respectively. The noise intensity D_1 is chosen as $D_1 = 0.25$ and the feedback gain is fixed at $K = 0.2$. Other parameters as in Fig. 6.42

$D_1 = 0.25$ and the other parameters remain unchanged. As in the previous figures, panels (a), (b), and (c) refer to a memory parameter of $R = 0$, 0.5, and 0.9, respectively.

One can observe in all panels that the two subsystems operate at a common frequency as the coupling strength increases indicate the merging of the two peaks. See, for instance, the red region in Fig. 6.52. For larger memory parameters the peak in the power spectrum becomes stronger for a time delay of $\tau = 2.7$ whereas it is reduced for $\tau = 4.7$ as in Fig. 6.53. This can be seen by a brighter peak in Fig. 6.52 and the reduced brightness of the main peak in Fig. 6.53 for larger R. As more information from integer multiples of a basic time delay is included for larger R, the feedback scheme acts against the system's dynamics in the case of $\tau = 4.7$ as in Fig. 6.53.

After this extensive discussion of the power spectrum in dependence on various system's and control parameters, the following figures present the interspike interval distributions. The provide further information concerning the question which timescales are enhanced or suppressed by time-delayed feedback control. Special emphasis is on the effects of the memory parameter R.

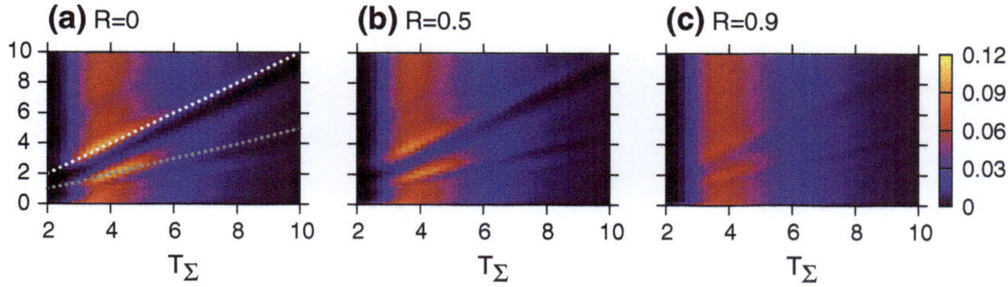

Fig. 6.54 Interspike interval (T_Σ) distribution of the summarized signal $u_\Sigma = u_1 + u_2$ for moderate synchronization ($C = 0.2$, $D_1 = 0.6$) in dependence on the time delay τ. The color coding denotes the probability of finding a certain interspike interval T_Σ. Panels (**a**), (**b**), and (**c**) correspond to a memory parameter of $R = 0$, $R = 0.5$, and $R = 0.9$, respectively. The feedback gain is fixed at $K = 1.5$. The *white* and *grey dotted lines* in panel (**a**) at $\tau = T_\Sigma$ and $\tau = T_\Sigma/2$, respectively, are guides to the eye. Other parameters as in Fig. 6.42

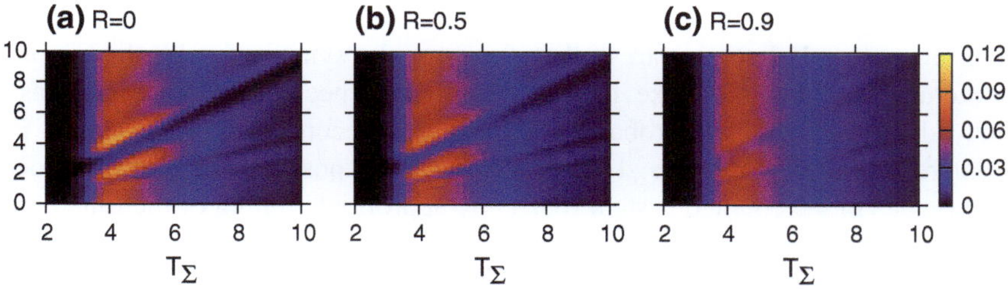

Fig. 6.55 Interspike interval (T_Σ) distribution of the summarized signal $u_\Sigma = u_1 + u_2$ for weak synchronization ($C = 0.1$, $D_1 = 0.6$) in dependence on the time delay τ. The color coding denotes the probability of finding a certain interspike interval T_Σ. Panels (**a**), (**b**), and (**c**) correspond to a memory parameter of $R = 0$, $R = 0.5$, and $R = 0.9$, respectively. The feedback gain is fixed at $K = 1.5$. Other parameters as in Fig. 6.42

Figures 6.54, 6.55, and 6.56 depict the dependence of the distributions of the interspike interval T_Σ for moderately ($C = 0.2$, $D_1 = 0.6$), weakly ($C = 0.1$, $D_1 = 0.6$), and strongly synchronized ($C = 0.2$, $D_1 = 0.15$) subsystems on the time delay, respectively. These are the three cases marked in Fig. 6.16 of Sect. 6.2. The corresponding power spectra are shown in Figs. 6.45, 6.46, and 6.47. Panels (a), (b), and (c) in each figure refer to a memory parameter of $R = 0$, 0.5, and 0.9, respectively. Note that all three figures 6.54, 6.55, and 6.56 display the same color code.

For increasing memory parameter R, the distribution exhibits a weaker dependence on the time delay τ. This can be seen, for instance, in panels (c) of Figs. 6.54 and 6.55. This is in agreement with the findings in the power spectra and the ratio of the average interspike intervals presented above.

For smaller R values, another effect becomes apparent. There are a minima in the distribution for interspike intervals slightly larger than τ, indicated by black stripes in the (T_Σ, τ) plane. These minima are located below the line $T_\Sigma = \tau$. See, for instance, the white line in Fig. 6.54a which is added as a guide to the eye.

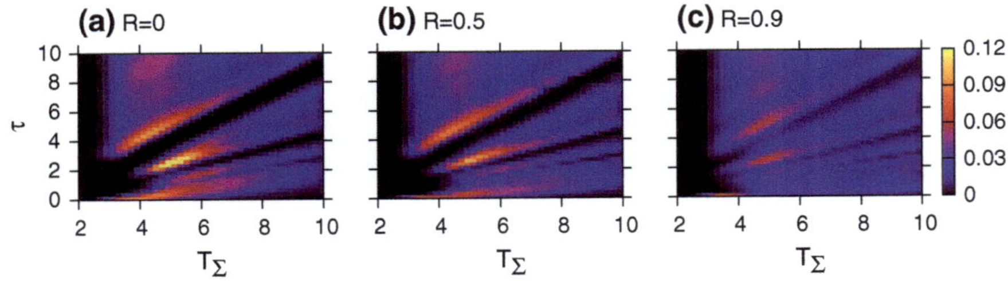

Fig. 6.56 Interspike interval (T_Σ) distribution of the summarized signal $u_\Sigma = u_1 + u_2$ for strong synchronization ($C = 0.2$, $D_1 = 0.15$) in dependence on the time delay τ. The color coding denotes the probability of finding a certain interspike interval T_Σ. Panels (**a**), (**b**), and (**c**) correspond to a memory parameter of $R = 0$, $R = 0.5$, and $R = 0.9$, respectively. The feedback gain is fixed at $K = 1.5$. Other parameters as in Fig. 6.42

A similar, but less pronounced effect can be observed if the interspike intervals match integer multiples of the time delay as shown by the grey line for $T_\Sigma = 2\tau$ in the same panel. Let me stress that for strongly synchronized subsystems this structuring of the interspike interval distribution becomes more visible as displayed in Fig. 6.56. The probability distribution becomes multimodal with peaks centered near $T_\Sigma = n\tau, n = 1, 2, \ldots$. Thus, it is possible to eliminate certain timescales in the coupled system (6.4.1) by appropriate tuning of the time delay in the controller.

To summarize the results on the interspike interval distribution in the presence of extended time-delayed feedback, the introduction of a large memory parameter renders the control method more robust against the specific choice of the time delay. This is depicted in Fig. 6.57 for a memory parameter of $R = 0.99$. Panels (a), (b), and (c) correspond to moderately, weakly, and strongly synchronized subsystems, respectively. For this large value of R, there is no structuring of the interspike interval distribution. Only panel (c) of already strongly synchronized subsystems shows some weakly pronounced modulation. For small memory

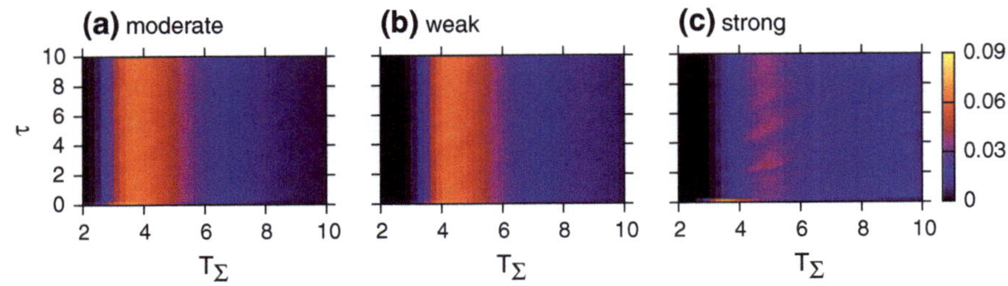

Fig. 6.57 Interspike interval (T_Σ) distribution of the summarized signal $u_\Sigma = u_1 + u_2$ in dependence on the time delay τ for a memory parameter $R = 0.99$ for moderate ($C = 0.2$, $D_1 = 0.6$), weak ($C = 0.1$, $D_1 = 0.6$), and strong ($C = 0.2$, $D_1 = 0.15$) synchronization in panels (**a**), (**b**), and (**c**), respectively. The color coding denotes the probability of finding a certain interspike interval T_Σ. The feedback gain is fixed at $K = 1.5$. Other parameters as in Fig. 6.42

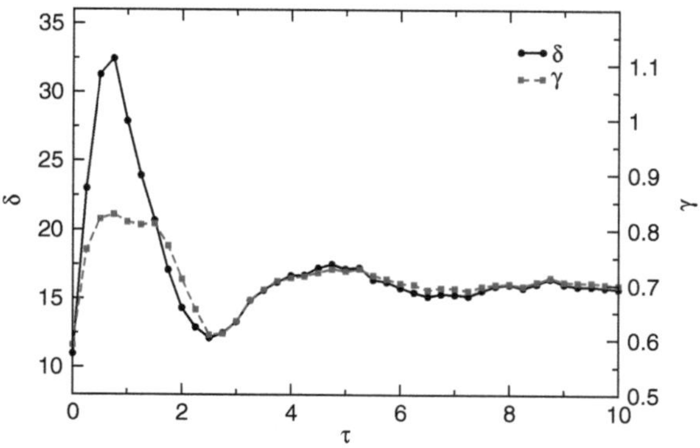

Fig. 6.58 Average phase synchronization interval δ (*black dots, solid curve*) and synchronization index γ (*red squares, dashed line*) in dependence on the time delay τ for the case of moderately synchronized subsystems ($C = 0.2$, $D_1 = 0.6$) and vanishing memory parameter R. Other parameters as in Fig. 6.42

parameters, there is a competing effect which structures the distribution in the sense that interspike intervals slightly larger than the time delay of the feedback are suppressed.

The measures of cooperative dynamics in this section are so far all related to frequency synchronization, but they are insensitive to phase synchronization. As introduced in Sect. 6.2, one can derive phase variables φ_1 and φ_2 from the time series of the respective subsystem. The corresponding formula is given by Eq. 6.2.2 which is reproduced here for convenience

$$\varphi_j(t) = 2\pi\frac{t - t_{i-1}}{t_i - t_{i-1}} + 2\pi(i-1), \tag{6.4.3}$$

with $j = 1, 2$. The times t_i denotes the events of the i-th spike. The difference $|\varphi_1 - \varphi_2|$ leads to a phase difference $\Delta\varphi$ from which a phase synchronization index can be calculated according to Eq. 6.2.4

$$\gamma = \sqrt{\langle\cos\Delta\varphi(t)\rangle^2 + \langle\sin\Delta\varphi(t)\rangle^2}. \tag{6.4.4}$$

An example for the time series of $\Delta\varphi$ is depicted in Fig. 6.22 of Sect. 6.2. There, it was shown the phase difference $\Delta\varphi$ is constant during time intervals of phase synchronization. The average length of these intervals denotes another measure of phase synchronization.

Figure 6.58 compares the sensitivity of the phase synchronization index γ and the average length of phase synchronization intervals δ. The values of γ are shown in red and the black curve corresponds to δ in dependence on the time delay τ. The other control parameters are fixed at $R = 0$ and $K = 1.5$ and the scale is chosen such that the points for $\tau = 0$ and $\tau = 10$ coincide.

Note that the average phase synchronization interval δ shows a stronger increase than the synchronization index γ. Therefore, the δ is more sensitive for the detection of phase synchronization as the control parameters K, τ, and R are varied than the synchronization index γ. Hence, I will restrict the investigations of phase synchronization to the discussion of δ in the following.

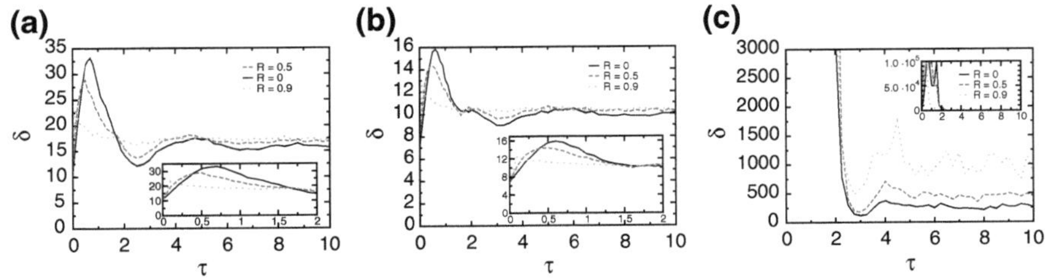

Fig. 6.59 Average phase synchronization interval δ in dependence on the time delay τ. The *black (solid)*, *red (dashed)*, and *green (dotted) curves* correspond to a memory parameter of $R = 0$, $R = 0.5$, and $R = 0.9$, respectively. The panels (**a**), (**b**), and (**c**) refer to the case of moderate, weak, and strong synchronization, respectively. The *insets* in panels (**a**) and (**b**) show an enlargement for small τ. The *inset* in panel (**c**) displays large δ. Other parameters as in Fig. 6.42

Figure 6.59 shows effects of extended time-delayed feedback in the average phase synchronization interval δ for varying time delay τ. The feedback gain K is fixed at $K = 1.5$. Panels (a), (b), and (c) display the case of moderately, weakly, and strongly synchronized subsystems, respectively. In all panels, the black, red, and green curves refer to a memory parameter of $R = 0$, $R = 0.5$, and $R = 0.9$, respectively. The insets in panels (a) and (b) are enlargements for small time delays and the inset in panel (c) display large values of δ.

In general, time-delayed feedback enlarges the average phase synchronization interval δ. Especially for small time delays, e.g., $\tau = 0.7$ for $R = 0$, δ becomes substantially larger. For $R = 0$, a modulation of δ can be seen for small delays, see insets in all panels of Fig. 6.59. These deviations are less pronounced for increasing R. Only panel (c) which refers to strong synchronization, shows larger values of δ with increasing memory parameter.

The sensitivity of δ, as discussed in Fig. 6.58, can also be seen in the case of strong synchronization. See panel (c) in Fig. 6.59. Since the already strong synchronization is further enhanced by the control force, the average phase synchronization interval rises by several orders of magnitude as shown in the inset. For perfect synchronization and simultaneous spiking, δ would be arbitrarily large and merely reflect the integration time.

Figure 6.60 depicts the correlation time, calculated from Eq. 6.1.17, for fixed coupling strength $C = 0.2$ in dependence on the noise intensity D_1, when no control is applied to the system. The blue (dotted), red (dashed), and black (solid) curves refer to the u_1-, the u_2-variable, and summarized global signal $u_\Sigma = u_1 + u_2$, respectively. Starting in the case where only small noise level in the second subsystem is present, i.e., $D_1 = 0$, the correlation time becomes larger for an increase of the noise intensity D_1. All three curves show coherence resonance with a maximum correlation time for $D_1 \approx 0.1$. For larger values of D_1, the correlation time decreases, since the dynamics is more noise-dominated. In the presence of control, care has to be taken in the definition of the correlation time because Eqs. 6.1.16 and 6.1.17 do not generally give the same result. As shown by Pomplun et al. [34] for the Van-der-Pol system with noise and extended

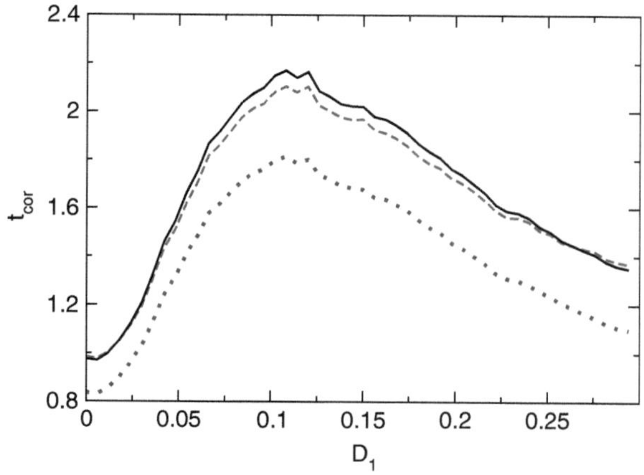

Fig. 6.60 Correlation time in dependence on the noise intensity D_1 for fixed coupling strength $C = 0.2$ in the absence of control. The blue (*dotted*), red (*dashed*), and black (*solid*) curves correspond to the u_1-, the u_2-variable, and summarized signal $u_\Sigma = u_1 + u_2$, respectively. No control is applied to the system. Other parameters as in Fig. 6.42

time-delayed feedback, two different timescales occur in the autocorrelation function $\Psi(s)$. The sharp peaks in the power spectrum correspond to long-term correlations and the broad background spectrum leads to an abrupt decrease of the autocorrelation function $\Psi(s)$ for small s. The long-term correlations are measured by the exponential slow decay of the autocorrelation function for larger s, and thus the correlation time can be determined by a fit of the envelope to an exponential function $\Psi(s) \propto \exp(-s/t_e)$ with $t_e = \frac{\pi t_{cor}}{2}$ as in Eq. 6.1.16, but with a different normalization constant. See Sect. 6.2 for details on the analytic derivations concerning the correlation time.

For two coupled FitzHugh–Nagumo systems given by Eqs. 6.4.1, the normalized autocorrelation function $\Psi(s)/\Psi(0)$ for $u_\Sigma = u_1 + u_2$ is depicted in Fig. 6.61 for the case of strongly synchronized subsystems with $C = 0.1$ and $D_1 = 0.15$. Panel (a) shows the uncontrolled system and the dashed curve is a fit of an exponential function of the local maxima through $\Psi(0)$, giving $t_e = 3.44$ and $t_{cor} = \frac{2t_e}{\pi} = 2.19$, in approximate agreement with Fig. 6.60. Panels (b) and (c) display $\Psi(s)/\Psi(0)$ in the presence of extended time-delayed feedback for memory parameters $R = 0$ and $R = 0.9$, respectively, and fixed time delay $\tau = 2$ and feedback gain $K = 1.5$. The dashed curves in panels (b) and (c) are fits of the envelope of $\Psi(s)$ by an exponential function in the range $s > 10$. One can see that for increasing memory parameter, the autocorrelation function decays more slowly. The corresponding exponential correlation times are $t_e = 9.03$ for vanishing memory parameter and $t_e = 26.5$ for $R = 0.9$, corresponding to $t_{cor} = 5.75$ and $t_{cor} = 16.9$, respectively. Thus, extended time-delayed feedback yields substantially longer correlations.

6.5 Coupling Effects of Time-Delayed Feedback

In the previous Sects. 6.3 and 6.4, time-delayed feedback was applied to a single and two coupled FitzHugh–Nagumo systems, where the feedback was generated most of

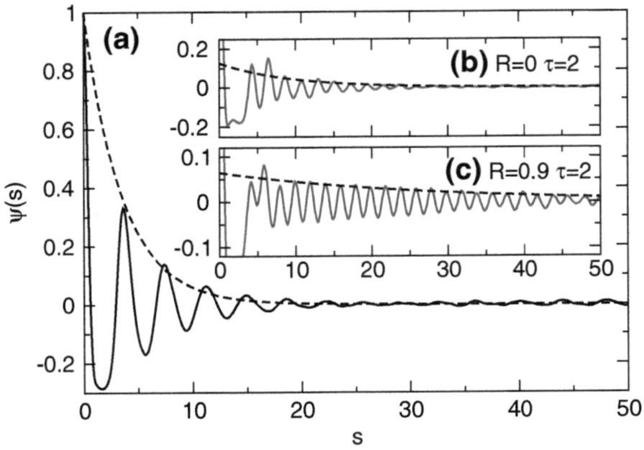

Fig. 6.61 Normalized autocorrelation function $\Psi(s)$ for $u_\Sigma = u_1 + u_2$ in dependence on R in the case of strong synchronization. No control is applied in panel (**a**), where the *dashed curve* is a fit to an exponential of the envelope through $\Psi(0) = 1$. Panels (**b**) and (**c**) show the autocorrelation function in the presence of multiple delayed feedback for memory parameters $R = 0$ and $R = 0.9$, respectively, where the dashed curve is an exponential fit to the envelope in the range $s > 10$. The other control parameters are fixed at $\tau = 2$ and $K = 1.5$. Other parameters as in Fig. 6.60 with $D_1 = 0.15$

the time from the inhibitor variable and applied to the same component. This section is concerned with other types of coupling and is mainly based in Ref. [71].

Before discussing the coupling effects, let me briefly review some aspects of neural systems and the cooperative dynamics of coupled elements. Neurons are excitable units which can emit spikes or bursts of electrical signals, i.e., the system rests in a stable steady state, but after it is excited beyond a threshold, it emits a pulse which is depicted in the beginning of this chapter in Fig. 6.3. In the following, I will consider electrically coupled neurons modelled by the FitzHugh–Nagumo system in the excitable regime similar to Sect. 6.4

$$\varepsilon_1 \frac{du_1(t)}{dt} = f(u_1(t), v_1(t)) + C[u_2(t) - u_1(t)] \tag{6.5.1a}$$

$$\frac{dv_1(t)}{dt} = g(u_1(t), v_1(t)) + D_1\xi_1(t) \tag{6.5.1b}$$

$$\varepsilon_2 \frac{du_2(t)}{dt} = f(u_2(t), v_2(t)) + C[u_1(t) - u_2(t)] \tag{6.5.2a}$$

$$\frac{dv_2(t)}{dt} = g(u_2(t), v_2(t)) + D_2\xi_2(t) \tag{6.5.2b}$$

with $f(u_i, v_i) = u_i - u_i^3/3 - v_i$ and $g(u_i, v_i) = u_i + a$ $(i = 1, 2)$. The fast activator variables u_i $(i = 1, 2)$ refer to the transmembrane voltage, and the slow inhibitor variables v_i are related to the electrical conductance of the relevant ion currents. The parameter a is the excitability parameter which is fixed at 1.05, such that there are no autonomous oscillations (excitable regime). C is the diffusive coupling strength between u_1 and u_2. To introduce different timescales for both systems, ε_1 is set to 0.005 and ε_2 is set to 0.1 as in the previous sections of this chapter. Both systems, when uncoupled, are driven entirely by independent noise sources which in the above equations are represented by ξ_i $(i = 1, 2)$ and realized by Gaussian white noise with zero mean and unity variance. See Fig. 6.2. D_i with $i = 1, 2$ is the noise intensity, and for the purposes of this paper, D_2 will be held fixed at 0.09 [26, 40].

It was discussed in Sect. 6.2 that the two subsystems exhibit cooperative dynamics for proper choice of the noise intensities and coupling strength. Sections 6.3 and 6.4 demonstrated that time-delayed feedback can influence this dynamics. So far, the control force was always implemented by inhibitor–inhibitor coupling in the first subsystem, i.e., the feedback signal was generated from the variable v_1 and applied to the same component.

Time-delayed feedback control force applied only to the first of the neural populations is schematically depicted in Fig. 6.41. As discussed in the previous sections and in detail in Chap. 2, the control method constructs a feedback F from the difference between the current value of a control signal s and the value for that quantity at time $t - \tau$. The difference is then multiplied by the feedback gain K

$$F(t) = K[s(t-\tau) - s(t)], \tag{6.5.3}$$

where s determines which components of the system enter the feedback as will be discussed in the following.

In this section, the variable w in the control force can be either the activator u_1 or the inhibitor v_1. Also, the control force can either be applied to the activator or the inhibitor differential equation. These possibilities lead to two self-coupling schemes (uu and vv) where either the activator is coupled to the activator equation or the inhibitor is coupled to the inhibitor equation, and two cross-coupling schemes (uv and vu). Thus, Eqs. 6.5.1 of the first subsystem can be rewritten including time-delayed feedback as

$$\begin{aligned}\begin{pmatrix} \varepsilon_1 \frac{du_1(t)}{dt} \\ \frac{dv_1(t)}{dt} \end{pmatrix} &= \begin{pmatrix} f(u_1(t), v_1(t)) + C[u_2(t) - u_1(t)] \\ g(u_1(t), v_1(t)) + D_1\xi_1(t) \end{pmatrix} \\ &\quad + K \begin{pmatrix} A_{uu} & A_{uv} \\ A_{vu} & A_{vv} \end{pmatrix} \begin{pmatrix} u_1(t-\tau) - u_1(t) \\ v_1(t-\tau) - v_1(t) \end{pmatrix},\end{aligned} \tag{6.5.4}$$

where the coupling matrix elements A_{ij} with $i, j \in \{u_1, v_1\}$ define the specific coupling scheme. The equations of the second subsystem (6.5.2a) remain unchanged. After the introduction of the model with general coupling of time-delayed feedback, I will discuss cooperative dynamics resulting in frequency synchronization and phase synchronization in the following.

A measure of frequency synchronization is the ratio of the average interspike intervals of the two neural populations [26, 40]. The respective average interspike interval of each neural population is denoted by $\langle T_1 \rangle$ and $\langle T_2 \rangle$. This measure was already used in previous sections of this chapter with and without feedback. See Sects. 6.2 and 6.4, respectively. The ratio $\langle T_1 \rangle / \langle T_2 \rangle$ compares the average timescales of both systems, where unity ratio describes two systems spiking at the same average frequency. It is for this reason that the interspike-interval ratio is often considered as a measure of frequency synchronization. Note that it does not contain information about the phase of synchronization, and a given interspike–interval ratio can also result from different interspike–interval distributions.

As a second type of synchronization, I discussed phase synchronization in the previous sections [45, 72]. Section 6.2 introduced a phase synchronization index γ to quantify this synchronization. It is derived from the time series of, e.g., the activator variables u_1 and u_2 which yields a phase according to Eq. 6.2.2

$$\varphi_j(t) = 2\pi \frac{t - t_{i-1}}{t_i - t_{i-1}} + 2\pi(i - 1), \tag{6.5.5}$$

where φ_j with $j = 1, 2$ corresponds to the phase of the subsystem j and t_i denotes the time of the ith spike in the respective subsystem. With this definition the phase increases by a value of 2π for each spike [26, 41, 43, 44]. The phase difference $\Delta\varphi = |\varphi_1(t) - \varphi_2(t)|$ leads to the desired measure γ by the following formula

$$\gamma = \sqrt{\langle \cos \Delta\varphi(t) \rangle^2 + \langle \sin \Delta\varphi(t) \rangle^2}. \tag{6.5.6}$$

The quantity γ varies between zero and unity corresponding to no phase synchronization and perfect phase synchronization, respectively.

After the introduction of the system, the coupling schemes, and measures of synchronization, I will present results on frequency and phase synchronization in the following. For this purpose, I will consider 16 different combinations of the noise intensity D_1 and the coupling strength C which are marked as green dots in Fig. 6.62. This figure is a reproduction of Figs. 6.16 and 6.23 added for convenience.

As discussed in Sect. 6.2, this figure displays both measures for stochastic synchronization described above in dependence on the coupling strength C and the noise intensity D_1 in the absence of of control. Panel (a) refers to the frequency synchronization characterized by the ratio of the average interspike intervals $\langle T_1 \rangle / \langle T_2 \rangle$ and panel (b) shows the phase synchronization index γ. The yellow color indicates configurations which are strongly synchronized, and dark blue and black regions refer to the case of weak synchronization. Thus, for a small value of D_1 and large coupling strength, the two subsystems display well synchronized behavior, $\langle T_1 \rangle / \langle T_2 \rangle \approx 1$ and $\gamma \approx 1$. The timescales in the interacting systems adjust themselves to 1:1 synchronization. On average, they show the same number of spikes and the two subsystems are in-phase which is indicated by yellow color. Since both panels in Fig. 6.62 share the same color code, one can see that frequency and phase synchronization coincide very well.

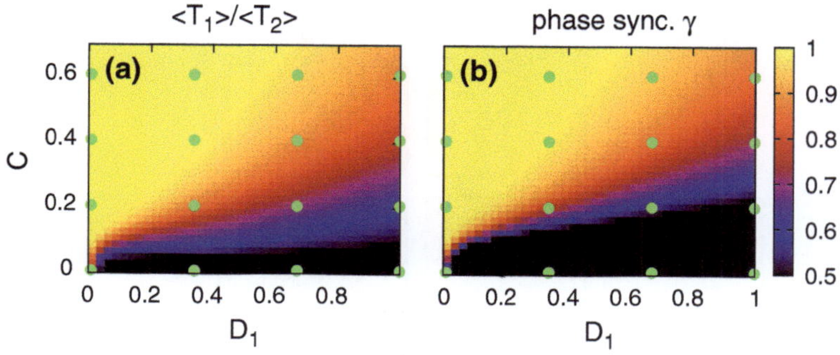

Fig. 6.62 Panels (**a**) and (**b**) show the ratio of interspike intervals $\langle T_1\rangle/\langle T_2\rangle$ and the phase synchronization index γ of the two subsystems as color code in dependence on the coupling strength C and noise intensity D_1, respectively. No control is applied to the system. The dots mark the parameter choice for different synchronization regimes used in the following. Other parameters: $\varepsilon_1 = 0.005$, $\varepsilon_2 = 0.1$, $a = 1.05$, and $D_2 = 0.09$

In the following, I will show the ratio of the average interspike interval $\langle T_1\rangle/\langle T_2\rangle$ and the phase synchronization index γ which are color coded in the (τ, K) plane for fixed combinations of D_1 and C. For each coupling scheme of time-delayed feedback control, i.e., cross-coupling schemes uv and vu and self-coupling schemes uu and vv, I will present a selection of (D_1, C) values. In all cases, only one element of the coupling matrix $\mathbf{A}$ is equal to unity and all other elements are zero.

The ordering of panels in Figs. 6.63–6.66 and 6.69–6.72 is the following: the rows correspond to fixed coupling strength chosen as $C = 0.01$, 0.21, 0.41, and 0.61 from bottom to top. The columns in each figure are calculated for constant noise intensity $D_1 = 0.01$, 0.34, 0.67, and 1.0 from left to right. The combinations (D_1, C) are summarized in Table 6.2.

Figures 6.63–6.66 show frequency synchronization measured by the ratio of average interspike intervals $\langle T_1\rangle/\langle T_2\rangle$ which is calculated from the summarized activator variable $u_\Sigma = u_1 + u_2$ as color code in dependence on the feedback gain K and the time delay τ. The system's parameters are fixed in each panel as described in Table 6.2. Figures 6.63 and 6.66 correspond to self-coupling (uu- and vv-coupling) and Figs. 6.64 and 6.65 depict the cross-coupling schemes (uv- and vu-coupling).

In terms of the coupling matrix $\mathbf{A}$, these four coupling schemes correspond to the following four matrices

$$\mathbf{A}^{(uu)} = \begin{pmatrix} 1 & 0 \\ 0 & 0 \end{pmatrix}, \quad \mathbf{A}^{(uv)} = \begin{pmatrix} 0 & 1 \\ 0 & 0 \end{pmatrix} \tag{6.5.7a}$$

$$\mathbf{A}^{(vu)} = \begin{pmatrix} 0 & 0 \\ 1 & 0 \end{pmatrix}, \quad \mathbf{A}^{(vv)} = \begin{pmatrix} 0 & 0 \\ 0 & 1 \end{pmatrix}. \tag{6.5.7b}$$

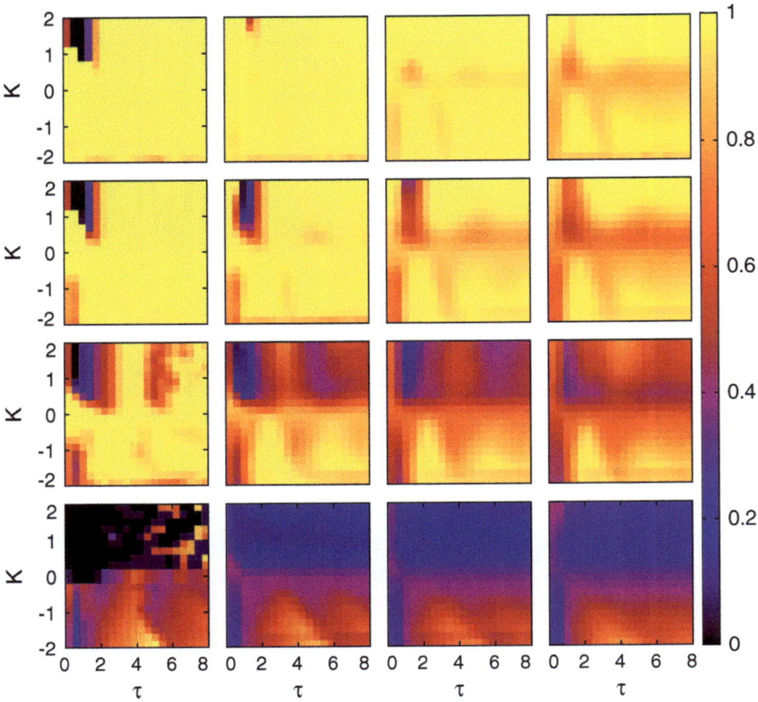

Fig. 6.63 *uu*-coupling: Ratio of average interspike intervals $\langle T_1\rangle/\langle T_2\rangle$ as color code. *Rows* and *columns* correspond to constant coupling strength C and noise intensity D_1, respectively. The parameter values of C and D_1 can be found in Table 6.2. Other parameters as in Fig. 6.62

Table 6.2 Combinations (C, D_1) used in Figs. 6.63 to 6.66 and Figs. 6.69 to 6.72

Row	Column			
	1	2	3	4
1	(0.61, 0.01)	(0.61, 0.34)	(0.61, 0.67)	(0.61, 1)
2	(0.41, 0.01)	(0.41, 0.34)	(0.41, 0.67)	(0.41, 1)
3	(0.21, 0.01)	(0.21, 0.34)	(0.21, 0.67)	(0.21, 1)
4	(0.01, 0.01)	(0.01, 0.34)	(0.01, 0.67)	(0.01, 1)

The dynamics in the white regions is outside the excitable regime and does not show noise-induced spiking, but rather the system exhibits large-amplitude self-sustained oscillations. This is shown for the coupling strength $C = 0.01$ and noise intensity $D_1 = 0.01$ for *uv*-coupling in Fig. 6.67 which depicts the time series of the inhibitor variable v_1 and the trajectory of the first subsystem in the phase space, where in addition the *u*- and *v*-nullclines are plotted for better orientation as blue (dashed) and red (dotted) curves, respectively. Panels (a)/(b), (c)/(d), and (e)/f) correspond to a feedback gain $K = -0.25, -1$, and -1.5, respectively, and the time delay is fixed at $\tau = 5$. These combinations K and τ are also marked as green

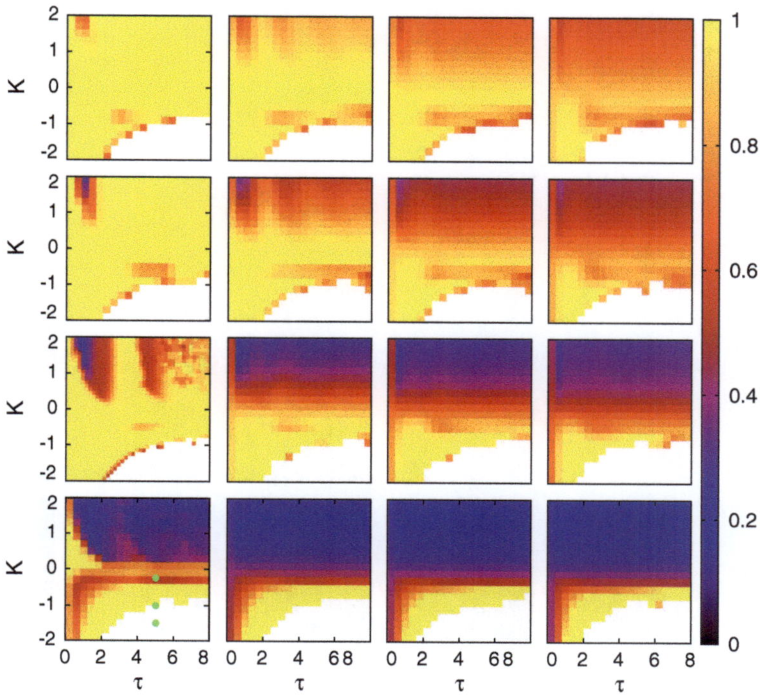

Fig. 6.64 *uv*-coupling: Ratio of average interspike intervals $\langle T_1\rangle/\langle T_2\rangle$ as color code. *Rows* and *columns* correspond to constant coupling strength C and noise intensity D_1, respectively. The parameter values of C and D_1 can be found in Table 6.2. The *green dots* for $C = 0.01$ and $D_1 = 0.01$ mark the choice of K and τ used in Fig. 6.67. Other parameters as in Fig. 6.62

dots in Fig. 6.64 of *uv*-coupling. While panels (a)/(b) are still in the yellow/orange range and show regular spiking, but no refractory phase, panels (c)/(d) depict the onset of large amplitude oscillations which can be clearly seen in panels (e)/(f) for $K = -1.5$. Note the different range of the inhibitor variable v_1 and activator variable u_1.

One can see that appropriate tuning of the control parameters leads to enhanced or deteriorated synchronization displayed by yellow and blue areas, respectively. In each figure, all panels show qualitatively similar features like a modulation of the ratio $\langle T_1\rangle/\langle T_2\rangle$ as the time delay increases, where range between maximum and minimum depends on D_1 and C. Comparing the rows, the systems are less (stronger) synchronized for small (large) values of C indicated by blue (yellow) color. As the noise intensity D_1 increases, the dynamics of the coupled subsystems is more and more noise-dominated and the dependence on the time-delay τ becomes less pronounced.

Note the symmetry in the cross-coupling schemes shown as Figs. 6.64 and 6.65 between K and its negative value $-K$ in the other cross-coupling. This symmetry is more visible in Fig. 6.68 which extracts the cross-coupling schemes (*uv*- and

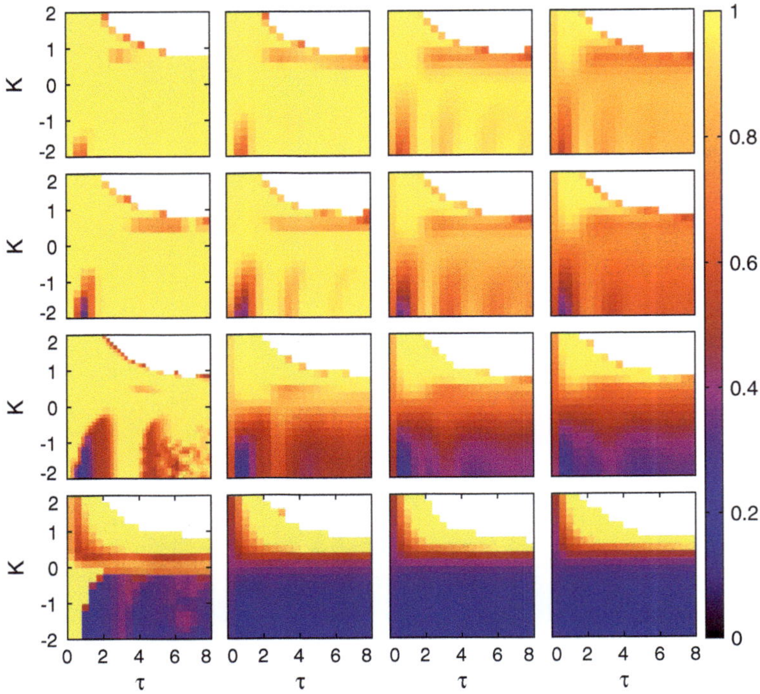

Fig. 6.65 *vu*-coupling: Ratio of average interspike intervals $\langle T_1 \rangle / \langle T_2 \rangle$ as color code. *Rows* and *columns* correspond to constant coupling strength C and noise intensity D_1, respectively. The parameter values of C and D_1 can be found in Table 6.2. Other parameters as in Fig. 6.62

vu-coupling) for a coupling strength $C = 0.21$ and noise intensity $D_1 = 0.01$ as in Figs. 6.64 and 6.65, respectively. Panels (a) and (c) show the ratio of average interspike intervals and panels (b) and (d) refer to the respective phase synchronization index. The reason for the symmetry is that enhancing the activator yields a similar effects on the dynamics as diminishing the inhibitor variable.

As for the discussion of ratio of average interspike intervals as a measure of frequency synchronization, I will present the phase synchronization index γ defined by Eq. 6.5.6 in the following for the same set of four coupling schemes as in Figs. 6.63–6.66. The respective coupling matrices are given by Eqs. 6.5.7.

Accordingly, Figs. 6.69, 6.70, 6.71, 6.72 depict the phase synchronization index γ as color code depending on the control parameters K and τ for *uu*-, *uv*-, *vu*-, and *vv*-coupling, respectively. The noise intensity D_1 and coupling strength C are fixed for each panel as described in Table 6.2.

Comparing Figs. 6.69–6.72 with the respective plot for frequency synchronization, i.e., Figs. 6.63–6.66, one can see that both types of synchronization coincide, but the phase synchronization index is more sensitive to the modulation features. Similar to the case of frequency synchronization, time delay feedback can

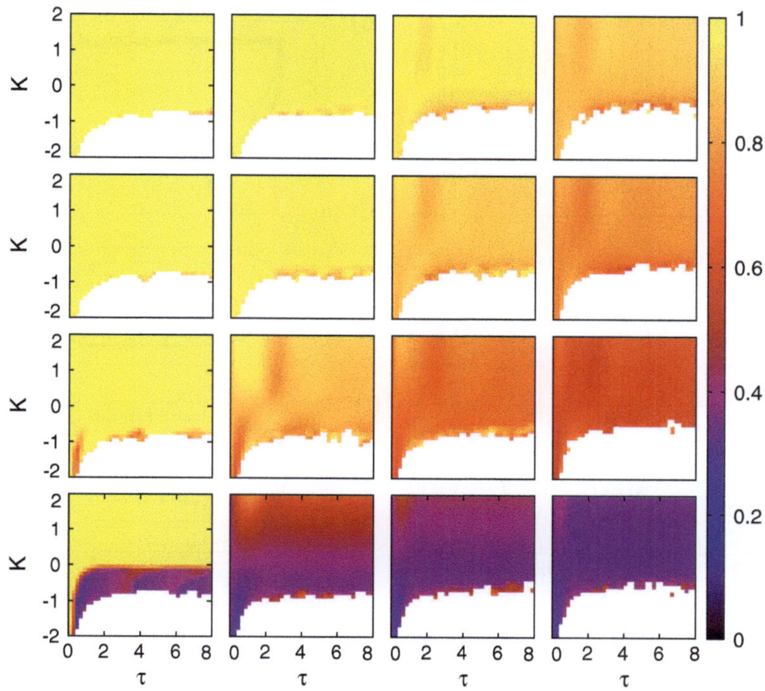

Fig. 6.66 *vv*-coupling: Ratio of average interspike intervals $\langle T_1 \rangle / \langle T_2 \rangle$ as color code. *Rows* and *columns* correspond to constant coupling strength C and noise intensity D_1, respectively. The parameter values of C and D_1 can be found in Table 6.2. Other parameters as in Fig. 6.62

lead to either support or suppression of phase synchronization depending on the specific choice of the feedback gain K and time delay τ indicated by yellow and blue regions. In general, these effects become less sensitive on the time delay as D_1 increases. For larger values of C, the two subsystems show enhanced phase synchronization.

To summarize this section, I have shown that stochastic synchronization in two coupled neural populations can be tuned by local time-delayed feedback control of one population. Synchronization can be either enhanced or suppressed, depending upon the specific values of the delay time and the coupling strength. The control depends crucially upon the coupling scheme of the control force. For inhibitor self-coupling (*vv*) synchronization is most strongly enhanced, whereas for activator self-coupling (*uu*) there exist distinct values of τ where the synchronization is strongly suppressed even in the strong synchronization regime. For cross-coupling (*uv*, *vu*) there is mixed behavior, and both schemes exhibit a strong symmetry with respect to inverting the sign of K. These observations might be important in the context of the deliberate application of control with the aim of suppressing synchronization, e.g., as therapeutic measures for Parkinson's disease.

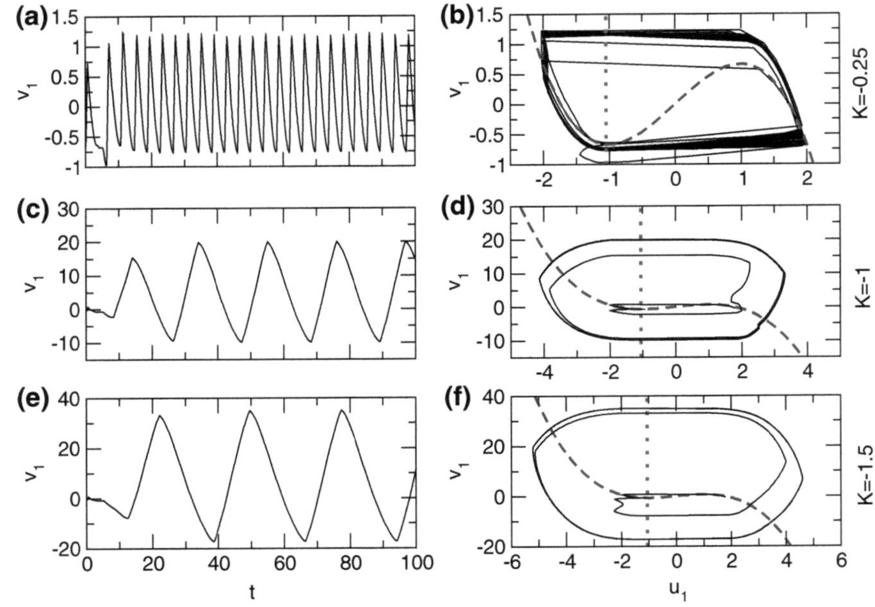

Fig. 6.67 Time series in the *uv*-coupling. Panels (**a**)/(**b**), (**c**)/(**d**), and (**e**)/(**f**) correspond to a feedback gain $K = -0.25$, -1, and -1.5, respectively. The time delay is fixed at $\tau = 5$. These values are marked by green dots in Fig. 6.64. The parameter values of C and D_1 are $C = 0.01$ and $D_1 = 0.01$. Other parameters as in Fig. 6.62

6.6 Towards Networks

In the previous sections of this chapter, I considered a single and two coupled FitzHugh–Nagumo systems with and without time-delayed feedback in various forms and coupling schemes. From the perspective of neurodynamics, this corresponds to the investigation of one and two neurons. Although the analysis of isolated neurons and their dynamics under the influence of feedback is important to understand underlying mechanisms, brain activity involves larger numbers of neurons and additional effects due to the network structure are expected. Therefore, I will discuss in this section the combination of three FitzHugh–Nagumo systems as a next step towards larger networks. At the end of the section, I will give an outlook for the development of a systematic way of grouping entities of a few neurons which serve as network motifs in order to construct a larger network. Then, the dynamics of a large number of neural oscillators can be analyzed on the basis of these notifs [73].

At first, I will consider the case of three FitzHugh–Nagumo systems which are diffusively coupled. As introduced in Sect. 6.1, this paradigmatic system provides a generic model of a two-dimensional activator–inhibitor type for neural activity. Similar to the case of two coupled neurons discussed Sect. 6.2, a combination of three elements can be written as follows

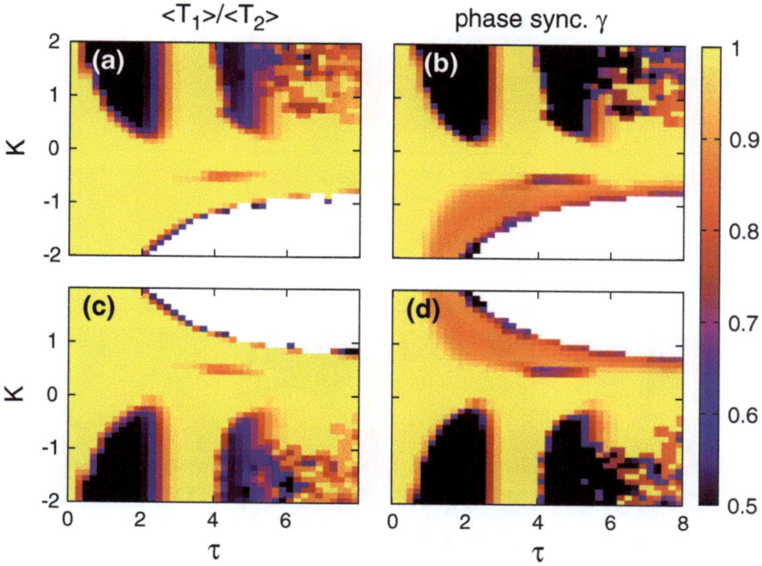

Fig. 6.68 Cross-coupling: panels (**a**) and (**c**) show the ratio of average interspike interval $\langle T_1\rangle/\langle T_2\rangle$, with time-delayed feedback in the *uv*- and *vu*-coupling, respectively. Panels (**b**) and (**d**) correspond to the respective phase synchronization index γ. The noise intensity and coupling strength are fixed at $D_1 = 0.01$ and $C = 0.21$. Other parameters as in Fig. 6.62

$$\begin{aligned} \varepsilon_1 \frac{du_1(t)}{dt} &= u_1(t) - \frac{u_1^3(t)}{3} - v_1(t) + C[-2u_1(t) + u_2(t) + u_3(t)] \\ \frac{dv_1(t)}{dt} &= u_1(t) + a \end{aligned} \tag{6.6.1a}$$

$$\begin{aligned} \varepsilon_2 \frac{du_2(t)}{dt} &= u_2(t) - \frac{u_2^3(t)}{3} - v_2(t) + C[u_1(t) - 2u_2(t) + u_3(t)] \\ \frac{dv_2(t)}{dt} &= u_2(t) + a \end{aligned} \tag{6.6.1b}$$

$$\begin{aligned} \varepsilon_3 \frac{du_3(t)}{dt} &= u_3(t) - \frac{u_3^3(t)}{3} - v_3(t) + C[u_1(t) + u_2(t) - 2u_3(t)] \\ \frac{dv_3(t)}{dt} &= u_3(t) + a, \end{aligned} \tag{6.6.1c}$$

where the variables u_i and v_i with $i = 1, 2, 3$ correspond to the activator and inhibitor whose timescale separation is given by the small parameter ε_i. The parameter a denotes the threshold parameter which is connected to the excitability of the subsystems. For $|a| < 1$, the subsystems are oscillatory and an absolute value greater than unity yields an excitable regime.

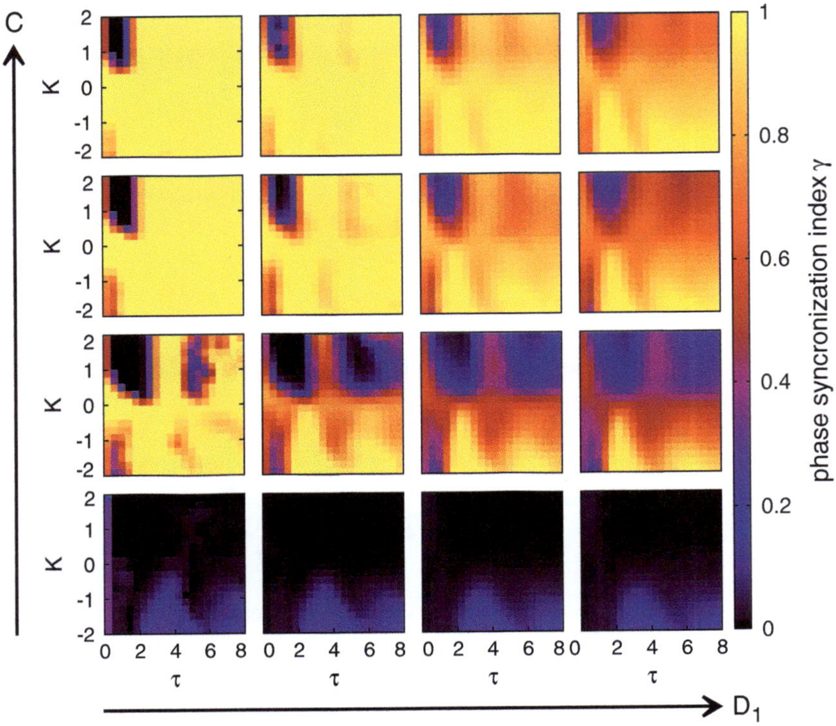

Fig. 6.69 *uu*-coupling: phase synchronization index γ. Noise intensity D_1 and coupling strength C chosen as described in Table 6.2. Other parameters as in Fig. 6.62

The three FitzHugh–Nagumo systems are diffusively coupled with a coupling strength C, where the notion of diffusion refers to a discrete version of the Laplace operator or the second space derivative of a function u

$$\frac{d^2u(x)}{dx^2} = \frac{d}{dx}\lim_{h\to 0}\frac{u(x+h)-u(x)}{h} \tag{6.6.2a}$$

$$= \lim_{h\to 0}\frac{\frac{u(x+2h)-u(x+h)}{h} - \frac{u(x+h)-u(x)}{h}}{h} \tag{6.6.2b}$$

$$= \lim_{h\to 0}\frac{u(x+2h)-2u(x+h)+u(x)}{h^2}, \tag{6.6.2c}$$

where the space is discretized in three positions $u(x+2h) = u_1$, $u(x+h) = u_2$, and $u(x) = u_3$ with periodic boundary conditions.

From the perspective of coupled systems, Eqs. 6.6.1a, b describe a specific realization of the coupling which corresponds to the following coupling matrix

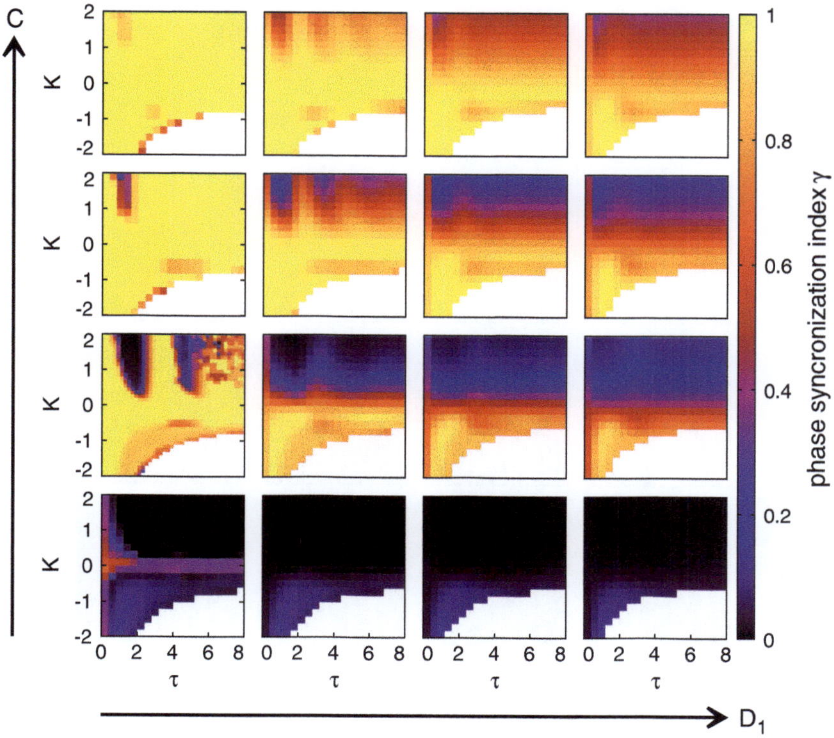

Fig. 6.70 *uv*-coupling: phase synchronization index γ. Noise intensity D_1 and coupling strength C chosen as described in Table 6.2. Other parameters as in Fig. 6.62

$$\mathbf{G} = \begin{pmatrix} -2 & 1 & 1 \\ 1 & -2 & 1 \\ 1 & 1 & -2 \end{pmatrix}. \tag{6.6.3}$$

Thus, the equations for the activator variables u_i are given in matrixform by

$$\begin{pmatrix} \varepsilon_1 \frac{du_1(t)}{dt} \\ \varepsilon_2 \frac{du_2(t)}{dt} \\ \varepsilon_3 \frac{du_3(t)}{dt} \end{pmatrix} = \begin{pmatrix} f(u_1(t), v_1(t)) \\ f(u_2(t), v_2(t)) \\ f(u_3(t), v_3(t)) \end{pmatrix} + C \begin{pmatrix} -2 & 1 & 1 \\ 1 & -2 & 1 \\ 1 & 1 & -2 \end{pmatrix} \begin{pmatrix} u_1(t) \\ u_2(t) \\ u_3(t) \end{pmatrix} \tag{6.6.4}$$

with the abbreviation $f(u_i, v_i) = u_i - u_i^3/3 - v_i$ $(i = 1, 2, 3)$.

The eigenvalues of this matrix are $\lambda_1 = 0$, $\lambda_2 = \lambda_3 = -3$, where the eigenvalue at zero reflects the zero-row sum of **G**. This corresponds to a conservation of the net flow between the elements, i.e., there is no capacitance in the connections and the coupling vanishes if all systems have identical temporal behavior, i.e., for complete synchronization. Figure 6.73 shows a schematic diagram of this configuration.

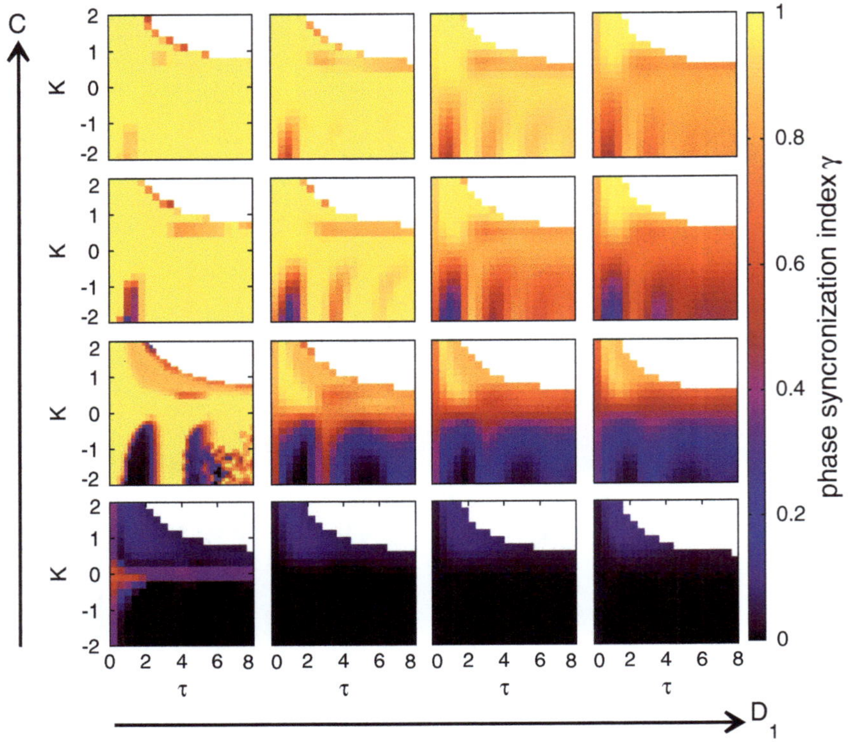

Fig. 6.71 *vu*-coupling: phase synchronization index γ. Noise intensity D_1 and coupling strength C chosen as described in Table 6.2. Other parameters as in Fig. 6.62

Figure 6.74 depicts exemplary time series of both the activator and inhibitor variables as well as the trajectory in the respective phase space according the Eqs. 6.6.1 for different coupling strengths $C = 0.0001, 0.01$, and 0.1 in the sets of panels (a), (b), and (c). The system's parameters are chosen as $\varepsilon_1 = \varepsilon_2 = \varepsilon_3 = 0.01$ and $a = 0.95$. Thus, each subsystem is prepared in the oscillatory regime. Thus, there is no need for stochastic input to excite spikes as in the previous sections of this chapter. This can also be seen from the nullclines which are added in the (u_i, v_i)-plane as dashed curves. The blue and red curves refer to the u- and v-nullclines, respectively, which intersect in the unstable part, i.e., where the cubic nullcline has positive slope. The initial conditions are chosen differently in each subsystem as follows: $(u_1(0), v_1(0)) = (2, 0)$, $(u_2(0), v_2(0)) = (-2, 0)$, and $(u_3(0), v_3(0)) = (-2, -2)$.

For a small coupling strength of $C = 0.0001$ as in panel (a), the subsystems are only weakly coupled and they oscillate independently. As the coupling becomes larger, two subsystems synchronize as shown in panel (b) for $C = 0.01$. This is the case of partial synchronization. Further increase of the coupling strength leads to complete synchronization depicted in panel (c) for $C = 0.1$.

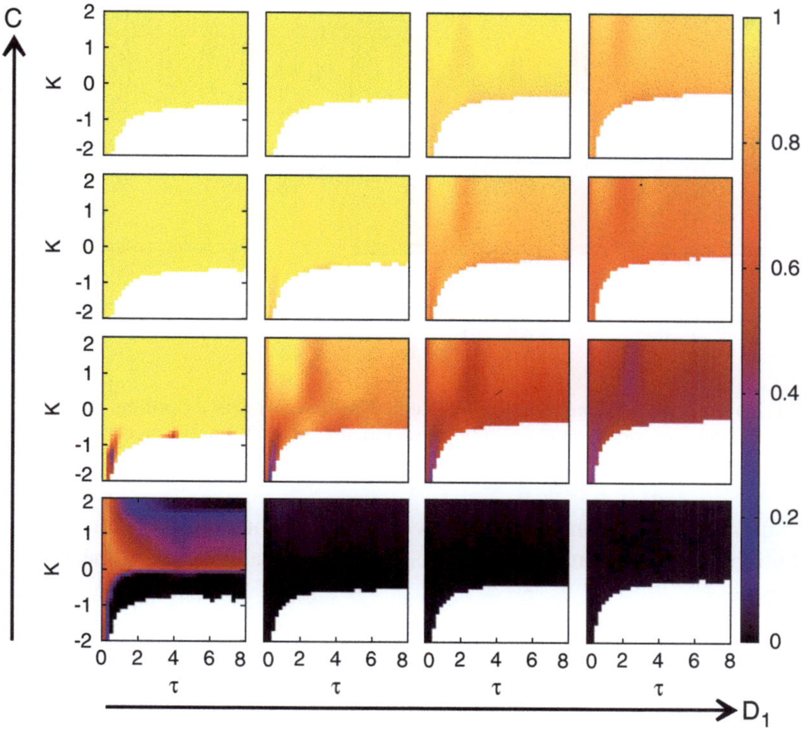

Fig. 6.72 *vv*-coupling: phase synchronization index γ. Noise intensity D_1 and coupling strength C chosen as described in Table 6.2. Other parameters as in Fig. 6.62

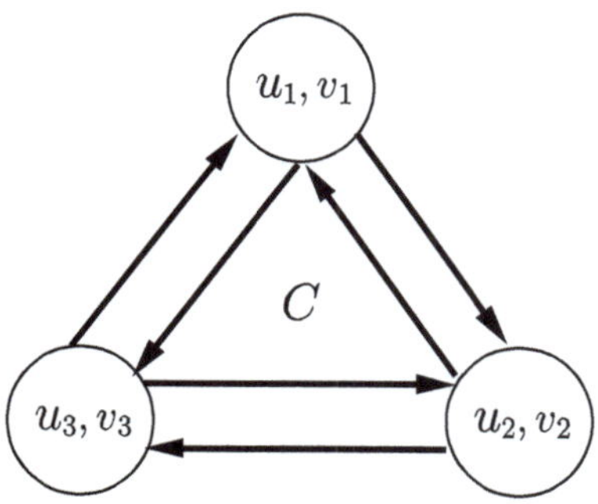

Fig. 6.73 Schematic diagram of three coupled FitzHugh–Nagumo systems including system's parameters according to Eqs. 6.6.1

The robustness of the synchronization can be investigated by the following protocol: prepare all compound system in the synchronous state as initial condition. Then, apply a random perturbation for some finite time interval, e.g., $t \in [20, 30]$. This random perturbations are chosen as independent Gaussian white noise $\xi_1(t)$, $\xi_2(t)$, and $\xi_3(t)$ with zero mean and unity variance which is added to the inhibitor variable v_i of all subsystems (6.6.1) with a fixed noise intensity D.

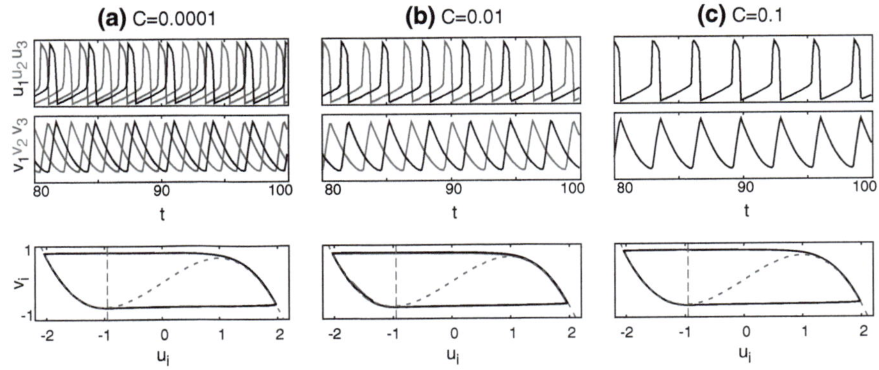

Fig. 6.74 Time series of three coupled FitzHugh–Nagumo systems according to Eqs. 6.6.1. Panels (**a**), (**b**), and (**c**) correspond to coupling strength of $C = 0.0001, 0.01$, and 0.1. Parameters: $\varepsilon_1 = \varepsilon_2 = \varepsilon_3 = 0.01$, $a = 0.95$

Since a complete synchronization results a vanishing coupling term, the averaged contribution of the coupling, for instance, in the u_i-variables gives a measures for the robustness

$$F_i = \langle | -2x_i(t) + x_j(t) + x_k(t)| \rangle \tag{6.6.5}$$

with $(i, j, k) = (1, 2, 3)$ or cyclic permutation. If all quantities F_1, F_2, and F_3 are below a threshold, e.g., $F_i < 0.01$, the subsystems are regarded as completely synchronized. For multiple iterations, the ratio of synchronous states to the total number of realizations gives a quantitative measure for the robustness of the synchronization against noise. By a similar argument, the system shows partial synchronization if only two of the quantities are below the threshold.

Figure 6.75 display the ratio of completely synchronized cases to the total number of realizations for 100 iterations as color code in dependence on the coupling strength C and noise intensity D of the perturbation. As can be seen in panel (a) and the enlargement in the range of small C depicted in panel (b), the synchronization becomes even more robust for large C.

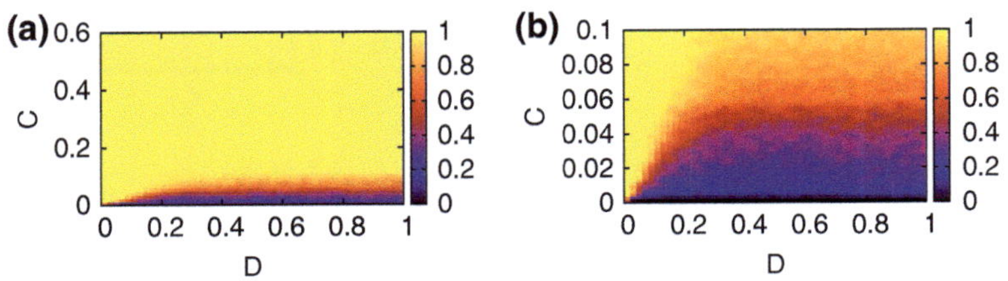

Fig. 6.75 Ratio of synchronous states to the number of iterations in dependence on the coupling strength C and noise intensity D. Each data point involves 100 iterations. Panel (**b**) shows an enlargement in the range of small C. Other parameters as in Fig. 6.74

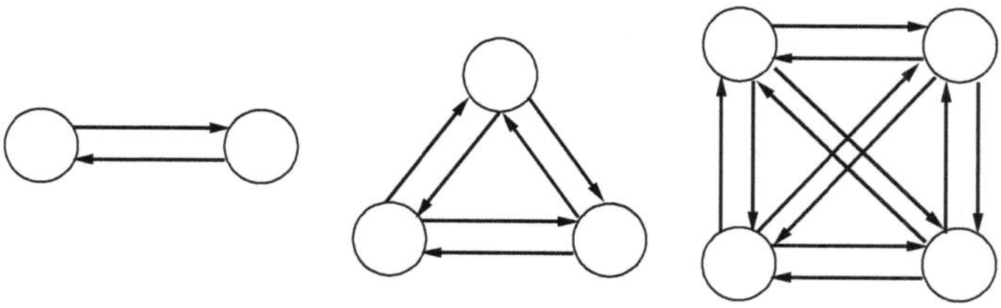

Fig. 6.76 Schematic diagram of network motifs with two, three, and four elements

Already for three coupled subsystems, one can see that the notion of synchronization needs to be revised. As a new scenario of the neural dynamics, only two of the three subsystems can exhibit the same temporal behavior while the third element is not synchronized. Thus, in addition to complete synchronization, there exists also partial synchronization. For a group of four coupled elements, there are even more scenarios possible as all four, only three, two, or no subsystem can synchronize. In the case, where only two subsystems synchronize, one needs to consider that the remaining two elements can also synchronize with each other. To make a long story short, as the number of coupled elements increases, the number of possible synchronization scenarios increases even stronger. On top of this, the number of choices for the coupling matrix **G** becomes unmanageable as its size increases quadratically. Thus, it is difficult, if not impossible, to keep track of all possible dynamics in a large network and one must think of a different approach for the analysis of network dynamics. For recent reviews concerning networks see Refs. [74–77].

One approach that has been successful in the context of complete synchronization of identical elements is called master stability function [78–80]. By this technique, the dynamics of the individual subsystems is separated from the topology of the network which enters only by the eigenvalues of the coupling matrix. For further details see the outlook at the end of Chap. 7. Let me mention that this formalism has been applied already in the context of neural dynamics [81–83].

Another approach to analyze the dynamics of large numbers of coupled elements is to break the network in smaller entities which consists of a few subsystems. These substructures might appear frequently in different parts of the network such that an understanding the dynamics of these network motifs can contribute to a systematic way of a bottom-up analysis of large collections of oscillators. Figure 6.76 depicts an example of network motifs containing two, three, and four elements in the case of all-to-all coupling.

6.7 Intermediate Conclusion

In this chapter, I investigated the influence of time-delayed feedback on the dynamics of two-variable, excitable systems of activator-inhibitor type representing neural dynamics. In all cases, the systems were prepared in the excitable regime such that stochastic external input created spiking behavior. I considered both single and coupled systems consisting of a few elements. In the beginning of the chapter, I introduced different measures to quantify the dynamics. These measures were based on the interspike interval characterized which I discussed be means of the average interspike intervals as well as the interspike interval distribution.

For coupled systems proper choice of the noise intensities and coupling strength led to cooperative dynamics like synchronization. This synchronization was investigated on the basis of the ratio of the average interspike intervals for frequency synchronization and based on the phase synchronization index and interval to check for phase synchronization. In general, both types of synchronization occurred simultaneously.

The application of time-delayed feedback control had various impacts on the dynamics of the neural systems: depending on the specific choices of the control parameters given by the feedback gain and time delay, the control method was able to enhance or deteriorate the synchronized behavior. As multiple time delays were incorporated in the feedback, the effect of the control was less sensitive on the time delay. In the single system as well as for coupled elements, time-delayed feedback control resulted in a suppression of timescales slightly larger than the time delay in the control.

In addition, I have investigated different coupling schemes of time-delayed feedback control, where I found a modulation of the synchronization in dependence on the time delay. For the cross-coupling schemes, where the activator couples via the control force to the inhibitor equation and vice versa, I observed a symmetry between the feedback gain and its negative value.

Finally, I have presented the case of three coupled FitzHugh–Nagumo systems without time-delayed feedback. The subsystems were bidirectionally coupled. Depending upon the coupling strength, the compound system exhibits no synchronization as well as partial and complete synchronization.

References

1. Koch C (1999) Biophysics of computation: information processing in single neurons. Oxford University Press, New York
2. Benucci A, Verschure PFMJ, König P (2004) High-order events in cortical networks: a lower bound. Phys Rev E 70:051909
3. Nijhawan R, Wu S (2009) Compensating time delays with neural predictions: are predictions sensory or motor? Phil Trans R Soc A 367:1063
4. Singer W (2007) Binding by synchrony. Scholarpedia 2:1657
5. Schiff SJ, Jerger K, Duong DH, Chang T, Spano ML, Ditto WL (1994) Controlling chaos in the brain. Nature (London) 370:615
6. Tass PA, Rosenblum MG, Weule J, Kurths J, Pikovsky AS, Volkmann J, Schnitzler A, Freund HJ (1998) Detection of n:m phase locking from noisy data: application to magnetoencephalography. Phys Rev Lett 81:3291

7. Grosse P, Cassidy MJ, Freund HJ (2002) EEG-EMG, MEG-EMG and EMG-EMG frequency analysis: physiological principles and clinical applications. Clin Neurophysiol 113:1523
8. Barnikol UB, Popovych OV, Hauptmann C, Sturm V, Freund HJ, Tass PA (2008) Tremor entrainment by patterned low-frequency stimulation. Phil Trans R Soc A 366:3545
9. Rosenblum MG, Pikovsky AS (2004) Delayed feedback control of collective synchrony: an approach to suppression of pathological brain rhythms. Phys Rev E 70:041904
10. Rosenblum MG, Pikovsky AS (2004) Controlling synchronization in an ensemble of globally coupled oscillators. Phys Rev Lett 92:114102
11. Popovych OV, Hauptmann C, Tass PA (2005) Effective desynchronization by nonlinear delayed feedback. Phys Rev Lett 94:164102
12. Popovych OV, Hauptmann C, Tass PA (2006) Control of neuronal synchrony by nonlinear delayed feedback. Biol Cybern 95:69
13. Gassel M, Glatt E, Kaiser F (2007) Time-delayed feedback in a net of neural elements: transitions from oscillatory to excitable dynamics. Fluct Noise Lett 7:L225
14. Schöll E, Schuster HG (Editors) (2008) Handbook of chaos control. Second completely revised and enlarged edition. Wiley-VCH, Weinheim
15. Gassel M, Glatt E, Kaiser F (2008) Delay-sustained pattern formation in subexcitable media. Phys Rev E 77:066220
16. Popovych OV, Hauptmann C, Tass PA (2005) Demand-controlled desynchronization of brain rhythms by means of nonlinear delayed feedback. In: 27th Annual conference proceedings of IEEE engineering in medicine and biology
17. Tass PA, Hauptmann C, Popovych OV (2006) Development of therapeutic brain stimulation techniques with methods from nonlinear dynamics and statistical physics. Int J Bif Chaos 16:1889
18. Hauptmann C, Tass PA (2007) Therapeutic rewiring by means of desynchronizing brain stimulation. Biosystems 89:173
19. FitzHugh R (1960) Thresholds and plateaus in the Hodgkin–Huxley nerve equations. J Gen Physiol 43:867
20. Nagumo J, Arimoto S, Yoshizawa S (1962) An active pulse transmission line simulating nerve axon. Proc IRE 50:2061
21. Hodgkin AL, Huxley AF (1952) A quantitative description of membrane current and its application to conduction and excitation in nerve. J Physiol 117:500
22. Socolar JES, Sukow DW, Gauthier DJ (1994) Stabilizing unstable periodic orbits in fast dynamical systems. Phys Rev E 50:3245
23. Erneux T (2008) Coupled FHN oscillators close to the singular opf bifurcation, unpublished notes
24. Prager T, Lerch HP, Schimansky-Geier L, Schöll E (2007) Increase of coherence in excitable systems by delayed feedback. J Phys A 40:11045
25. Hizanidis J, Balanov AG, Amann A, Schöll E (2006) Noise-induced front motion: signature of a global bifurcation. Phys Rev Lett 96:244104
26. Hauschildt B, Janson NB, Balanov AG, Schöll E (2006) Noise-induced cooperative dynamics and its control in coupled neuron models. Phys Rev E 74:051906
27. Stratonovich RL (1963) Topics in the theory of random noise, vol 1. Gordon and Breach, New York
28. Gardiner CW (2002) Handbook of stochastic methods for physics, chemistry and the natural sciences. Springer, Berlin
29. van Kampen NG (2003) Stochastic processes in physics and chemistry. North-Holland, Amsterdam
30. Schöll E, Balanov AG, Janson NB, Neiman A (2005) Controlling stochastic oscillations close to a Hopf bifurcation by time-delayed feedback. Stoch Dyn 5:281
31. Pomplun J, Amann A, Schöll E (2005) Mean field approximation of time-delayed feedback control of noise-induced oscillations in the Van der Pol system. Europhys Lett 71:366
32. Hu G, Ditzinger T, Ning CZ, Haken H (1993) Stochastic resonance without external periodic force. Phys Rev Lett 71:807

33. Pikovsky AS, Kurths J (1997) Coherence resonance in a noise-driven excitable system. Phys Rev Lett 78:775
34. Pomplun J, Balanov AG, Schöll E (2007) Long-term correlations in stochastic systems with extended time-delayed feedback. Phys Rev E 75:040101(R)
35. Hiller G (2008) Synchronization and control of delay-coupled excitable systems. Master's thesis, Technische Universität Berlin
36. Dahlem MA, Hiller G, Panchuk A, Schöll E (2009) Dynamics of delay-coupled excitable neural systems. Int J Bifur Chaos 19:745
37. Schöll E, Hiller G, Hövel P, Dahlem MA (2009) Time-delayed feedback in neurosystems. Phil Trans R Soc A 367:1079
38. Hövel P, Dahlem MA, Dahms T, Hiller G, Schöll E (2009) Time-delayed feedback control of delay-coupled neurosystems and lasers. In: Preprints of the second IFAC meeting related to analysis and control of chaotic systems (CHAOS09). World Scientific, Singapore. arXiv:0912.3395
39. Erneux T (2008) Echo waves for two FHN excitable systems coupled with delay. unpublished notes
40. Hövel P, Dahlem MA, Schöll E (2010) Control of synchronization in coupled neural systems by time-delayed feedback. Int J Bifur Chaos 20:813
41. Pikovsky AS, Rosenblum MG, Kurths J (2001) Synchronization, A universal concept in nonlinear sciences. Cambridge University Press, Cambridge
42. Mosekilde E, Maistrenko Y, Postnov D (2002) Chaotic synchronization: applications to living systems. World Scientific, Singapore
43. Pikovsky AS, Rosenblum MG, Kurths J (1996) Synchronisation in a population of globally coupled chaotic oscillators. Europhys Lett 34:165
44. Hauschildt B (2005) Control of noise-induced multimode oscillations in coupled neural systems. Master's thesis, Technische Universität Berlin
45. Rosenblum MG, Pikovsky AS, Kurths J, Schäfer C, Tass PA (2001) Phase synchronization: from theory to data analysis. In: Handbook of biological physics. Elsevier Science, Amsterdam, vol 4, chap 9, 1st edn, pp 279–321
46. Park K, Lai YC (2005) Characterization of stochastic resonance. Europhys Lett 70:432
47. Lai YC, Frei MG, Osorio I (2006) Detecting and characterizing phase synchronization in nonstationary dynamical systems. Phys Rev E 73:26214
48. Park K, Lai YC, Krishnamoorthy S (2007) Noise sensitivity of phase-synchronization time in stochastic resonance: theory and experiment. Phys Rev E 75:46205
49. Pyragas K (1992) Continuous control of chaos by self-controlling feedback. Phys Lett A 170:421
50. Janson NB, Balanov AG, Schöll E (2004) Delayed feedback as a means of control of noise-induced motion. Phys Rev Lett 93:010601
51. Balanov AG, Janson NB, Schöll E (2004) Control of noise-induced oscillations by delayed feedback. Phys D 199:1
52. Pototsky A, Janson NB (2008) Excitable systems with noise and delay, with applications to control: renewal theory approach. Phys Rev E 77:031113
53. Schöll E, Hövel P, Flunkert V, Dahlem MA (2010) Time-delayed feedback control: from simple models to lasers and neural systems. In: Atay FM (ed) Complex time-delay systems: theory and applications. Springer, Berlin
54. Terry JR, Thornburg KS, DeShazer DJ, VanWiggeren GD, Zhu S, Ashwin P, Roy R (1999) Synchronization of chaos in an array of three lasers. Phys Rev E 59:4036
55. Wünsche HJ, Bauer S, Kreissl J, Ushakov O, Korneyev N, Henneberger F, Wille E, Erzgräber H, Peil M, Elsäßer W, Fischer I (2005) Synchronization of delay-coupled oscillators: a study of semiconductor lasers. Phys Rev Lett 94:163901
56. Shaw LB, Schwartz IB, Rogers EA, Roy R (2006) Synchronization and time shifts of dynamical patterns for mutually delay-coupled fiber ring lasers. Chaos 16:015111
57. Schwartz IB, Shaw LB (2007) Isochronal synchronization of delay-coupled systems. Phys Rev E 75:046207

58. de Sousa Vieira M (2007) Properties of zero-lag long-range synchronization via dynamical relaying. arXiv:0705.1807
59. Vicente R, Pipa G, Fischer I, Mirasso CR (2007) Zero-lag long range synchronization of neurons is enhanced by dynamical relaying. In: Artificial Neural Networks ICANN 2007. Lecture notes in computer science. Springer, Berlin, vol 4668, pp 904–913
60. Ashwin P, Buescu J, Stewart I (1994) Bubbling of attractors and synchronisation of chaotic oscillators. Phys Lett A 193:126
61. Ott E, Sommerer JC (1994) Blowout bufurcations: the occurrence of riddled basins and on-off intermittency. Phys Lett A 188:39
62. Venkataramani SC, Hunt BR, Ott E (1996) Bubbling transition. Phys Rev E 54:1346
63. Flunkert V, D'Huys O, Danckaert J, Fischer I, Schöll E (2009) Bubbling in delay-coupled lasers. Phys Rev E 79:065201(R)
64. Hövel P, Dahlem MA, Schöll E (2007) Synchronization of noise-induced oscillations by time-delayed feedback. In: Proceedings of 19th international conference on noise and fluctuations (ICNF-2007). American Institute of Physics, College Park, Maryland 20740-3843, vol 922, pp 595–598. ISBN 0-7354-0432-8
65. Unkelbach J, Amann A, Just W, Schöll E (2003) Time–delay autosynchronization of the spatiotemporal dynamics in resonant tunneling diodes. Phys Rev E 68:026204
66. Schlesner J, Amann A, Janson NB, Just W, Schöll E (2003) Self-stabilization of high frequency oscillations in semiconductor superlattices by time–delay autosynchronization. Phys Rev E 68:066208
67. Dahms T, Hövel P, Schöll E (2007) Control of unstable steady states by extended time-delayed feedback. Phys Rev E 76:056201
68. Dahms T, Hövel P, Schöll E (2008) Stabilizing continuous-wave output in semiconductor lasers by time-delayed feedback. Phys Rev E 78:056213
69. Schöll E, Majer N, Stegemann G (2008) Extended time delayed feedback control of stochastic dynamics in a resonant tunneling diode. Phys Stat Sol (c) 5:194
70. Majer N, Schöll E (2009) Resonant control of stochastic spatio-temporal dynamics in a tunnel diode by multiple time delayed feedback. Phys Rev E 79:011109
71. Hövel P, Shah SA, Dahlem MA, Schöll E (2009) Feedback-dependent control of stochastic synchronization in coupled neural systems. In: Fortuna L, Frasca M (eds) Proceedings 4th international scientific conference on physics and control (PhysCon 09). IPACS Open Access Library. http://lib.physcon.ru (e-Library of the International Physics and Control Society). arxiv:0911.2334v1
72. Rosenblum MG, Pikovsky AS, Kurths J (2001) Synchronization—a universal concept in nonlinear sciences. Cambridge University Press, Cambridge
73. D'Huys O, Vicente R, Erneux T, Danckaert J, Fischer I (2008) Synchronization properties of network motifs: influence of coupling delay and symmetry. Chaos 18:037116
74. Watts DJ, Strogatz SH (1998) Collective dynamics of 'small-world' networks. Nature 393:440
75. Strogatz SH (2001) Exploring complex networks. Nature 410:268
76. Albert R, Barabási A-L (2002) Statistical mechanics of complex networks. Rev Mod Phys 74:47
77. Newman MEJ (2003) The structure and function of complex networks. SIAM Rev 45:167
78. Pecora LM, Carroll TL (1998) Master stability functions for synchronized coupled systems. Phys Rev Lett 80:2109
79. Pecora LM (1998) Synchronization conditions and desynchronizing patterns in coupled limit-cycle and chaotic systems. Phys Rev E 58:347
80. Fink KS, Johnson G, Carroll TL, Mar D, Pecora LM (2000) Three coupled oscillators as a universal probe of synchronization stability in coupled oscillator arrays. Phys Rev E 61:5080
81. Dhamala M, Jirsa VK, Ding M (2004) Enhancement of neural synchrony by time delay. Phys Rev Lett 92:074104
82. Dhamala M, Jirsa VK, Ding M (2004) Transitions to synchrony in coupled bursting neurons. Phys Rev Lett 92:028101

83. Rossoni E, Chen Y, Ding M, Feng J (2005) Stability of synchronous oscillations in a system of Hodgkin–Huxley neurons with delayed diffusive and pulsed coupling. Phys Rev E 71:061904

Chapter 7
Summary and Outlook

You have to allow a certain amount of time in which you are doing nothing in order to have things occur to you, to let your mind think.

(Mortimer J. Adler)

In this thesis, I have investigated effects of time-delayed feedback on different classes of dynamical systems. Originally, this control method was invented for the stabilization of periodic orbits that are embedded in strange attractors of deterministic, chaotic systems. Stabilization is achieved by a control force constructed from the difference between the time-delayed and current control signal which is generated from the system's variables. Thus, there is no need for a reference system. It was previously shown that the period of the target orbit provides an optimal choice for the delay in the controller. This choice renders the control method noninvasive in the event of successful stabilization because the control force vanishes.

In my work, I have extended the application of time-delayed feedback control to the stabilization of steady states. I have demonstrated that fixed points of focus type can be stabilized while the method fails for the stabilization of saddle points. In the case of an unstable focus, the intrinsic period given by the rotation around the fixed point yields a timescale that can be employed for the choice of the time delay. It turns out that the domain of control becomes largest if the time delay is chosen as one half of this intrinsic period. In fact, I have analytically derived that the time-delayed feedback control fails if the time delay is equal to integer multiples of this period.

Furthermore, I have discussed several modifications of the original control scheme which are experimentally relevant. Among these are additional filters, multiple delays, and latency times in the control loop. I have explored the domains of control, i.e., the areas of stability in the control parameter space, on the basis of transcendental characteristic equations for the eigenvalues of the system with control.

Besides linear systems with steady states, I have demonstrated that time-delayed feedback control can be used to stabilize periodic orbits generated in a

P. Hövel, *Control of Complex Nonlinear Systems with Delay*,
Springer Theses, DOI: 10.1007/978-3-642-14110-2_7,

subcritical Hopf bifurcation. Previously, it was a common belief that these orbits cannot be stabilized because they have an odd number of real Floquet multipliers greater than unity. Thus, the odd number limitation theorem should apply. I have shown that this theorem does not hold for autonomous systems. The control mechanism is the following: The feedback scheme generates a pair of additional, delay-induced periodic orbits in a saddle-node bifurcation. Beyond this bifurcation point, the stable orbit exchanges its stability with the subcritical orbit in a transcritical bifurcation and subsequently disappears in another subcritical Hopf bifurcation with the fixed point at the origin. In addition to this counterexample of the odd number limitation theorem, I have presented similar results in the context of periodic orbits generated in a fold bifurcation.

As another class of dynamical systems, I have considered neutral delay-differential equations which have an intrinsic time delay in the highest derivative. These systems occur, for instance, in the substructuring technique, where part of an experiment is transferred to numerical simulations. These calculations lead to supplemental forces which are applied to the system via actuators in real time. An intrinsic time delay occurs in this configuration due to the latency between the remaining parts of the experiment and the computer.

I have analyzed the eigenvalue spectrum of this delayed compound system and investigated the asymptotics for large time delays. I have found that there is an upper bound of the coupling strength of the delayed signal beyond which the considered system becomes unstable. Furthermore, I have demonstrated that time-delayed feedback control shifts this maximum coupling strength to large values. In addition, I have discussed the interplay of the intrinsic and controller time delay and derived analytical expressions for the parameterization of the stability boundaries.

The last type of dynamical systems that I have considered exhibits excitable dynamics, where the excitations are due to stochastic input of random fluctuations. Excitable systems can be found, for instance, on neurodynamics and accordingly, I have chosen the FitzHugh–Nagumo system for my investigations. The FitzHugh–Nagumo system is a two-dimensional activator–inhibitor model and is paradigmatic for neural dynamics.

After the discussion of a single element, where I found coherence resonance, I have considered two symmetrically coupled FitzHugh–Nagumo systems. I have shown that the two subsystems exhibit cooperative dynamics depending upon the coupling strength and noise intensities. Beyond this previously established results, I have applied time-delayed feedback control in its original form as well as in the extension involving multiple time delays.

In general, time-delayed feedback is able to suppress and enhance the synchronization of the two subsystems depending on the specific choices of the feedback gain and time delay. In the case of multiple time delays, I have found that the controller is less sensitive to the choice of control parameters as more and more information from the past is included.

I have also analyzed different coupling schemes of the control method. These involve construction of the control force from the activator or inhibitor variable and application to either component. For different coupling strengths and noise intensities, I have investigated frequency and phase synchronization in dependence on the control parameters. I have found a modulation depending on the time delay in all coupling schemes and a symmetry between the value of the feedback gain and its negative value for the cross-coupling schemes, where the control force constructed from the activator is coupled to the inhibitor equation and vice versa.

To conclude this summary, I have elaborated in detail that time-delayed feedback control provides a powerful tool for the stabilization of unstable states in a large variety of different dynamical systems. I will finish this thesis with an outlook towards future work.

An open problem which has recently become of increasing interest is the topic of synchronization of individual elements in networks, especially, in the presence of unavoidable delays. In networks, time delays arise naturally due to the propagation of signals between distant nodes. For this, the topology described, for instance, by the adjacency or coupling matrix plays a crucial role for the establishment of synchronization in the network. Prominent topologies include scale-free, all-to-all, and regular networks. The simulation of networks is often computationally expensive and involves a large number of parameters which can be independently varied. Thus, a systematic analysis is not obvious.

For the investigation of complete synchronization, a powerful technique called *master stability function* has been recently developed. Complete synchronization refers to a scenario, where all elements of the network show identical temporal dynamics without a phase shift. The main steps of this techniques are briefly summarized in the following: Starting from a given network of identical elements, the individual dynamics of each node is decoupled from the network structure. For this, a linearization of the coupled subsystems at the synchronous state is performed. This is followed by diagonalization of the matrix that determines the coupling of the elements. Under the assumption of constant row sum of the coupling matrix, the linearized equations of each node are not affected by this diagonalization. The constant row sum reflects that each element of the network is subject to the same coupling signal in the case of complete synchrony. Finally, one arrives at a single set of equations for each eigenvalue of the coupling matrix and it is sufficient to calculate the Lyapunov exponent of this reduced system to check for linear stability. This is an elegant way to determine the stability of the completely synchronous state for a given network topology.

Another procedure to analyze delay effects in networks could be a bottom-up approach involving network motifs. By this strategy, identical parts of the network containing a few elements are investigated separately before their mutual interaction is studied. From this perspective, the analysis of two or three coupled subsystems presented, for instance, in Chap. 6 is a first step towards larger network structures.

About the Author

Philipp Hövel was born in Berlin, Germany, in 1980. He received the Diploma in physics from the *Technische Universität Berlin* in 2004, for which he was awarded the *Physik-Studienpreis* by the *Wilhelm und Else Heraeus-Stiftung*. In 2006, he received the Diploma in mathematics from the *Technische Universität Berlin*. For both studies he was supported on a scholarship for academic excellence provided by the *Bischöfliche Studienförderung Cusanuswerk*. From 2002 to 2003, he was a Visiting Scholar with Duke University, Durham, NC, on a Fulbright grant. He visited the University of Tokyo in 2007 supported by the *Deutsche Akademische Austauschdienst*. Philipp Hövel received the Dr. rer. nat. degree from *Technische Universität Berlin* in 2009. His work was performed in the group of Prof. Dr. Eckehard Schöll, Ph.D. at the *Institut für Theoretische Physik*. His research interests include nonlinear dynamics, control theory, stabilization of peridoc orbits and steady states by time-delayed feedback, neural and excitable systems, and dynamics in networks and network motifs.

Index

MIX
Papier aus verantwortungsvollen Quellen
Paper from responsible sources
FSC® C105338

If you have any concerns about our products, you can contact us on
ProductSafety@springernature.com

In case Publisher is established outside the EU, the EU authorized representative is:
Springer Nature Customer Service Center GmbH
Europaplatz 3, 69115 Heidelberg, Germany

Printed by Libri Plureos GmbH
in Hamburg, Germany